Schrödinger Operators, Markov Semigroups, Wavelet Analysis, Operator Algebras

Mathematical Topics

Volume 11

Advances in Partial Differential Equations

Schrödinger Operators, Markov Semigroups, Wavelet Analysis, Operator Algebras

edited by
Michael Demuth
Elmar Schrohe
Bert-Wolfgang Schulze (editor-in-chief)
Johannes Sjöstrand

Akademie Verlag

Editors:

Prof. Dr. Michael Demuth, Technical University of Clausthal
Dr. Elmar Schrohe, Max Planck Research Group for Partial Differential Equations and Complex Analysis, University of Potsdam
Prof. Dr. Bert-Wolfgang Schulze, University of Potsdam and Max Planck Research Group for Partial Differential Equations and Complex Analysis (editor-in-chief)
Prof. Dr. Johannes Sjöstrand, Ecole Polytechnique, Palaiseau

1st edition
Editorial Director: Dipl.-Math. Gesine Reiher

Library of Congress Card Number pending

Die Deutsche Bibliothek – CIP-Einheitsaufnahme

Schrödinger operators, Markov semigroups, wavelet analysis, operator algebras / ed. by Michael Demuth ... –
1. ed. – Berlin : Akad. Verl., 1996
(Mathematical topics ; Vol. 11 : Advances in partial differential equations)
ISBN 3-05-501710-2
NE: Demuth, Michael [Hrsg.]; GT

ISSN 0946-3844

Akademie Verlag is a member of the VCH Publishing Group.

Printed on non-acid paper.
The paper used corresponds to both the U.S. standard ANSI Z.39.48 – 1984 and the European standard ISO TC 46.

Printing: GAM Media GmbH, Berlin
Bookbinding: Verlagsbuchbinderei Mikolai GmbH, Berlin

Printed in the Federal Republic of Germany

Akademie Verlag GmbH
Mühlenstraße 33–34 · D-13187 Berlin
Federal Republic of Germany

VCH Publishers, Inc.
333 7th Avenue
New York, NY 10001

Preface

Partial differential equations govern many phenomena in nature, and their analysis has been an indispensable tool in science and engineering for more than two centuries. Moreover, partial differential equations lie at the interface of several branches of mathematics, areas as diverse as differential geometry, functional analysis, the theory of Lie groups, and algebraic topology. During the past decades, the discovery of microlocal analysis and pseudodifferential operators has greatly advanced the theory of partial differential equations. It has led to major breakthroughs and still offers fascinating prospects for the future.

This book is the third title of "Advances in Partial Differential Equations", a series of books originating from the work of the Max Planck Research Group for Partial Differential Equations and Complex Analysis at the University of Potsdam. It is our intention to promote expositions of a systematic character in the field of partial differential equations. The articles will be largely self-contained and written for a wide readership by specialists, who will report on recent progress and give specific insight into their own work.

The analysis of partial differential equations has stimulated large areas of research in mathematical physics, harmonic analysis, and operator theory. The present volume illuminates the depth and variety of these interactions.

It begins with a survey on the use of semiclassical analysis and maximum-principle techniques in statistical mechanics. There follows an article presenting the perturbation theory for generators of Markov semigroups acting on L^p.

The third contribution provides a self-contained introduction to continuous wavelet analysis, including its relations to function spaces and microlocal regularity; this is particularly topical, as wavelet methods have been applied with great success in the past decade to problems in harmonic and numerical analysis as well as in diverse fields of engineering.

The final section explores pseudodifferential analysis on singular configurations, with special emphasis on C^*-algebra techniques, Mellin operators, and analytical index formulas.

Potsdam, June 1996 M. Demuth, E. Schrohe, B.-W. Schulze, J. Sjöstrand

Contents

Recent Results and Open Problems on Schrödinger Operators, Laplace Integrals, and Transfer Operators in Large Dimension

Bernard Helffer

Abstract

In this survey, we intend to present how the semiclassical analysis or the maximum principle techniques can be useful in statistical mechanics. After some presentation of the origin of the problem in statistical mechanics, which is partly inspired by a course by M. Kac, and we shall analyze in parallel three situations where an asymptotic behavior with respect to the dimension or to a small parameter has to be considered:

- Laplace integrals,
- Schrödinger operators,
- transfer operators.

In these three cases the starting point is a family $V^{(m)}$ of C^∞ potentials on $\mathbb{R}^m$ parametrized by m which appears as the sum of a quadratic term $\sum_j x_j^2$ and of an interaction which is in the simplest case of the form $\sum_j I(x_j, x_{j+1})$.

The main techniques presented here were introduced in the context of large dimension by J. Sjöstrand and then developed recently in different papers mainly by B. Helffer or J. Sjöstrand.

Acknowledgements. I would like at first to thank J. Sjöstrand whose collaboration was very important for me. This survey gives me in particular the opportunity to present his decisive contribution in the subject. I thank him also for useful comments during the elaboration of this manuscript.
I thank also the other colleagues who helped me to clarify my ideas or to complete my knowledge of some of the questions presented here and especially J. Bellissard, M. Brunaud, F. Daumer, B. Derrida, I. Herbst, A. Jensen, P. Kerdelhue, S.T. Kuroda, M. Schatzman, R. Seiler, V. Sordoni, V. Tchoulaevski and W. M. Wang.
This research would not have been possible without the good working conditions I found at the Wissenschaftskolleg zu Berlin in 1990-1991 and at the Mittag Leffler Institute in 1992-1993.
Finally I would like to thank M. Demuth who convinced me to write this survey.

1 Introduction

1.1 On the way of a program by M. Kac

In this survey we intend to present results concerning semiclassical techniques and maximum principle techniques in statistical mechanics. After some presentation of the origin of the problem in statistical mechanics which is mainly inspired by a course by M. Kac [Ka66] we analyze in parallel three situations concerning asymptotic behavior with respect to the dimension or to a small parameter: Laplace integrals, Schrödinger operators and Kac operators (which are more commonly called transfer operators).
In these three cases, the starting point is a family $V^{(m)}$ of C^∞ functions on $\mathbb{R}^m$ parametrized by m which play the role of the potential in mechanics. We plan to analyze in the survey the multiple links between the different problems and will compare the different approachs. These links can be at the technical level (pseudodifferential techniques) or at a deeper level: link between quantum mechanics and statistical mechanics. Although we have tried to present a wide variety of techniques, the probabilistic methods and some techniques coming from quantum field theory are not presented due to the ignorance of the author. We apologize for this lack of completeness.

1.2 Presentation of the mathematical problems

Although the problems we are considering are strongly motivated by statistical mechanics, as we shall see in Section 2, the problems, we are interested in, appear in other areas in mathematics and can be formulated independently of the language of statistical mechanics. We have selected the following problems.

(I) Study of the asymptotic behavior of the Laplace integral

$$\frac{1}{m}\left[\ln((1/h\pi)^{\frac{m}{2}}\int \exp\,(-\frac{V^{(m)}(x)}{h})\,dx)\right] \tag{1.2.1}$$

as m tends to ∞, and control of the limit with respect to h as h tends to $+0$.

(II) If $\mu_1(m,h)$ is the largest eigenvalue of the operator

$$K^{(m)}(h) = \exp\left(-\frac{V^{(m)}(x)}{2}\right)\cdot\exp\,(h^2\Delta^{(m)})\cdot\exp\left(-\frac{V^{(m)}(x)}{2}\right), \tag{1.2.2}$$

consider the asymptotic behavior of the quantity $-\dfrac{\ln\mu_1(m,h)}{m}$ as m tends to ∞ and control this limit with respect to h as h tends to $+0$. Here $\Delta^{(m)}$ is the Laplacian on $\mathbb{R}^m$.

(III) If $\lambda_1(m,h)$ is the smallest eigenvalue of the Schrödinger operator

$$S^{(m)}(h) = -h^2\Delta^{(m)} + V^{(m)}(x), \tag{1.2.3}$$

what is the asymptotic behavior of the quantity $\lambda_1(m,h)/m$ as m tends to ∞ and control this limit with respect to h as h tends to $+0$.

These three questions are of course strongly related and we shall analyze these links carefully. If we think of a potential which is invariant by circular permutation of the variables and "near" in a suitable sense of the harmonic oscillator, all these questions are well analyzed for fixed m as h tends to zero. (I) can be treated by application of the stationary phase theorem (or more correctly of the Laplace method), (III) corresponds with a semiclassical analysis of the Schrödinger operator at the bottom (see [HeSj84-87] and [Si83]) and the study of (III) can be considered as a pseudo-differential extension of (II) (see [BruHe91] or [He92b]). In particular this study gives for example that

$$-\ln(\mu_1(m,h)) = \lambda_1(m,h) + O_m(h^2). \tag{1.2.4}$$

Another link is that, by Golden Thompson inequality (cf. [Len71]), we have

$$\begin{aligned}\mathrm{Tr}\,(\exp\;-(-\Delta+W)) &\leq \mathrm{Tr}\,(\exp(-\tfrac{W}{2})\cdot\exp\;\Delta\cdot\exp(-\tfrac{W}{2}))\\ &= C_m\textstyle\int\exp\;-W(x)dx^{(m)}\;.\end{aligned} \tag{1.2.5}$$

Let us recall also Segal's Lemma which says under suitable assumptions on the self-adjoint semibounded operators A and B on an Hilbert space H, (see Lemma 4.6 in [Si74], [ReSi75] (p.333) or [ReSi78] (p.320)) that

$$\|\exp\;-(A+B)\|_{\mathcal{L}(H)} \leq \left\|\exp\;-\frac{A}{2}\cdot\exp\;-B\cdot\exp\;-\frac{A}{2}\right\|_{\mathcal{L}(H)}. \tag{1.2.6}$$

It is by the way also interesting (cf. [Sj92b]) to look at the asymptotic behavior of the quantity $\left[\ln \operatorname{Tr}\left(\exp -t(-h^2\Delta + V^{(m)})\right)\right]/m$ as m tends to ∞ and control (for fixed $t > 0$) this limit with respect to h as h tends to $+0$.

The difficult new problem is of course a good control with respect to m as m is large. But other interesting (and sometimes more difficult) problems appear in the same context.

(IV) In the context of the Laplace integrals, the connected problem is to analyze the behavior as $|i - j| \to \infty$ of the pair correlation

$$\lim_{m \to \infty} \operatorname{Cor}^{(m)}(i, j), \tag{1.2.7}$$

where

$$\operatorname{Cor}^{(m)}(i, j) = \langle x_i \cdot x_j \rangle_m - \langle x_i \rangle_m \cdot \langle x_j \rangle_m \tag{1.2.8}$$

and $\langle f \rangle_m$ is the mean value with respect to the measure $\exp -V^{(m)}(x)dx^{(m)}$ of the function f depending on a finite fixed number of variables:

$$\langle f \rangle_m = \frac{\int f(x_1, \cdots, x_k) \exp\, -V^{(m)}(x)\, dx^{(m)}}{\int \exp\, -V^{(m)}(x)\, dx^{(m)}}\,. \tag{1.2.9}$$

(V) Study the lim inf and the lim sup of

$$m \mapsto \mu_2(m, h)/\mu_1(m, h)$$

as $m \to \infty$, where $\mu_2(m, h)$ is the second eigenvalue of $K^{(m)}(h)$.

(VI) Study the lim inf and the lim sup of

$$m \mapsto (\lambda_2(m, h) - \lambda_1(m, h))$$

as $m \to \infty$, where $\lambda_2(m, h)$ is the second eigenvalue of $S^{(m)}(h)$.

The problem (VI) corresponds with the well known problem of the study of the splitting between the two first eigenvalues.

(VII) A similar problem occurs in geometry. In his article "Open problems in geometry" S.T Yau [Yau93] proposes the following 44th problem. Consider in the manifold M_ϵ defined in $(S^3)^n$ by:

$$d(x_i, x_{i+1}) < \epsilon = \frac{1}{n}$$

the operator

$$- \Delta + c \sum_{i=1}^{n} d(x_i, x_{i+1})^2 \tag{1.2.10}$$

and give a lower bound for the splitting between the two first eigenvalues [1].

1.3 Our main toy models

Our initial motivation comes from questions by M. Brunaud in 1989-1990 whose origin was a course of M. Kac [Ka66] in which he develops partially heuristic ideas in order to prove the existence of phase transition by semiclassical techniques. Another motivation was that J. Sjöstrand found at this time powerful methods in [Sj93a], [Sj93b] and [Sj92b] which permit to control with respect to the dimension a large part of the semiclassical study of the multiple wells problem which we developed in collaboration with him in the 80's in [HeSj84-87] (see also B. Simon [Si83], [Si84] and partially rigorous contribution of V.P. Maslov [Mas84]). The results we shall present here are mainly obtained by J. Sjöstrand, B. Helffer or J. Sjöstrand and B. Helffer in collaboration. They correspond to a class containing the model "Kac" potential (K):

$$V^{(m)}(x;\nu) = \frac{1}{4}\sum_{k=1}^{m}(x_k)^2 - \sum_{k=1}^{m} \ln\cosh\left(\left(\frac{\nu}{2}\right)^{\frac{1}{2}}(x_k + x_{k+1})\right), \tag{1.3.1}$$

with the convention $(x_{m+1} = x_1)$.
But other connected potentials like

$$\Phi(x) = \sum_{j=1}^{p} V^{(m)}(x_j;\nu) + \frac{1}{2}|x_j - x_{j+1}|^2, \tag{1.3.2}$$

with $x_j \in \mathbb{R}^m$, which depend on $n = mp$ variables arise very naturally in the lattice field theory and will be considered.
The Kac model was our initial[2] toy model because we meet in this model all the difficulties of the problem of analyzing phase transition but we can also analyze as toy models other natural models coming from the field theory:

- Model of Dobrokhotov, Kolokoltsov [DoKo94] (DK)

$$V^{(m)}(x) = \frac{a}{2}\sum_{k=1}^{m}(x_k - x_{k+1})^2 + \frac{b}{4}\sum_{k=1}^{m}(x_k - y)^2(x_k + y)^2 \tag{1.3.3}$$

  where $a > 0$, $b > 0$ and $y > 0$.

[1] The author does not indicate which selfadjoint realization is chosen.
[2] in collaboration with J. Sjöstrand,

- More generally, we can consider:

$$V^{(m)}(x) = \frac{a}{2}\sum_{k=1}^{m}(x_k - x_{k+1})^2 + \sum_{k=1}^{m} v(x_k) \tag{1.3.4}$$

where v is a C^∞ potential with one or two symmetric non degenerate wells satisfying the inequality:

$$v(t) \geq \frac{1}{C}t^2 - C$$

for some strictly positive constant C. This model was also considered by C. Fefferman (Theorem 7 in [Fef83]) and is a standard model occuring in the ϕ_2^4-field theory.

- Model of E. Brézin, S. Hikami, A. V. Koudinov, M. A. Smondyrev (BH) (See [BreHi79], [KoSm82], [Sor93a] and [Sor93b])

$$V^{(m)}(x) = \frac{a}{2}\sum_{k=1}^{m} x_k^2 \, + \, m^{1-\ell}\left(\sum_{k=1}^{m} x_k^2\right)^{\ell} \tag{1.3.5}$$

for some $\ell \in \mathbb{N}$, $\ell > 1$.

1.4 Contents of the survey

This survey is (unfortunately?) not completely selfcontained. But I have tried to avoid to transform it in a (surely incomplete) list of results. I have sketched most of the proofs, giving more details when the presented result or point of view was new. For the "standard" results in semiclassical analysis, I refer to my study with J. Sjöstrand starting in 1984 [HeSj84-87] and to more recent pedagogical presentations like [CFKS87], [He88] and [CDV93].
I hope that this survey will be interesting for at least two types of readers:

- the specialist in semiclassical mechanics or quantum mechanics which will find a motivation for applying its techniques with the idea of controling with respect to the dimension,
- the specialist in statistical mechanics or quantum field theory who will find, I hope, a new insight in the theory mainly brought by J. Sjöstrand.

This survey is organized as follows.

- In Section 2, we present different problems coming from statistical mechanics and motivate the study of the problems given in this introduction. This chapter can be considered as an extension of some sections of the beautiful course in Brandeis by M. Kac [Ka66].

- In Section 3, we shall analyze the different problems concerning Laplace integrals: Laplace's method, thermodynamic limit, decay of the correlations. We present here the results of Sjöstrand [Sj93a], [Sj93b], [Sj93c] and [Sj94], Helffer [He94a] and Helffer-Sjöstrand [HeSj93]. This subject is also well developed in probability theory (see for example R. Ellis [El85], G. Ben Arous-M. Brunaud [BeBru90] and Albeverio-Kondratiev-Röckner [AKR94]) but we are not sufficiently expert in this direction to compare deeply the connection between the different approaches. In particular, all the concepts relative to the keywords: Wiener integrals, large deviations, Feynman-Kac formula are surely not sufficiently analyzed in this survey.
  Although we work essentially with Laplace integrals, the results have of course connections at least formally with the stationary phase theorems. We will not analyze this point in detail. We think that the control of the stationary phase expansion is more difficult. We refer to the work of Fujiwara [Fu91] where some control of the remainder term is given with respect to the dimension in the same spirit as here for the Laplace integrals.

- In Section 4, we study the Schrödinger equation in large dimension mainly in the case when the potential V is convex and present universal estimates.

- In Section 5, we present the semiclassical approach to the Schrödinger operator in large dimension. We study in more detail the model of Dobrokhotov-Kolokol'tsov which was not analyzed before in the rigorous literature. This is done by adapting the techniques of [HeSj92a] and [HeSj92a], using more recent improvements of [Sj92b].

- Section 6 is devoted to precize some of the links between the Schrödinger operator, the Laplace integrals and the Kac operator. This will make more precise some of the links observed already in the section devoted to the statistical mechanics.

- Section 7 is devoted to a presentation of the results on the Kac operator. We will be rather short on this part because we write recently a survey on this subject (see [He93b]) but we have tried to include the recent improvements of [He94b], [He94c] and [He94d] and recall older semiclassical results obtained with M. Brunaud [BruHe91] or alone [He92b].

- The last section is devoted to emphasize some of the open problems remaining in this theory. The program of Kac is indeed far to be done and many questions remain which seem to me good challenge to analysts, sometimes independently of its direct application to statistical mechanics.

2 Problems in statistical mechanics

Our aim is here to give strong motivations for the study which will be presented in the next sections. In particular, we shall present various situations in statistical mechanics, where Laplace integrals and spectral properties of matrices or more generally unbounded selfadjoint operators are present.

2.1 The Curie-Weiss model

The Curie-Weiss model is one of the simple models in statistical mechanics which plays an important role in the mean field theory. We consider a N-spins lattice $\sigma = (\sigma_1, \sigma_2, \cdots, \sigma_N)$ with $\sigma_j \in \{-1, +1\}$. In order to take account of the magnetic field, we associate to any σ the energy $E_B(\sigma)$ defined by

$$E_B(\sigma) = -\frac{2J}{N} \sum_{1 \leq i < j \leq N} \sigma_i \cdot \sigma_j - B \sum_{1 \leq i \leq N} \sigma_i \tag{2.1.1}$$

The partition function Z_N is given by

$$\begin{aligned} Z_N(B) &= 2^{-N} \left[\sum_{\sigma \in \{-1,+1\}^N} \exp -E_B(\sigma)/kT \right] \\ &= 2^{-N} \exp -\nu \cdot \left[\sum_{\sigma \in \{-1,+1\}^N} \exp \left(\nu \left(\sum_{i=1}^{N} \frac{\sigma_i}{\sqrt{N}} \right)^2 - (kT)^{-1} B \sum_{i=1}^{N} \sigma_i \right) \right] \end{aligned} \tag{2.1.2}$$

with $\nu = J/kT$.
A simple computation based on the easy identity

$$\exp\, a^2 = \frac{1}{\sqrt{2\pi}} \int_{\mathbb{R}} \exp \left(-\frac{\xi^2}{2} + \sqrt{2} a \xi \right) d\xi \tag{2.1.3}$$

gives the formula

$$Z_N(B) = \exp -\nu \cdot \sqrt{\frac{N}{2\pi}} \cdot \int_{\mathbb{R}} \exp\, N \frac{\eta^2}{2} \left(\cosh \left(\eta \sqrt{2\nu} + \frac{B}{kT} \right) \right)^N d\eta \tag{2.1.4}$$

We rewrite this integral in the form:

$$Z_N(B) = \exp -\nu \cdot \sqrt{\frac{N}{2\pi}} \cdot \int_{\mathbb{R}} \exp\, -N \Phi_{\nu,B}(\eta)\, d\eta \tag{2.1.5}$$

with

$$\Phi_{\nu,B}(\eta) = \frac{\eta^2}{2} - \ln\cosh\left[\eta\sqrt{2\nu} + \frac{B}{kT}\right] . \tag{2.1.6}$$

We are first interested in the so called thermodynamic limit

$$-\lim_{N\to\infty} N^{-1}\ln Z_N$$

and according to formula (2.1.5), this is easily analyzed as a Laplace integral. In particular, an easy computation gives:

$$-\lim_{N\to\infty} N^{-1}\ln Z_N = \inf_{\eta} \Phi_{\nu,B}(\eta) . \tag{2.1.7}$$

More precise information (asymptotic analysis of $N^{-1}\ln Z_N$ with respect to N as $N\to\infty$) can also be obtained by standard methods. We shall recall these results in Subsection 3.2. In particular we consider the minima of the phase $\Phi_{\nu,B}(\eta)$ and observe that the critical points are given by the equation

$$\eta = \sqrt{2\nu}\cdot\tanh\left[\eta\sqrt{2\nu} + \frac{B}{kT}\right] . \tag{2.1.8}$$

In the case $B = 0$, the value $\nu = \frac{1}{2}$ of the parameter appears critical. We observe indeed:

- If $\nu < \frac{1}{2}$, the phase has two symmetric minima (double well situation),
- If $\nu = \frac{1}{2}$, the phase has a unique degenerate minimum,
- If $\nu > \frac{1}{2}$, the phase is strictly convex (single well situation) and has a unique nondegenerate minimum.

2.2 The 1-d Ising model

We consider now the following model

$$H_n(\sigma) = -J\sum_{1\le r\le n}\sigma_r\cdot\sigma_{r+1} - h\sum_{1\le r\le n}\sigma_r \tag{2.2.1}$$

where we identify σ_1 and σ_{1+n}. If one introduces

$$T(\sigma,\sigma') = \exp\,\beta\left(J\sigma\cdot\sigma' + \frac{h}{2}(\sigma+\sigma')\right) , \tag{2.2.2}$$

for σ, $\sigma' = \pm 1$, we see immediately that we can write:

$$\exp -\beta H_n(\sigma) = T(\sigma_1,\sigma_2)\cdot T(\sigma_2,\sigma_3)\cdots T(\sigma_{n-1},\sigma_n)\cdot T(\sigma_n,\sigma_1) \tag{2.2.3}$$

We observe now that $T(\sigma_1, \sigma_1')$ with $\sigma_1, \sigma_1' = \pm 1$ can be considered as a 2×2-matrix

$$T = \begin{pmatrix} \exp\ \beta(J+h) & \exp\ -\beta J \\ \exp\ -\beta J & \exp\ \beta(J-h) \end{pmatrix} \tag{2.2.4}$$

which is usually called transfer matrix and will appear very useful in order to express statistical quantities. Let us first consider the partition function

$$Z_n =_{def} \sum_{\sigma \in \{-1,+1\}^n} \exp\ -\beta H_n(\sigma)\ . \tag{2.2.5}$$

An immediate computation gives

$$Z_n = \sum_{\sigma_1} T^n(\sigma_1, \sigma_1)\ , \tag{2.2.6}$$

or

$$Z_n = \mathrm{Tr}\ T^n = \lambda_1^n + \lambda_2^n \tag{2.2.7}$$

where λ_1 and λ_2 are the two eigenvalues of T

$$\begin{aligned} \lambda_1 &= \exp\ \beta J \cosh(\beta h) + \sqrt{\sinh^2(\beta h) \exp\ 2\beta J\ +\ \exp\ -2\beta J)}\ , \\ \lambda_2 &= \exp\ \beta J \cosh(\beta h) - \sqrt{\sinh^2(\beta h) \exp\ 2\beta J\ +\ \exp\ -2\beta J)}\ . \end{aligned} \tag{2.2.8}$$

We then compute the free energy

$$F_n = -kT \ln\ Z_n = -kT \ln\ (\lambda_1^n + \lambda_2^n) = -nkT \ln\ (\lambda_1) - kT \ln\ (1 + \lambda_2^n/\lambda_1^n)\ . \tag{2.2.9}$$

At the thermodynamic limit when $n \to \infty$, observing that $\lambda_1 > \lambda_2$ we get

$$f = \lim_{n \to \infty} F_n/n = -kT \ln \lambda_1\ . \tag{2.2.10}$$

We observe that the speed of convergence to the thermodynamic limit is immediately related to the comparison of λ_2/λ_1 with 1. More precisely, the strict inequality $\lambda_2/\lambda_1 < 1$ permits to get exponentially rapid convergence in (2.2.10). This will explain why we shall later be interested in estimates on the splitting in more general situations. We shall also compute in this very simple situation the correlation pair

$$\langle \sigma_s \cdot \sigma_{r+s} \rangle_n =_{def} Z_n^{-1} \sum_{\sigma \in \{-1,+1\}^n} \sigma_r \cdot \sigma_{s+r} \exp\ -\beta H_n(\sigma)\ . \tag{2.2.11}$$

An easy computation gives again:

$$\langle \sigma_s \cdot \sigma_{s+r} \rangle_n = Z_n^{-1} \sum_{\sigma, \sigma' = \pm 1} \sigma T^r(\sigma, \sigma') \sigma' T^{n-r}(\sigma', \sigma)\ . \tag{2.2.12}$$

If we now observe that the third Pauli matix

$$\tau_3(\sigma,\sigma') = \sigma\delta_{\sigma,\sigma'}$$

satisfies:

$$(\tau_3 \circ A)(\sigma,\sigma') = \sigma A(\sigma,\sigma') \ ,$$

we can rewrite (2.2.12) in the form

$$\langle \sigma_s \cdot \sigma_{s+r} \rangle_n = Z^{-1}\, \mathrm{Tr}\, (\tau_3 \circ T^r \circ \tau_3 \circ T^{n-r}) \ . \tag{2.2.13}$$

If we now denote by $|1\rangle$ and $|2\rangle$ the eigenvectors corresponding to λ_1 and λ_2, we finally obtain

$$\lim_{n\to\infty} \langle \sigma_s \cdot \sigma_{s+r} \rangle_n = |\langle 1|\tau_3|1\rangle|^2 + (\lambda_2/\lambda_1)^r |\langle 1|\tau_3|2\rangle|^2 \ . \tag{2.2.14}$$

We observe that if $h = 0$, then $|\langle 1|\tau_3|1\rangle| = 0$ and $|\langle 1|\tau_3|2\rangle| \neq 0$. In this case the correlation function decays exponentially with r:

$$\langle \sigma_s \cdot \sigma_{s+r} \rangle_\infty \approx \exp\ -r/\xi \tag{2.2.15}$$

with

$$\xi = (\ln(\lambda_1/\lambda_2))^{-1} \tag{2.2.16}$$

which is called the correlation length. In any case we observe that

$$\langle (\sigma_s - \langle \sigma_s \rangle_\infty) \cdot (\sigma_{s+r} - \langle \sigma_{s+r} \rangle_\infty) \rangle_\infty \approx \exp\ -r/\xi \ .$$

Again we observe that the splitting plays a very important role in the asymptotic behavior of the correlations.

2.3 The 2-d Ising model

Following M. Kac [Ka66], we now review some facts concerning the two-dimensional Ising model. We consider a two-dimensional $n \times m$ lattice of m rows and n columns, with spins $\sigma_P = \pm 1$ occupying the vertices of the lattice. The interaction energy in a given configuration of spins is now given by

$$H_{n,m}(\sigma) = -J \sum^{*} \sigma_P \cdot \sigma_Q \ , \tag{2.3.1}$$

where the star on the sum denotes a sum over nearest-neighbor lattice points P and Q, and we have assumed that the coupling J between nearest neighbors is the same for spins in rows and columns. We assume also that $J > 0$, i.e. that the interaction i The partition function is now:

$$Z_{n,m} = \sum_{\sigma\in\{-1,+1\}^{\{\mathbb{Z}/n\mathbb{Z}\times\mathbb{Z}/m\mathbb{Z}\}}} \exp\ -H_{n,m}(\sigma)/kT = \sum_{(\sigma)} \exp\ \nu \sum^{*} \sigma_P \cdot \sigma_Q \ , \tag{2.3.2}$$

with

$$\nu = J/kT\,. \tag{2.3.3}$$

The free energy per spin ψ in the thermodynamic limit is defined[3] by

$$-\psi/kT = \lim_{m,n\to\infty} (mn)^{-1} \ln Z_{n,m}\,. \tag{2.3.4}$$

In order to evaluate $Z_{n,m}$, Kramers and Wannier relate this quantity to a spectral problem for the $2^m \times 2^m$ transfer matrix $\mathbf{T}$ defined by

$$T(\sigma,\sigma') = \exp\left(\frac{\nu}{2}\sum_{k=1}^{m}\sigma_k\cdot\sigma_{k+1}\right)\cdot\exp\left(\nu\sum_{k=1}^{m}\sigma_k\cdot\sigma'_k\right)\cdot\exp\left(\frac{\nu}{2}\sum_{k=1}^{m}\sigma'_k\cdot\sigma'_{k+1}\right)\,. \tag{2.3.5}$$

It is then a simple matter to get the following expression for the partition function:

$$Z_{n,m} = \mathrm{Tr}\ T^n = \sum_{j=1}^{2^m}\mu_j^n\,, \tag{2.3.6}$$

where $\mu_1 > \mu_2 \geq \mu_3 \cdots \mu_{2^m}$ are the eigenvalues of the matrix T. If we first take the limit as $n \to \infty$ we have immediately

$$\lim_{n\to\infty} n^{-1}\ln Z_{n,m} = \ln\mu_1(m) \tag{2.3.7}$$

and hence from (2.3.4)

$$-\psi/kT = \lim_{m\to\infty} m^{-1}\ln\mu_1(m) \tag{2.3.8}$$

The explicit computation of $\mu_1(m)$ was done by Onsager in 1944. This explicit computation permits to show that as $m \to \infty$ the eigenvalue becomes asymptotically degenerate for T smaller than some critical temperature T_c. More precisely, for $T > T_c$, the spectrum of T is simple (this point is always true due to the Perron-Frobenius Theorem) and the spacing of the $(\ln\mu_j(m))/m$ is of order $1/m$, in particular

$$\ln\mu_1(m) - \ln\mu_2(m) \geq 1/C \tag{2.3.9}$$

and below T_c there are two such spectra with spacings of $\frac{1}{m}\ln\mu_j$ of order $\frac{1}{m}$ but shifted with respect to each other by an amount of order $\exp\ -cm$ for some $c > 0$, in particular:

$$\ln\mu_1 - \ln\mu_2 \leq \frac{1}{c}\exp\ -cm\,, \tag{2.3.10}$$

for some $c > 0$.

The basic idea suggested by M. Kac and illustrated by heuristic computations was to relate these properties to tunneling phenomena observed in quantum mechanics.

[3] A precise definition of this limit is surely needed but we do not enter in the details (See for example [Ru69]). Here we shall choose an order by taking successively the limit $n \to \infty$ and then the limit $m \to \infty$.

2.4 A simple quantum model

We now study another standard example following a course by S.H. Shenker ([Sh82]). Let us consider a simple quantum mechanical system: a two dimensional Hilbert space with observables described by Pauli-matrices. Let the Hamiltonian be

$$H = \frac{1}{\ell}(\sigma_1 - \kappa\sigma_3)\ , \tag{2.4.1}$$

where ℓ is some length, $\kappa \geq 0$ and

$$\sigma_1 = \begin{pmatrix} 0 & 1 \\ 1 & 0 \end{pmatrix}\ ,\ \sigma_3 = \begin{pmatrix} 1 & 0 \\ 0 & -1 \end{pmatrix}\ . \tag{2.4.2}$$

The matrix H has two eigenvalues given by

$$\lambda_1 = -\frac{1}{\ell}\sqrt{1+\kappa^2}\ ,\ \lambda_2 = +\frac{1}{\ell}\sqrt{1+\kappa^2}\ . \tag{2.4.3}$$

The corresponding eigenvectors are:

$$\begin{aligned} |e_1\rangle &= (1, -\sqrt{\kappa^2+1}+\kappa)/\sqrt{2(\kappa^2+1-\kappa\sqrt{\kappa^2+1})} \\ |e_2\rangle &= (1, \sqrt{\kappa^2+1}+\kappa)/\sqrt{2(\kappa^2+1+\kappa\sqrt{\kappa^2+1})} \end{aligned} \tag{2.4.4}$$

Let us now consider the operator exp $-HT$ (which is a well defined matrix). Using the semi-group relation, we divide the time interval T into N segments of width ΔT, with $N\Delta T = T$. We have (inserting a basis of eigenstates of say σ_3 denoted by ± 1)

$$\begin{aligned} &\langle b|\exp\ -HT|a\rangle \\ &= \textstyle\sum_{s_1=\pm 1}\sum_{s_2=\pm 1}\cdots\sum_{s_{N-1}=\pm 1}\langle b|\exp\ -H\Delta T|s_1\rangle\cdot\langle s_1|\exp\ -H\Delta T|s_2\rangle\cdots \\ &\quad\cdots\langle s_{N-1}|\exp\ -H\Delta T|a\rangle\ . \end{aligned} \tag{2.4.5}$$

Note that these sums over intermediate states look like the sums that appear in statistical mechanical partition functions. To further press this analogy define

$$\begin{aligned} \langle b|\exp\ -HT|a\rangle &= \sum_{s_1=\pm 1}\sum_{s_2=\pm 1}\cdots \\ \cdots\sum_{s_{N-1}=\pm 1}&\exp\ -\left(V(b,s_1)+V(s_1,s_2)+\cdots+V(s_{N-2},s_{N-1})+V(s_{N-1},a)\right)\ . \end{aligned} \tag{2.4.6}$$

This really does look like a partition function, that of a 1-d Ising model with nearest neighbor couplings. Here we have by definition:

$$\exp\ -V(s,s') =^{def} \langle s|\exp\ -\Delta T\cdot H\,|s'\rangle\ . \tag{2.4.7}$$

The physicists usually do the approximate computation in the limit ΔT small.

$$\exp\ -\Delta T H \approx \begin{pmatrix} 1-\kappa\frac{\Delta T}{\ell} & \frac{\Delta T}{\ell} \\ \frac{\Delta T}{\ell} & 1+\kappa\frac{\Delta T}{\ell} \end{pmatrix} . \tag{2.4.8}$$

The form for V is then

$$V(s,s') \approx \frac{K}{2}(s-s')^2 + \frac{h}{2}(s+s') . \tag{2.4.9}$$

We find

$$\exp\ -2K = \frac{\Delta T}{\ell}\ ,\ \exp\ h = 1+\kappa\frac{\Delta T}{\ell} . \tag{2.4.10}$$

This representation is called a path integral. We now proceed to identify various parts of this mapping. Consider

$$Z_T = \mathrm{Tr}\ \exp\ -HT .$$

Inserting states we see this is just the 1-d Ising model with periodic boundary conditions. The free energy per site of the system in the infinite volume limit is

$$\begin{aligned} F_T &= \lim_{N\to\infty}(-\frac{1}{N}\ln Z_T) \\ &= \Delta T\cdot\lim_{T\to\infty}\left(-\frac{1}{T}\ln\ \mathrm{Tr}\ \exp\ -HT\right) \\ &= \Delta T\cdot\lim_{T\to\infty}\left(-\frac{1}{T}\ln(\exp\ -\lambda_1 T + \exp\ -\lambda_2 T)\right) \\ &= \Delta T\cdot\lambda_1 . \end{aligned} \tag{2.4.11}$$

So the statistical mechanical free energy maps onto the quantum ground state energy. Correlation functions are a standard diagnostic of statistical mechanical systems. The two spin correlation function[4] is

$$\langle s_0 s_r\rangle_T = \left(\sum_{s_1,\cdots,s_N=\pm1}\exp\left(-\sum_i V(s_i,s_{i+1})\right)s_0\cdot s_r\right)\Big/ Z_T\ , \tag{2.4.12}$$

with $s_{N+1}=s_1$.
What does it mean in quantum mechanics? We can rewrite (2.4.12) as

$$\langle s_0\ s_r\rangle_T = \mathrm{Tr}\ (\sigma_3\cdot\exp\ -H\,r\cdot\sigma_3\cdot\exp\ -H(T-r))\,/Z_T \tag{2.4.13}$$

[4]We shall later take the limit $T\to\infty$.

Inserting eigenstates of H

$$\begin{aligned}\langle s_0\ s_r\rangle_T &= \left(\frac{1}{(\exp\ -\lambda_1 T + \exp\ -\lambda_2 T)}\right) \times \\ &\quad \times \left(\langle e_1|\sigma_3|e_1\rangle|^2 \exp\ -\lambda_1 T + \langle e_2|\sigma_3|e_2\rangle|^2 \exp\ -\lambda_2 T \right. \\ &\quad \left. + \langle e_1|\sigma_3|e_2\rangle|^2 \left[\exp -(\lambda_2 r + \lambda_1 (T-r)) + \exp -(\lambda_1 r + \lambda_2 (T-r))\right]\right) .\end{aligned} \tag{2.4.14}$$

Taking $T \to \infty$ we find

$$\langle s_0\ s_r\rangle_\infty = |\langle e_1|\sigma_3|e_1\rangle|^2 + |\langle e_1|\sigma_3|e_2\rangle|^2 \cdot \exp\ -(\lambda_2 - \lambda_1) r\ . \tag{2.4.15}$$

The correlation function decays to its asymptotic value exponentially. The dimensionless decay length, often called the correlation length ξ is related to the gap between the ground and excited states of H

$$\Delta T \cdot \xi = \frac{1}{\Delta}\ ,\ \Delta = \lambda_2 - \lambda_1\ . \tag{2.4.16}$$

If $\langle e_1|\sigma_3|e_1\rangle \neq 0$ (i.e. $\kappa \neq 0$) then (see (2.4.15))

$$\lim_{r\to\infty} \langle s_0\ s_r\rangle \neq 0\ . \tag{2.4.17}$$

There is long range in the system. On the contrary we observe (compare with (2.2.14)) that

$$\langle s_0\ s_r\rangle - |\langle e_1|\sigma_3|e_1\rangle|^2 \tag{2.4.18}$$

has always exponential decay as $r \to \infty$

2.5 The Kac model A

The model we want to consider is called Model A in Section 7 of [Ka66]. We consider the following Hamiltonian:

$$E_{\Lambda(n,m)}(\sigma) = - \sum_{(P,Q)\in\Lambda(n,m)\times\Lambda(n,m)} v_{P,Q}\ \sigma_P\ \sigma_Q \tag{2.5.1}$$

with

$$\Lambda(n,m) = [1,...,n] \times (\mathbb{Z}/m\mathbb{Z})\ , \sigma_P \in \{-1,+1\}, J \in \mathbb{R}_+^\star\ , \gamma \in \mathbb{R}_+^\star \tag{2.5.2}$$

with

$$v_{P,P} = 0$$

the Schrödinger operator.
Let us now consider the correlation of two spins in a row separated by a distance t:

$$\mathrm{Cor}_{\mathrm{col}}(t) = \lim_{m\to\infty} \mathrm{Cor}_{\mathrm{col}}^{(m)}(t) \tag{2.5.14}$$

where

$$\mathrm{Cor}_{\mathrm{col}}^{(m)}(t) = \lim_{n\to\infty} \langle \sigma_{k,\ell} \cdot \sigma_{(k,\ell+t)} \rangle_{n,m}. \tag{2.5.15}$$

The computation of $\mathrm{Cor}_{\mathrm{col}}^{(m)}(t)$ (cf. formula (7.36) in [Ka66]) gives:

$$\mathrm{Cor}_{\mathrm{col}}^{(m)}(t) = \int_{\mathbb{R}^m} u_1^{(m)}(x)^2 . \tanh(\sqrt{\nu\gamma/2}(x_1 + x_2)). \tanh(\sqrt{\nu\gamma/2}(x_{1+t} + x_{2+t}))dx. \tag{2.5.16}$$

This appears as the mean value of the function:

$$x \to \tanh(\sqrt{\nu\gamma/2}(x_1 + x_2)) \cdot \tanh(\sqrt{\nu\gamma/2}(x_{1+t} + x_{2+t}))$$

with respect to the probability measure $u_1^{(m)}(x)^2\, dx$.

2.6 The multiexponential model

This model corresponds to the following potential on $\mathbb{R}^m$

$$V^{(m)}(x) = \frac{1}{4}\sum_{k=1}^{m} \sigma_{km}^2 x_k^2 - \ln\cosh\left(\sqrt{\nu}\sum_k \sqrt{a_{km}\sigma_{km}} \cdot x_k\right), \tag{2.6.1}$$

where as typical example

$$\sigma_{km} = k/m, k = 1, ..., m \tag{2.6.2}$$

and

$$a_{km} = (1/m) \cdot (k/m)^a , \tag{2.6.3}$$

for a strictly positive real a. The computation of the critical points gives

$$\frac{1}{2}\sigma_{km}^2 = \sqrt{\nu} \cdot \sqrt{a_{km}\sigma_{km}} \tanh\left(\sqrt{\nu}\sum_k \sqrt{a_{km}\sigma_{km}}\right) \tag{2.6.4}$$

Let us introduce

$$\alpha_{0m} = \sum_k \frac{a_{km}}{\sigma_{km}} \tag{2.6.5}$$

and

$$\tau_m(x) = \sum_k \sqrt{a_k \sigma_k}\, x_k \,. \tag{2.6.6}$$

A necessary condition for having a critical point is:

$$\tau_m(x) = 2\sqrt{\nu} \cdot \alpha_{0m} \cdot \tanh(\sqrt{\nu} \cdot \tau_m(x)) \tag{2.6.7}$$

This last equation determines a critical value for ν

$$\nu_c(m) = \frac{1}{2}\alpha_{0m} \tag{2.6.8}$$

which will play an important role in the discussion.

- For $\nu > \nu_c(m)$, there are three solutions for the equation

  $$\tau = 2\sqrt{\nu} \cdot \alpha_{0m} \cdot \tanh(\sqrt{\nu} \cdot \tau) \tag{2.6.9}$$

  which we denote by

  $$\pm\,\tau^+(m)\;,\;0 \tag{2.6.10}$$

  and this corresponds to two minima and a maximum for $V^{(m)}$. We note for later that $\tau^+(m)$ is actually a function of ν and α_{0m}. In particular, if α_{0m} tends to a limit $\alpha_{0\infty}$ as $m \to \infty$, $\tau^+(m)$ tends to a limit $\tau^+(\infty)$.

- For $\nu < \nu_c(m)$, 0 corresponds to a unique minimum.

Let us analyze the situation more carefully. We first remark that if x is a critical value then we have

$$x_{km} = 2a_{km}^{1/2} \cdot \sigma_{km}^{-3/2} \cdot \sqrt{\nu} \tanh(\sqrt{\nu}\tau_m(x)) \,. \tag{2.6.11}$$

The minimum is equal to

$$\min V^{(m)} = \nu\alpha_{0m} \tanh^2(\sqrt{\nu} \cdot \tau^+(m)) - \ln\cosh(\sqrt{\nu} \cdot \tau^+(m)) \tag{2.6.12}$$

Let us also observe that

$$\nu_c(\infty) = \lim_{m\to\infty} \nu_c(m) = 1/(2\alpha_{0\infty}) = a/2 \tag{2.6.13}$$

This model was introduced by M. Kac [Ka66] in order to approximate a one dimensional spin model whose interaction was given by

$$\gamma\rho(\gamma|i-j|) = \gamma \int_0^1 t^a \exp(-t\gamma|i-j|)\, dt \,. \tag{2.6.14}$$

In order to understand why a transition of phase should exist for $0 < a < 1$ and not in the case $a > 1$, condition which corresponds to the finiteness of $\int_0^\infty t\rho(t)dt$, M. Kac approximates the interaction (2.6.14) by taking the corresponding Riemann sum

$$\rho_m(t) = \frac{1}{m}\sum_{k=1}^{m}(\frac{k}{m})^a \cdot \exp -\frac{k}{m}|t|$$

and by solving explicitely this new approximate model whose interaction energy is given by

$$E_{n,m} = -J\gamma \sum_{1\leq i<j\leq n} \rho_m(\gamma|i-j|)\sigma_i \cdot \sigma_j$$

he finds the formulas we introduced above by taking first the limit $n \to \infty$. Interesting computations using fine analysis are given by M. Kac in [Ka66] Section 6 but a rigorous analysis has not been completed. In particular one can compute in the case where 0 is the unique minimum of the potential the harmonic approximation at the origin. The different modes of this quadratic approximation are obtained by considering the m solutions ω_{sm} $(s = 1, \cdots, m)$ of the equation

$$\frac{1}{2\nu} = \sum_{k=1}^{m} a_{km}\sigma_{km}/(\sigma_{km}^2 - \omega)\ . \tag{2.6.15}$$

The condition that the smallest solution ω_{1m} is strictly positive corresponds exactly to the condition that $\nu < \nu_c(m)$. Moreover the potential appears in this case to be convex

$$\text{Hess}\, V^{(m)}(x) \geq \frac{1}{2}\omega_{1m}. \tag{2.6.16}$$

This last inequality will be important to measure the splitting between the two first eigenvalues.

2.7 Comments

In all these models we have met the different problems which we mentioned in the introduction through various ways. For two models, it appears a semiclassical problem associated with the introduction of a semiclassical parameter γ. We emphasize that the semiclassical limit

$$\gamma \to 0$$

is considered **after** the thermodynamic limit

$$m \text{ or } n \to \infty$$

and this is actually the crucial difficulty. Anyway the semiclassical study for fixed m can be considered as a first step to understand the situation.

3 Laplace integrals

This section is devoted to the study of statistical quantities associated with a family (indexed by the dimension m) of measures

$$\exp -\frac{1}{h}\Phi^{(m)}(x)dx^{(m)}$$

on $\mathbb{R}^m$, where $\Phi^{(m)}(x)$ is a phase (or potential) and h is as possibly small parameter. Although the case when the space $\mathbb{R}^m$ is replaced by a more general measure space and the corresponding measure by a more general one (this permits in some case to treat simultaneously spin systems and classical integrals) is quite interesting, the techniques we shall present in this survey are adapted only to this special form and we shall limit our presentation to this situation (or to cases leading to this situation as it is the case in the next subsection). We refer to the book of Ellis [El85] for a more general point of view.

3.1 From spin models to Laplace integrals

This subsection is in some sense a transition. We follow here the presentation given in [ItzDr89] and [He94a]. As a consequence we shall obtain expansions in the large temperature regime, but other techniques (where one looks directly to the spin systems) as it said in Chapter 4 in [Ru69] seem to give the way to obtain a convergent expansion in power of β for the thermodynamic limit (cf. also [GaMiRo67]). We will only get a weaker result by justifying the approach which follows. Let us look at:

$$Z_n(\beta) = \sum_{\sigma\in\{-1,+1\}^n} \exp\left(\frac{\beta}{2}\sum_{\substack{i=1,\cdots,n\\ j=1,\cdots,n}} \sigma_i\cdot J_{ij}\cdot\sigma_j\right) \tag{3.1.1}$$

and at its behavior as $\beta \to 0$ with control with respect to n (we are actually working with a family of matrices $J^{(n)}$, $n \in \mathbb{N}$). We use the Laplace transform and get:

$$\begin{aligned} & Z_n(\beta) && (3.1.2)\\ = \ & (2\pi)^{-n/2}(\det\beta J)^{-1/2}\int_{\mathbb{R}^n} \exp\left(-\frac{1}{2\beta}\sum_{ij} x_i\cdot(J^{-1})_{ij}\cdot x_j + \sum_i \ln\cosh x_i\right) dx. \end{aligned}$$

This is only true if the $n\times n$ matrix J_{ij} is positive definite (and consequently its inverse). If this property is not satisfied, one can use the following classical trick (see formula (2.1.3)). We rewrite:

$$\begin{aligned} &\exp\left(\tfrac{\beta}{2}\sum\ {i,j}\ =\ 1,\cdots,n\ \ \sigma_i\cdot J_{ij}\cdot\sigma_j\right)\\ &\quad = \exp\left(\tfrac{\beta}{2}\sum\ {i,j}\ = 1,\cdots,n\ \ \sigma_i\cdot \tilde{J}_{ij}\cdot\sigma_j\right)\cdot\exp(-\beta\lambda n/2) \end{aligned} \tag{3.1.3}$$

with

$$\tilde{J}_{ij} = J_{ij} + \lambda\delta_{ij} \ .$$

If one knows for example that $\sup_j \sum_i |J_{ij}|$ is finite (as it is the case when for example $J_{ij} = J(i-j)$ with J in $\ell^1(\mathbb{Z})$, J symmetric), the existence of λ is evident. But we always put this type of condition in order to obtain the existence of the thermodymamic limit.
In the case of the exponential interaction, we have

$$J_{ij} = -\tau \exp -\tau|i-j|$$

and its inverse is given (see [Ka66] p.254) by

$$\sum_{ij}(J_\tau^{-1})_{ij}\xi_i \cdot \xi_j = (1/\tau)\left(\xi_1^2 + \left(\sum_{j=1}^{n-1}(\xi_{j+1} - \xi_j \exp -\tau)^2/(1-\exp -2\tau)\right)\right) \ .$$

In this last case, we do not need the trick. In the Ising case, we have

$$J_{ij} = \delta_{i,j\pm 1} \ .$$

The eigenvalues are symmetric around 0 and we need the trick. We observe also that this choice of λ has to be independent of n if we want to take later the thermodynamic limit $n \to \infty$. Hence we have to calculate

$$\begin{aligned} Z_n(\beta) &= \exp(-(\beta\lambda n/2).(2\pi)^{-n/2}(\det \beta\tilde{J}^{(n)})^{-1/2}\times \\ &\quad \times \textstyle\int_{\mathbb{R}^n} \exp\{(-1/2\beta)\sum_{ij} x_i((\tilde{J}^{(n)})^{-1})_{ij}x_j + \sum_i \ln\cosh x_i\}dx \end{aligned}$$

with

$$\tilde{J}_{ij} = J_{ij} + \lambda\delta_{ij}$$

We find again a Laplace integral and we shall see later that it enters inside the scope of the general theory of Sjöstrand [Sj93a]. We possibly have to play with λ in order to apply the theory. In this direction, we remark that

$$\tilde{J}_{ij} = J_{ij} + \lambda\delta_{ij} = \lambda(I + (J/\lambda))$$

and that consequently

$$\tilde{J}_{ij}^{-1} = \lambda^{-1}(I + (J/\lambda))^{-1} \ .$$

If we introduce $h = \beta\lambda$, we then have to analyze the behavior as h tends to 0 of:

$$\begin{aligned} Z_n(\beta) &= \exp(-(\beta\lambda n/2) \cdot (2\pi h)^{-n/2}(\det A^{(n)})^{1/2} \times \\ &\quad \times \int_{\mathbb{R}^n} \exp\left\{-\frac{1}{2h}\sum_{ij} x_i \cdot A_{ij}^{(n)} \cdot x_j + \sum_i \ln\cosh x_i\right\} dx \ , \end{aligned}$$

with

$$A^{(n)} = (I + (J^{(n)}/\lambda))^{-1} .$$

When λ is large enough, $||J^{(n)}/\lambda||_{\mathcal{L}(\ell^2)}$ can be made as small [5] as we need.

Remark 3.1.1 Although the standard Laplace method can be generally applied (we shall recall the main statement in the next subsection), it is more difficult to control the dependence on n of the coefficients of the expansion in h. In some cases the result will be only obtained under an additional condition $n \leq h^{-k_0}$.
But sometimes (see [HeSj92a] or [He94a]), we can use the existence of the thermodynamic limit in order to eliminate this type of conditions. The theory will give then the existence of an expansion in powers of h with coefficients controlled with respect to n. We can hope that the convergence of the coefficients (divided by n) as n tends to ∞ is sufficiently rapid and hope a justification of the interversion of the limits. In the domain of the possible λ, the coefficients have to be independent of λ (and we may have to play with the choice of λ in order to improve the results). But at least at the formal level we have just to compute for $\lambda = 0$ in order to get the formal results. This is what is described in [ItzDr89]. Our approach seems to give a good way to justify these computations (at least under strong assumptions on the interaction).

Remark 3.1.2 In some cases results concerning our continuous Laplace integrals can be deduced from the results on spin systems. This is for example in the case when

$$\Phi^{(m)}(x) = \sum_{ij} J_{ij} x_i \cdot x_j + a \sum_i x_i^2 .$$

This is presented in [SiGr73] (See [Si74] Chapter IX and also Remark 2, p.79 in [Si79]). Conversely some continuous results can have (through a limit procedure) application to spin systems. For example it is some times useful to observe properties like:
The normalized measure associated to $\exp -a^2(x^2-1)^2\, dx$ tends to

$$\frac{1}{2}(\delta(x-1) + \delta(x+1))$$

as $a \to \infty$.

3.2 Classical Laplace methods

Before to enter in more general questions it is perhaps better to recall very simple results concerning Laplace integrals. The first one is modelled on the stationary phase theorem:

Theorem 3.2.1 *Let Φ a real C^∞ phase defined in a neighborhood $\mathcal{V}$ of the closure of the ball $B(0,1)$ in $\mathbb{R}^m$ such that*

[5] Note here that the choice of the norm which is not important if we do not want to control with respect to n can be very important if we want to pass to the thermodynamic limit.

- *$\Phi \geq 0$ on $B(0,1)$; $\Phi > 0$ on $\partial B(0,1)$,*
- *$\Phi(0) = \nabla\Phi(0) = 0$,*
- *Φ has a unique non degenerate minimum at 0.*

Let a be a C^∞ phase defined in $\mathcal{V}$ and let us consider the Laplace integral

$$I(a,\Phi;h) = \int_{B(0,1)} a(x)\ \exp -\Phi(x)/h\ dx\ ,$$

where $h \in]0, h_0]$. Then as h tends to 0, $I(a,\Phi;h)$ has the following asymptotic behavior

$$I(a,\Phi;h) \sim h^{\frac{m}{2}} \sum_j \alpha_j \cdot h^j\ , \tag{3.2.1}$$

with

$$\alpha_0 = (2\pi)^{\frac{m}{2}} \cdot a(0) \cdot (\det \mathrm{Hess}\,\Phi(0))^{-\frac{1}{2}}\ . \tag{3.2.2}$$

The proof is rather simple (and actually simpler than for an oscillatory integral). The assumptions permit to reduce modulo exponentially small contributions (in $\mathcal{O}(\exp -\epsilon_0/h)$) to an arbitrarily small neighborhood of 0. We can then use the Morse Lemma in order to write in new coordinates: $\Phi(x) = \sum y_j^2 =: \hat{\Phi}(y)$. We are then reduced to the study of $I(b, \hat{\Phi}; h)$ which is easy by taking the Taylor expansion of b at 0. We are then reduced to the computation of very explicit integrals associated with gaussian measures on $\mathbb{R}^m$. One of our basic problems in this section will be to control all this procedure with respect to the dimension. There is also an analog of the above theorem for a phase with more than one minimum and under suitable assumptions at ∞ one can also consider Laplace integrals over $\mathbb{R}^m$. The assumption that Φ is real can be made weaker if one assumes holomorphy of a and Φ and we then enter in the world of the saddle point method (see J. Sjöstrand [Sj82] Chapter 2 for a sophisticated and general presentation of the method).
We shall also have to discuss problems with parameter when the phase is deformed from a convex phase into a double well. The most simple case of such a situation (see for example [He92c]) is the phase defined on $\mathbb{R}$

$$x \mapsto \phi(x,\theta) = (1-\theta)x^2 + x^4\ ,$$

and, for $a \in C_0^\infty(\mathbb{R})$, one can consider the integral $I(a, \phi(\cdot,\theta); h)$ over $\mathbb{R}$ and it is interesting to follow the behavior of this integral with respect to $\theta \in]0, \infty[$. For fixed θ, we can distinguish three cases

- For $\theta = 1$, $\quad I(\theta;h) \sim a(0)\, h^{\frac{1}{4}} \left(\int_{\mathbb{R}} \exp -x^4\, dx + \mathcal{O}(h^{\frac{1}{2}})\right)$
- For $\theta < 1$, $\quad I(\theta;h) \sim \pi^{\frac{1}{2}}\, a(0)\, h^{\frac{1}{2}}(1 + \mathcal{O}(h))$,

- For $\theta > 1$, $I(\theta;h)$
$\sim h^{\frac{1}{2}}\cdot(2(\theta-1))^{-\frac{1}{2}}\cdot\exp\frac{(\theta-1)^2}{4h}\times\Big(a(+((\theta-1)/2)^{\frac{1}{2}})+a(-((\theta-1)/2)^{\frac{1}{2}})+\mathcal{O}(h)\Big).$

As $\theta \sim 1$, one has to be more precise in order to get a behavior in h uniformly with respect to θ. This can be done by using as a reference $J(\alpha) = \int_{\mathbb{R}} \exp -(\alpha x^2 + x^4)\, dx$ where α is a new parameter $\alpha = h^{-\frac{1}{2}}(\theta - 1)$. We observe that

- For $\alpha \to \infty$, $J(\alpha) \sim \alpha^{-\frac{1}{2}}$,
- For $\alpha = 0$, $J(0) = \int_{\mathbb{R}} \exp -x^4 dx$,
- For $\alpha = -\infty$, $J(\alpha) \sim \sqrt{2\pi}\cdot \exp\frac{\alpha^2}{4}\cdot\alpha^{-\frac{1}{2}}$.

It is then possible to prove that

$$I(\theta;h) = \frac{1}{2}h^{\frac{1}{4}}\cdot J(\alpha)\cdot\Big(a(+((\theta-1)_+/2)^{\frac{1}{2}})+a(-((\theta-1)_+/2)^{\frac{1}{2}})+\mathcal{O}(h^{\frac{1}{2}})\Big)\ .$$

In a very simple situation this explains what is going on near the transition of phase. We refer also to the survey of J.Zinn-Justin [Z-J84] (Section 1) which presents similar examples in connection with the instanton method.

3.3 Laplace integrals in large dimension: the thermodynamic limit

In this section we shall analyze more precisely the existence of the thermodynamic limit and control the speed of convergence of the limit in the spirit of Sjöstrand [Sj92a].

Thermodynamic limit

Let us first give rather general results on the existence of the thermodynamic limit. The first proposition is rather standard (see for example [Ru69]) but permits to understand what is implied in the proof of the existence of the thermodynamic limit.

Proposition 3.3.1 *Let C be positive constant and let us consider a family of C^∞ real potentials, $\mathbb{R}^m \ni x \mapsto \Phi^{(m)}(x)$, indexed by $m \in \mathbb{N}$ such that*

$$\int_{\mathbb{R}^m} \exp -\Phi^{(m)}(x)\, dx^{(m)} < \infty \tag{3.3.1}$$

and

$$\left|\left((\Phi^{(m+n)} - (\Phi^{(m)}\tilde{\oplus}\Phi^{(n)}))\right)(x)\right| \leq C\ , \tag{3.3.2}$$

Then the thermodynamic limit

$$\lambda_\infty =: -\lim_{m\to\infty}\left(\ln\int_{\mathbb{R}^m}\exp -\Phi^{(m)}(x)\, dx^{(m)}\right)/m$$

exists and the speed of the convergence to the limit is controlled in the following manner

$$\left|\lambda_\infty + \left(\ln \int_{\mathbb{R}^m} \exp -\frac{1}{h}\,\Phi^{(m)}(x)\,dx^{(m)}\right)/m\right| \leq C/(mh)\;. \tag{3.3.3}$$

Here we have used the notation $x = (x', x'') \in \mathbb{R}^{n+m}$ and

$$(\Phi^{(m)}\tilde{\oplus}\Phi^{(n)})(x) = \Phi^{(m)}(x') + \Phi^{(n)}(x")\;.$$

Proof of Proposition 3.3.1. In order to get the existence of the thermodynamic limit and (3.3.10), it is sufficient to prove a result of "approximative additivity". It is sufficient (exercise!) to prove that

$$|a(m+n) - a(m) - a(n)| < D\;, \tag{3.3.4}$$

for some D where we recall that:

$$a(m) = -\ln\left(\int \exp(-\Phi^{(m)}(x)/h)dx\right)\;.$$

But we have immediately the inequality

$$\exp -C/h \leq \frac{\int_{\mathbb{R}^{(m+n)}} \exp -\frac{1}{h}\,\Phi^{(m+n)}(x)\,dx^{(m+n)}}{\int_{\mathbb{R}^m} \exp -\frac{1}{h}\,\Phi^{(m)}(x)\,dx^{(m)} \cdot \int_{\mathbb{R}^n} \exp -\frac{1}{h}\,\Phi^{(n)}(x)\,dx^{(n)}} \leq \exp C/h\;.$$

This gives (3.3.4) and (3.3.2) is a corollary. ∎
This simple criterion is not sufficient for the examples we are interested in because it is assumed that the interaction is bounded. This is also the case for other examples given in [Ru69] or [Sj92b]. The next theorem treats examples where the interaction can be unbounded.

Proposition 3.3.2 *Let C_1, C_2, C_3 be positive constants and let us consider the following family of C^2 real potentials on $\mathbb{R}^m$, indexed by $m \in \mathbb{N}$,*

$$\Phi^{(m)}(x) = \frac{x^2}{2} + \Psi^{(m)}(x) \tag{3.3.5}$$

satisfying the following properties,

$$\Psi^{(m)}(0) = 0\;, \tag{3.3.6}$$

$$|\nabla\Psi^{(m)}(x)|_{\ell^\infty} \leq C_1\;, \tag{3.3.7}$$

$$\left|\left(\nabla\left(\Psi^{(m+n)} - (\Psi^{(m)}\tilde{\oplus}\Psi^{(n)})\right)\right)(x)\right|_{\ell^1} \leq C_2\;, \tag{3.3.8}$$

$$\left|\left(\Delta(\Psi^{(m+n)} - (\Psi^{(m)}\tilde{\oplus}\Psi^{(n)}))\right)(x)\right| \leq C_3 \; . \tag{3.3.9}$$

Then the thermodynamic limit

$$\lambda_\infty =: - \lim_{m\to\infty} \left(\ln \int_{\mathbb{R}^m} \exp -\frac{1}{h}\, \Phi^{(m)}(x)\, dx^{(m)}\right) /m \; ,$$

exists and the speed of the convergence to the limit is controlled in the following manner

$$\left|\lambda_\infty + \left(\ln \int_{\mathbb{R}^m} \exp -\Phi^{(m)}(x)\, dx^{(m)}\right) /m\right| \leq D/m \; , \tag{3.3.10}$$

with $D = C_3 \; + \; C_1 \cdot C_2/h$ *.*

Corollary 3.3.3 Let W be a C^2-function on $\mathbb{R}^2$ such that ∇W and ΔW are bounded. Then the assumptions of Proposition 3.3.2 are satisfied for

$$\Psi^{(m)}(x) = \Sigma_{i=1}^m \; W(x_i, x_{i+1}) \tag{3.3.11}$$

with the convention that $m+1 = 1$.

This corollary was obtained in [HeSj92a] or [HeSj93] but similar results are of course older (see for example [Ru69]). It can also be obtained by the technique of the transfer operator (See Section 6). The thermodynamic limit will then appear as the largest eigenvalue (assume $W(s,t) = W(t,s)$) of the operator $L^2(\mathbb{R}) \ni u \mapsto Ku$ where K is the compact operator with distribution kernel

$$K(s,t) = \exp -\frac{t^2}{4} \cdot \exp -W(t,s) \cdot \exp -\frac{s^2}{4} \; .$$

Example 3.3.4 The model of Kac.

$$W(u,v) = -\ln\cosh(\sqrt{\nu}(u+v))$$

This example was one of the motivating examples for the recent studies in [Sj91], [Sj93a], [Sj93b], [Sj92a], [HeSj92a], [HeSj92b] and comes from M. Kac [Ka66]. The corresponding family of potentials is convex for $\nu \leq \frac{1}{4}$.

Proof of Proposition 3.3.2. As before it is sufficient to prove approximate additivity. We introduce the interpolating family:

$$\Phi^{(m,n)}(x,s) = (1-s)\Phi^{(m+n)}(x) + s(\Phi^{(m)}(x') + \Phi^{(n)}(x")) \tag{3.3.12}$$

with $x = (x', x") \in \mathbb{R}^{n+m}$; $s \in [0,1]$ and the corresponding $a(m,n;s)$. We observe that

$$a(m,n;1) = a(m) + a(n) \; ; \; a(m,n;0) = a(m+n) \; .$$

Remark 3.3.8 The model of Dobrokhotov-Kolokoltsov. It is possible to replace in (3.3.5) the quadratic term x^2 by more general terms

$$\Phi^{(m)}(x) = \sum_{j=1}^{m} v(x_j) + \Psi^{(m)}(x) ,$$

where v is a C^∞ even potential on $\mathbb{R}$ such that $|v'(t)| \geq \frac{1}{C}|t| - C$. The important property seems here that

$$\exp v(t) \int_t^{+\infty} u \exp -v(u)/h \, du \leq C \cdot h \ , \quad \text{for } t \geq 0 \ .$$

This permits in particular to cover the case when $v(t) = (t^2 - 1)^2$.

Remark 3.3.9 The models in Ben Arous-Brunaud. Other classical examples appear in [BeBru90](Section I) where the potential has the more general form

$$\phi^{(m)}(x) = \sum_{j=1}^{m} v(x_j) + \sum_{k=2}^{r} m^{-(k-1)} \sum_{1 \leq i_1, \cdots, i_k \leq m} V_k(x_{i_1}, \cdots, x_{i_k}) \tag{3.3.18}$$

where

- $\exp -v(t)dt$ is a probability measure on $\mathbb{R}$,
- r is an integer ≥ 2,
- V_k is a continuous bounded symmetric function from $\mathbb{R}^k$ into $\mathbb{R}$ for $k \in \{2, \cdots, r\}$.

These models correspond to extensions of the model of Curie-Weiss.

Exponential convergence

In order to get more precise results where we have exponential convergence, we introduce first a new parameter t and consider the family of potentials:

$$\Phi^{(m)}(x,t) = \frac{x^2}{2} + t\Psi^m(x) \ , \ t \in [0,1]$$

and we are interested in the behavior with respect to m of the quantity

$$a(m,t) = -\ln \left(\int \exp -\Phi^{(m)}(x,t)/h \, dx \right)$$

for $t = 1$.
Because $a(m, 0)$ is well known and we are interested in $a(m, 1)$ it is natural to study the derivative with respect to t $a'(m, t)$:

$$a'(m,t) = -\frac{1}{h}\left(\int_{\mathbb{R}^m} (\partial_t \Phi^{(m)}(x,t)) \exp -\Phi^{(m)}(x,t)/h \, dx\right), \tag{3.3.19}$$

and we get the information on $a(m, 1)$ by integration over $[0, 1]$ with respect to t. We are then reduced to the proof that $a'(m, t)/m$ converges exponentially rapidly with respect to m to a "thermodynamic limit" and uniformly with respect to t. We forget now an explicit reference to m and introduce

$$b(t) = a'(m,t) .$$

Following an idea of [Sj92b] as developed in [HeSj93] that we follow here, we are looking for a vector field $v(x, t)$ with temperate coefficients such that:

$$(\partial_t \Phi)(x,t) = b(t) + v(x,t) \cdot \nabla_x \Phi(x,t) - \operatorname{div}_x v(x,t), \tag{3.3.20}$$

and observe that if Φ and v satisfy this equation, then

$$b = -a'(m,t)/h .$$

If we find such a vector field $v(x, t)$, then we get an expression for $b(t)$ by observing that at the (unique) critical point $x(t)$ of Φ we have:

$$b(t) = (\partial_t \Phi)(x(t),t) - (\operatorname{div}_x v)(x(t),t) \tag{3.3.21}$$

Good estimates on $\operatorname{div}_x v$ and on $(\partial_t \Phi)(x(t), t)$ will give good estimates on $b(t)$. If we define:

$$g(x,t) = (\partial_t \Phi)(x,t) ,$$

and differentiate the equation (3.3.20) with respect to x, we get the "basic" equation

$$\nabla_x g(x,t) = (-\Delta + (\nabla_x \Phi)(x,t).\nabla_x) v(x,t) + \operatorname{Hess}_x \Phi(x,t) v(x,t) \tag{3.3.22}$$

(assuming v to be a gradient) which was studied in [HeSj93]. We shall recall some of the results in the next subsection. Let us also observe that in our example:

$$(\nabla_x g)(x,t) = (\nabla_x \Psi)(x) ,$$

and that we shall assume in Theorem 3.3.11 that $\nabla \Psi^{(m)}(0) = 0$ which implies using the convexity of Φ:

$$x(t) = 0.$$

We then get in this case $b(t)$ by the formula:

$$b(t) = -(\operatorname{div}_x v)(t,0) .$$

We are looking at the "basic" equation (see (3.3.22))

$$\begin{aligned}\nabla_x \Psi^{(m,n)}(x,s_2) &= \nabla_x g^{(m,n)}(x,s_2)\\ &= \left(-\Delta + \nabla_x \Phi^{(m,n)} \cdot \nabla_x\right) v^{(m,n)}(x,s_1,s_2) \\ &\quad + \mathrm{Hess}_x \Phi^{(m,n)} v^{(m,n)}(x,s_1,s_2)\ ,\end{aligned} \tag{3.3.41}$$

with

$$v^{(m,n)}(x,s_1,s_2) = \nabla_x f^{(m,n)}(x,s_1,s_2)\ .$$

The basic idea developed in [Sj92b], [HeSj92a] and [HeSj93] was to study the basic equation

$$w = (-\Delta + \nabla\Phi \cdot \nabla)v + \mathrm{Hess}\,\Phi\ v \tag{3.3.42}$$

in different weighted spaces and to use the Maximum Principle. Later J. Sjöstrand improved the technique in [Sj93c], [Sj93d] and [Sj94] making in particular a natural connection with the Witten's Laplacian ([Wi82] and [HeSj84-87], Part IV). We shall present some of these techniques in the next subsection. We define suitable weights in the following manner. The class $\mathcal{R}$ of weights is the class of constant multiple of weights satisfying (3.3.29) and (3.3.30), but in this application we shall only use: $\rho_{m,n} = 1$. But we shall associate to the two extra parameters (s_1, s_2) the following family of weights

$$\hat{\rho}_1 = 1\ ; \tag{3.3.43}$$

and

$$\begin{aligned}&\hat{\rho}_2(j) = \hat{\rho}_{m,n}(j) = 1 && \text{for } m \leq j \leq m+n\\ &\hat{\rho}_2(j) = \hat{\rho}_{m,n}(j) = \exp \kappa \inf(j-1, m-j) && \text{for } 1 \leq j \leq m\ .\end{aligned} \tag{3.3.44}$$

Now according to the assumption (3.3.24) and due to the uniqueness of the solution v, we get (with $t = s_1$)

$$\mathrm{div}\, v^{(m)}(0,t)/m = (\partial_i v_i^{(m)})(0,t)\ ;\ \forall i = 1, \ldots\ldots, m\ . \tag{3.3.45}$$

This permits in particular, in order to estimate

$$b(m,t) = (\partial_i v_i^{(m)})(0,t)\ ,$$

to choose i depending on m.
In order to get a control of the convergence, we shall bound $(b(m+n,t) - b(m,t))$ observing that

$$b(m+n,s_1) - b(m,s_1) = \int_0^1 (\partial^2_{i,s_2} v_i^{(m,n)})(0,s_1,s_2)\, ds_2$$

for any $i \in \mathbb{N}$ s.t. $1 \leq i \leq m$.
One can then obtain[6] , as a consequence of a similar property for $w = \nabla_x \Psi^{(m,n)}$, the following inequality for $v^{(m,n)}$:

$$|(\partial^2_{i,s_2} v_i^{(m,n)})(0, s_1, s_2)| \leq$$

$$\leq (\exp -\kappa \inf(i, m-i)) \left(\sup_{(x,s_1,s_2)} |(\partial^2_{i,s_2} v_i^{(m,n)}))(x, s_1, s_2)|_{\mathcal{L}(\ell^\infty, \ell^\infty_{\tilde{\rho}_2})} \right).$$

The choice of $i = [m/2]$ gives finally:

$$|(b(m+n,t)/(m+n)) - (b(m,t)/m)| \leq D \exp -(\kappa m/2). \tag{3.3.46}$$

This proves that $(b(m,t)/m)$ is a Cauchy sequence (uniformly with respect to t) and taking the limit $n \to \infty$ in (3.3.46), we get finally:

$$\left| \left(\lim_{p \to \infty} b(p,t)/p \right) - (b(m,t)/m) \right| \leq D \exp -(\kappa m/2). \tag{3.3.47}$$

∎

Remark 3.3.13 Theorem 3.3.11 is valid without assumption (3.3.23). It is indeed possible to control the terms

$$\partial_t \Phi^{(m)}(x^{(m)}(t), t)/m = \Psi^{(m)}(x^{(m)}(t))/m$$

and

$$(\mathrm{div}_x v^{(m)})(x^{(m)}(t), t)/m .$$

This is analyzed in detail in [HeSj93].

3.4 Around the Maximum Principle

We follow here the recent paper by J. Sjöstrand [Sj93b] where a generalization of arguments previously given in [Sj93a] and [HeSj93] is presented. Our main object is the study of regularity properties of the "basic equation" which was the essential point in the proof sketched above as well as in the next subsection. Let us start from

$$w = (-\Delta + \nabla\Phi \cdot \nabla)v + \mathrm{Hess}\,\Phi\; v \tag{3.4.1}$$

where $v \in C^1_b(\mathbb{R}^m; \mathbb{R}^m)$ and $w \in C^0_b(\mathbb{R}^m; \mathbb{R}^m)$. We recall that this equation appears naturally by the derivation of the scalar equation

$$f = (-\Delta + \nabla\Phi \cdot \nabla)u + \text{ const.} \tag{3.4.2}$$

[6] This is a little more sophisticated version of what we shall explain in the next subsection, particularly for obtaining (3.4.4). We have to apply a Maximum Principle argument to an equation deduced from (3.3.42) by differentiation.

The constant appears actually (if a solution u exists in a suitable space) as the mean value of f

$$\langle f\rangle_\Phi = \int f \exp -\Phi dx \Big/ \int \exp -\Phi dx \,.$$

$f - \langle f\rangle_\Phi$ appears to be orthogonal to the first eigenfunction $x \to 1$ of the selfadjoint operator

$$u \mapsto (-\Delta + \nabla\Phi\cdot\nabla)u$$

on $L^2(\mathbb{R}^m, \exp -\Phi\cdot dx)$.
This appears as the Witten's Laplacian on the functions and the operator

$$v \mapsto (-\Delta + \nabla\Phi\cdot\nabla)v + \text{Hess}\,\Phi\; v$$

can be considered as the Witten's Laplacian on the 1-forms. In order to study this new Maximum Principle we consider $\mathbb{R}^m$ as a normed space B for a given norm and we denote by B^* the dual space. We observe that B is reflexive (B is finite dimensional). Let

$$A : B \mapsto B$$

be a linear map and let $\delta \geq 0$. We say that A satisfies $mp(\delta)$ (with respect to the space B) if we have the property,

$$\text{if } t \in B, s \in B^* \text{ and } \langle t|s\rangle_{B\times B^*} = ||t||_B \cdot ||s||_{B^*} \text{then } \langle At|s\rangle \geq \delta\, ||t||_B \cdot ||s||_{B^*}\,. \tag{3.4.3}$$

Example 3.4.1 $B = B^* = \ell^2(\mathbb{Z}/m\mathbb{Z})$ and A is a symmetric matrix such that $A \geq \delta \geq 0$. Then A satisfies $mp(\delta)$.

Example 3.4.2 $B = \ell^p_\rho$ for some $p \in [1, +\infty]$, $\rho : \{1, \cdots, m\} \mapsto]0, \infty[$, where ℓ^p_ρ is equipped with the norm $||x||_{\ell^p_\rho} = ||\rho x||_{\ell^p}$ and $(\rho x)_j = \rho(j)x_j$.

Assume that $A = D + W$, where D is diagonal and satisfies $D \geq r_0 \geq r_1 \geq 0$ and where $||W||_{\mathcal{L}(\ell^p_\rho)} \leq r_1$. Then A satisfies $mp(r_0 - r_1)$. In fact, we have $B^* = \ell^q_{\frac{1}{\rho}}$ with $1 = \frac{1}{p} + \frac{1}{q}$. If we have $\langle t|s\rangle = ||t||_B \cdot ||s||_{B^*}$, then $t_j\, s_j \geq 0$, for every j. Consequently,

$$\langle Dt|s\rangle \geq r_0\, ||t||_B \cdot ||s||_{B^*}\,,$$

while

$$|\langle Wt|s\rangle| \leq r_1\, ||t||_B \cdot ||s||_{B^*}\,.$$

Let us see now how it can be applied in our context. We neglect technicalities like regularization and our operator A will be

$$A = \text{Hess}\,\Phi(x_0)$$

for some x_0 in $\mathbb{R}^m$ and we assume that there exists $\delta > 0$ such that

$$\text{Hess}\,\Phi(x) \text{ satisfies } mp(\delta)$$

for all $x \in \mathbb{R}^m$. We also assume [7] that $v(x) \to 0$ as $|x| \to +\infty$. Let $x_0 \in \mathbb{R}^m$ be a point where $||v(x)||_B$ is maximal:

$$||v(x_0)|| = \sup_{x \in B} ||v(x)|| =: m_1 \,.$$

Let $s \in B^*$ be a unit vector such that

$$\langle v(x_0)|s\rangle = m_1 \,.$$

Then, considering the function $x \mapsto \langle v(x)|s\rangle$, we get

- $-\Delta(\langle v|s\rangle)(x_0) = \langle -(\Delta v)(x_0)|s\rangle \geq 0$,
- $\nabla\phi \cdot \nabla(\langle v|s\rangle)(x_0) = (\langle \nabla\phi \cdot \nabla v(x_0)|s\rangle) = 0$, and
- $\langle \text{Hess}\,\Phi(x_0)v(x_0)|s\rangle \geq \delta \cdot m_1$, using $mp(\delta)$ and the property that $\langle v(x_0)|s\rangle = ||v(x_0)|| \cdot ||s|| = m_1$.

If now v is a solution of the basic equation, we can take the scalar product with s and we get at the point x_0

$$\delta \cdot m_1 \leq \langle w(x_0)|s\rangle \leq ||w(x_0)||_B$$

and so we obtain, for every $x \in \mathbb{R}^m$,

$$||v(x)||_B \leq \delta^{-1} \sup_{x \in \mathbb{R}^m} ||w(x)||_B \,. \tag{3.4.4}$$

Depending on the different results we want to prove, we use different choices of norms on $\mathbb{R}^m$. We can also obtain in the same way the control of derivative to higher order. We refer to [HeSj93] or [Sj93b] for this type of results which can be useful for various results like the control of correlations of high order. The main idea is to differentiate the basic equation and to apply the Maximum Principle to the new equation. The assumption permitting to verify $mp(\delta)$ appears for example in (3.4.5) (for the case when $B = \ell^2$) or in (3.8.32).

In particular we have given the main steps for obtaining the following result (corresponding to $B = \ell^2$):

[7] In our applications, a regularization argument permits to reduce the study to this case, see for example [HeSj93].

Theorem 3.4.3 *Let Φ in $C^\infty(\mathbb{R}^m;\mathbb{R})$ satisfy*

$$\mathrm{Hess}\,\Phi(x) \geq \sigma\ ,\ \forall x \in \mathbb{R}^m \tag{3.4.5}$$

$$|\partial^\alpha \Phi(x)| \leq C\ ,\ \forall \alpha\ s.t.\ |\alpha| \geq 2\ . \tag{3.4.6}$$

Let g in $C^1(\mathbb{R}^m;\mathbb{R})$ s.t. $x \to \nabla g(x)$ is bounded. Then the equation (3.4.2) in $\mathbb{R}^m$,

$$g - \langle g \rangle = \nabla_x \Phi \cdot \nabla u - \Delta u\,,$$

admits a unique temperate solution u such that

$$u \in L^2(\mathbb{R}^m, \exp -\Phi \cdot dx)$$

and

$$\langle u \rangle = 0$$

The gradient of this solution is bounded, and we have the estimate

$$\sup_x ||\nabla u(x)||_{\ell^2} \leq \frac{1}{\sigma} \sup_x ||\nabla g(x)||_{\ell^2} \tag{3.4.7}$$

Complements for the sketch of the proof. We come back to the operator $\mathcal{A}$:

$$u \to \nabla_x \Phi \cdot \nabla u - \Delta u$$

which appears to be essentially selfadjoint (starting from C_0^∞) on the weighted space $L^2(\mathbb{R}^m, \exp -\Phi \cdot dx)$. Using the conjugation $f \mapsto f \exp -\Phi/2$, the problem of existence of solutions is reduced to the study of a "pseudo" harmonic oscillator

$$h \to -\Delta h + \frac{1}{4}|\nabla \Phi|^2\, h - \Delta \Phi\, h\ . \tag{3.4.8}$$

Under the weaker condition (condition at ∞) that for some $C > 0$ and some $\sigma > 0$:

$$\mathrm{Hess}\,\Phi(x) \geq \sigma\ ,\ \forall x \in \mathbb{R}^m \setminus B(O, C)\ , \tag{3.4.9}$$

and condition (3.4.6), we observe using standard techniques that this new operator $\tilde{A}$ admits a unique selfadjoint extension which defines a positive operator whose domain is characterized as:

$$B^2(\mathbb{R}^m) = \{u \in L^2(\mathbb{R}^m) | x^\beta \partial_x^\alpha u \in L^2(\mathbb{R}^m)\ ,\ \forall \alpha\,, \beta\ \text{s.t.} |\alpha| + |\beta| \leq 2\}$$

and consequently with compact resolvent. The smallest eigenvalue is 0 and the first eigenfunction is the function $x \to \exp -\Phi(x)/2$. The first part of the statement can consequently seen as a consequence of the Fredhom alternative.

We need also a regularization argument (reducing first to the case when Φ is of the form $x^2/2 + \Psi$ with Ψ compactly supported) in order to work with the Maximum Principle. As proven in [Sj93b] (see also [HeSj93] and [Sj93b]), we observe that if in equation (3.4.2) g belongs to C_0^∞ then the solution u has the property that $\nabla u(x)$ tends to 0 as $|x|$ tends to ∞ and that the same property is true for the higher derivatives of u. One way to prove this property, which will be useful also in order to apply the Maximum Principle, is in the following remark:

Remark 3.4.4 If u is a solution of (3.4.2), then $v =: \nabla u$ is a solution of (3.4.1) with

$$w = \nabla g \,. \tag{3.4.10}$$

Remark 3.4.5 As in [Sj92b], [HeSj93], we can also obtain, if g is C^∞, similar estimates for the higher derivatives and get for example the control of $||\mathrm{Hess}\, u(x)||_{\mathcal{L}(B)}$ using the control of $||\mathrm{Hess}\, g(x)||_{\mathcal{L}(B)}$, $||\nabla g(x)||_{B'}$ in a suitable space B' for which $mp(\sigma)$ is also satisfied, and an additional assumption on the derivatives of order 3 on Φ. We refer for example to the estimate (4.3) in [Sj93c].

3.5 Laplace integrals in large dimension: decay of the correlations

We now study more precisely the family of measures on cylindrical functions. This is the object of

Proposition 3.5.1 *Let c be a C^∞ function defined on $\mathbb{R}^k$ s.t. $\nabla c \in C_b^\infty(\mathbb{R}^k)$ that we identify* [8] *with a function on $\mathbb{R}^m$ by:*

$$c^{(m)}(x) - c(x_1, x_2, \ldots\ldots, x_k),$$

then, under the assumptions of Theorem 3.3.11, *the limit as m tends to ∞ of $\langle c^{(m)}\rangle_m$ exists and the convergence is exponentially fast as in (*3.3.38*).*

There is a slightly more general result permitting to treat the important case where $c(x) = x_1 \cdot x_j$. Let us say only here that the theorem above is also true for this case by a simple integration by parts argument which works easily in the situation of Corollary 3.3.3 and also in the general case.

This theorem of convergence permits actually to define a limit measure whose properties seem connected to what is for example considered in [AKR94].

[8] This choice is arbitrary. We could take also

$$c_n^{(m)}(x) = c(x_{1+n}, x_{2+n}, \ldots\ldots, x_{k+n}),$$

but the limit will not depend on this choice.

Theorem 3.5.2 *Let W be a C^3 function on $\mathbb{R}^2$ such that $\nabla^p W$ is bounded (for $p = 1, 2, 3$). If*

$$\Psi^{(m)}(x) = \sum_{i=1}^{m} W(x_i, x_{i+1})$$

and if $||\text{Hess}\, W||$ is small enough, then the correlation

$$\text{Cor}^{(m)}(i,j) = \langle x_i x_j \rangle_m - \langle x_i \rangle_m \langle x_j \rangle_m$$

admits a limit and there exists $\kappa_0 > 0$ and C_0 s.t.

$$|\text{Cor}^{(m)}(i,j)| \leq C_0 \exp -(\kappa_0 \text{ dist }_{\mathbb{Z}/m\mathbb{Z}}(i,j)) \tag{3.5.1}$$

for all m and all pairs (i,j) s.t $|i-j| \geq 2$.

The computation of the best κ_0 is a quite interesting problem which may have various applications (see for example [He94b] and [He94c]).

Remark 3.5.3 The theorem can actually be proved more easily by the technique of the transfer matrix which will be explained in Section 6. But we have to emphasize two points. We can treat by this method more general interactions where $\Psi^{(m)}(x)$ has not this nearest neighbor one dimensional structure and the second point is that this approach gives informations, for example on κ_0, which are not necessarily given by the other method. We shall play with this property in Sections 6 and 7.

Let us sketch the proof of this theorem. If $v^{(i)}$ is the solution in $C_b^1(\mathbb{R}^m; \mathbb{R}^m)$ of the basic equation for $w^{(i)} = \delta_i$, that is $w_j^{(i)}(x) = \delta_{ij}$, we shall apply the Maximum Principle by choosing the following Banach space: $B = \ell^2_{\rho^{(i)}}(\mathbb{Z}/m\mathbb{Z})$ where $\rho^{(i)}$ is chosen such that:

$$\rho^{(i)}(j) = \exp \kappa_0 \text{ dist }_{\mathbb{Z}/m\mathbb{Z}}(i,j) \; .$$

With this choice we observe that

$$\sup_x ||w^{(i)}(x)||_B \leq 1 \; ,$$

and if B (that is κ_0) is chosen such that $\text{Hess}\, \Phi(x)$ satisfies $(mp(\delta))$ for some $\delta > 0$, we get

$$\sup_x ||v^{(i)}(x)||_B \leq \frac{1}{\delta} \; .$$

In particular we obtain that,

$$\sup_x |v_j^{(i)}(x)| \leq \frac{1}{\delta} \cdot \exp -(\kappa_0 \text{ dist }_{\mathbb{Z}/m\mathbb{Z}}(i,j)) \; . \tag{3.5.2}$$

In order to finish the proof we now relate the correlation $\mathrm{Cor}^{(m)}(i,j)$ and $v_j^{(i)}(x)$ by observing

$$\mathrm{Cor}^{(m)}(i,j) = \langle v_j^{(i)} \rangle_{\Phi^{(m)}} . \tag{3.5.3}$$

The proof of this last formula is as follows. If $u^{(i)}$ is the solution orthogonal in $L^2(\mathbb{R}^m; \exp -\Phi \cdot dx)$ to the constant of the equation

$$x_i - \langle x_i \rangle = (-\Delta + \nabla\Phi \cdot \nabla) u^{(i)} ,$$

it is possible to prove that

$$\nabla u^{(i)} = v^{(i)} ,$$

and this gives in particular

$$x_i - \langle x_i \rangle = -\mathrm{div}\, v^{(i)} + \nabla\Phi \cdot v^{(i)} . \tag{3.5.4}$$

An integration by part in the definition of the correlation using also (3.5.4) gives (3.5.3). The combination of (3.5.3) and of (3.5.2) gives the theorem.∎

Another application is the study of the magnetization attached to the following family of potentials

$$\Phi^{(m)}(x,B) = \frac{1}{2}\sum_{\ell} (x_\ell - B)^2 + \Psi^{(m)}(x), \tag{3.5.5}$$

and defined by

$$\mathcal{M}(m,B) = (1/m)(\partial a/\partial B)(m,B) \tag{3.5.6}$$

or

$$\mathcal{M}(m,B) = \langle x_1 - B \rangle_m .$$

It is for example proved in [HeSj93] the following theorem

Theorem 3.5.4 *Under the same assumptions as for Theorem 3.5.2, the magnetization $\mathcal{M}(m,B)$ is convergent (exponentially rapidly) to a continuous limit. In particular, if $\Psi^{(m)}$ is even, then the limit as B tends to 0 is zero.*

Around contributions of A. Sokal and J. Sjöstrand

Let us finish this subsection by reporting on the contributions of A.D. Sokal [Sok82] and more recently J. Sjöstrand [Sj93c]. A. Sokal considers as starting point the probability measure on $\mathbb{R}^m$

$$d\mu(x) = Z^{-1} \exp\left[\frac{1}{2}\sum_{i,j} J_{ij} x_i x_j\right] \prod_i d\nu_i(x_i) , \tag{3.5.7}$$

with

$$J_{ij} = J_{ji} \geq 0 \, .$$

This measure is sometimes called (see [Si74] (p.273)) a general ferromagnetic Ising magnet.
A. Sokal assumes the rather implicit assumption that each $d\nu_i$ is an even probability measure satisfying the assumptions of the GHS and Lebowitz inequalities, that is

$$\text{(GHS)} \qquad \langle x_j x_k x_\ell \rangle \leq 0 \, , \tag{3.5.8}$$

and

$$\text{(Leb)} \qquad \langle x_j x_k x_\ell x_m \rangle - \langle x_j x_k \rangle \langle x_\ell x_m \rangle - \langle x_j x_\ell \rangle \langle x_k x_m \rangle - \langle x_j x_m \rangle \langle x_k x_\ell \rangle \leq 0 \, . \tag{3.5.9}$$

Examples are the spin $\frac{1}{2}$-model

$$d\nu_i(x) = \delta(x^2 - 1) dx \tag{3.5.10}$$

and the ϕ^4 lattice theory

$$d\nu_i(x) = \text{const.} \cdot \exp(-ax^2 - bx^4) \, dx \, , \tag{3.5.11}$$

with $b > 0$. More generally it is known from [BFS82] that (Leb) is true for the measure

$$d\nu_i(x) = \text{const.} \cdot \exp -v(x) \, dx \, , \tag{3.5.12}$$

where $v(x) = f(x^2)$ with f convex on $(0, \infty)$. This is for example the case when v is even with v' convex. We met also natural examples of this type in Subsection 3.1 (in particular (3.1.2)) where $v(x) = -\ln \cosh x + ax^2$.
Then A.Sokal compares the correlation functions of this model with the corresponding correlations of the Gaussian model $d\mu'$ having the same pair interaction J_{ij} but with $d\nu_i$ replaced by the new measure

$$d\nu_i'(x) = (2\pi \, c_i)^{-\frac{1}{2}} \cdot \exp(-x^2/2c_i) \, dx \, , \tag{3.5.13}$$

where

$$c_i = \int x^2 d\nu_i(x) \, . \tag{3.5.14}$$

The theorem is then the following

Theorem 3.5.5 *Assume that each $d\nu_i$ satisfies the hypothesis of the Lebowitz inequality. Then*

$$0 \leq \langle x_i x_j \rangle_\mu \leq \langle x_i x_j \rangle_{\mu'} \, , \tag{3.5.15}$$

whenever the measure μ' is well defined.

Let us just sketch some elements of the proof. We consider the probability measure $d\mu_\lambda$ defined as in (3.5.7) but with λJ_{ij} replacing J_{ij}, with $0 \leq \lambda \leq 1$. Then

$$\frac{d}{d\lambda}\langle x_i x_j\rangle_{\mu_\lambda} \leq \sum_{k,\ell}\langle x_i x_k\rangle_{\mu_\lambda} J_{k\ell}\langle x_\ell x_j\rangle_{\mu_\lambda} , \tag{3.5.16}$$

using the Lebowitz inequality. Now (3.5.16) is a system of differential inequalities for the functions

$$G_{ij}(\lambda) \equiv \langle x_i x_j\rangle_{\mu_\lambda} .$$

Since the right side of (3.5.16) is an increasing function of each element of the matrix G (in the relevant region $G_{ij} \geq 0$), due to the ferromagnetic assumption on the J_{ij}, it follows from a well known lemma that the solution of the differential inequality is bounded above by the solution of the corresponding equation with the same condition at $\lambda = 0$. This initial condition is

$$G_{ij}(0) = c_i \delta_{ij} ,$$

and the solution of the differential equation is

$$\bar{G}(\lambda) = [G(0)^{-1} - \lambda J]^{-1}$$

as long as the matrix $G(0)^{-1} - \lambda J$ is strictly positive definite. We thus have

$$G_{ij}(\lambda) \leq \bar{G}_{ij}(\lambda) ,$$

for all i, j. But notice now that $\bar{G}_{ij}(\lambda)$ is nothing other than the two-point function of the Gaussian model μ'_λ (because the Lebowitz inequality is an equality obtained simply by integration by parts in the case of the Gaussian model). In particular we have

$$\bar{G}_{ij}(1) = \langle x_i x_j\rangle_{\mu'}$$

and this proves the theorem. Theorem 3.5.5 can now be applied to a translation-invariant model with all $c_i = c$ and $J_{ij} = J(i-j)$. We can introduce

$$\mathcal{J} = \sum_j J_{0j} . \tag{3.5.17}$$

Then the norm of the matrix J, considered as an operator on ℓ^2 is at most $\mathcal{J}$. So, if

$$c\mathcal{J} < 1 \tag{3.5.18}$$

$[G(0)^{-1} - \lambda J]$ is invertible for $\lambda \in [0,1]$ and the theory above applies. This condition is precisely the "mean field" condition. Moreover, it is easy to see that $\bar{G}(1)$ then decays exponentially if J does (we already used or will use this type of property when playing with weighted spaces ℓ^p_ρ). For example, in the nearest-neighbor model with interaction strength β on the lattice $\mathbb{Z}^d$ (so that the corresponding $\mathcal{J}$ satisfies $\mathcal{J} = 2d\beta$, we have

$$\bar{G}(1)_{ij} \leq \text{const.} \cdot \exp(-m_0|i-j|) \; , \tag{3.5.19}$$

with

$$m_0 = \cosh^{-1}[(2\beta c)^{-1} - (d-1)] \; . \tag{3.5.20}$$

m_0 is called the mass gap. In order to be precise, the above analysis should be carried first for a finite box Λ, and then the inequalities carried over to the infinite-volume limit. Although the assumptions are not exactly the same, this work of Sokal was further developped by Bricmont, Fontaine, Lebowitz, Spencer [BFLS81] and quite recently by J. Sjöstrand [Sj93c] (Sections 1,2,3). J. Sjöstrand analyzes for example the case when $c\mathcal{J} = 1$ (when $d \geq 3$) and the case when $c\mathcal{J} = 1 - \epsilon$ (with $\epsilon > 0$) with a control as $\epsilon \to 0$. That is the critical situation. Actually it seems that the two results can be combined. A. Sokal has indeed implemented the use of the Lebowitz inequality which leads to Theorem 3.5.5 and J. Sjöstrand is in some case more precise for the control of $[G(0)^{-1} - J]^{-1}$. It is clear for example that, for the model (3.5.11) with $a > 0$, the approach based on the Maximum Principle does not take account of the gain due to b. But as a function of b, the variance $\langle x^2 \rangle$ which determines c is strictly decreasing for $b > 0$ (b small) as a consequence of the strict inequality:

$$\begin{aligned} &-(\textstyle\int x^6 \exp -ax^2 \, dx) \cdot (\int \exp -ax^2 \, dx) \\ &\quad +(\textstyle\int x^4 \exp -ax^2 \, dx) \cdot (\int x^2 \exp -ax^2 \, dx) \; < \; 0 \; , \end{aligned}$$

which can be explicitely verified.

Finally J. Sjöstrand analyzes "non strictly convex cases" in [Sj94] where $\langle \text{Hess}\, \Phi \rangle$ appears to play a role in connection with the Witten Laplacian. This has maybe a connection with what A. Sokal did. This paper contains also a new analysis of the family of convolution equations which appears above. As motivating reference, J. Sjöstrand mentioned also contributions by R. Minlos (see [ZM88]).

3.6 Around Brascamp-Lieb inequalities

H.J. Brascamp and E. Lieb [BraLi76] have proved the following celebrated inequality which plays an important role in different contexts in the study of the Schrödinger equation.

Theorem 3.6.1 *Let $F(x) = \exp(-\Phi(x)), x \in \mathbb{R}^m$, with Φ in C^2 and strictly convex. We assume that Φ has a minimum and consequently F decays exponentially in all directions. Let $h \in C^1(\mathbb{R}^m)$, and let us assume that* var $h < \infty$. *Then*

$$\text{var}\; h \leq \langle \nabla h, (\text{Hess}\, \Phi\, (x))^{-1} \nabla h \rangle \tag{3.6.1}$$

where ∇h is the gradient of h.

We want to give a new proof of this theorem inspired by [HeSj93], [He93b] and [Sj94]. For h in C^1, such that ∇h is bounded, we have seen in Section 3.4 that there exists a gradient vector field v s.t.:

$$h - \langle h \rangle = \nabla\Phi \cdot v - \operatorname{div} v, \tag{3.6.2}$$

where $\langle h \rangle = \int h \exp -\Phi\ dx / \int \exp -\Phi\ dx$ is the mean value of h. If we take the formula giving the variance, we get

$$\begin{aligned} \operatorname{var}(h) &\stackrel{\text{def}}{=} \int (h - \langle h \rangle)^2 \exp -\Phi\ dx \Big/ \int \exp -\Phi\ dx \\ &= \int (h - \langle h \rangle)(v \cdot \nabla\Phi - \operatorname{div} v) \exp -\Phi\ dx \Big/ \int \exp -\Phi\ dx, \end{aligned} \tag{3.6.3}$$

and finally

$$\operatorname{var}(h) = \left(\int v \cdot \nabla h \exp -\Phi\ dx \right) \Big/ \int \exp -\Phi\ dx \tag{3.6.4}$$

As in [HeSj93], we get by differentiation of (3.6.2)

$$\nabla h = A_1 v\ , \tag{3.6.5}$$

where A_1 is the operator

$$v \mapsto A_1 v = (\nabla\Phi \cdot \nabla - \Delta) v + \operatorname{Hess} \Phi \cdot v\ ,$$

which can be interpreted as a Witten's Laplacian. This operator is positive and actually strictly positive (as first observed in [Sj94]) *without any assumption of strict convexity*.

Let us sketch the proof.

If v satisfies $A_1 v = 0$, then taking the scalar product in $L^2(\mathbb{R}^m; \exp -\Phi \cdot dx)$ we get:

$$\int |dv|^2 \exp -\Phi \cdot dx + \int |\nabla\Phi \cdot v - \operatorname{div} v|^2 \exp -\Phi\ dx = 0\ . \tag{3.6.6}$$

This implies $dv = 0$, and consequently $v = \nabla u$. Finally we get also:

$$\nabla\Phi \cdot \nabla u - \Delta u = 0$$

which implies modulo the verification that u can be chosen (see [HeSj93] or (1.7) in [Sj94]) in the domain of A_0:

$$A_0\, u = 0\ .$$

But this implies that u is a collinear to the first eigenfunction of A_0: $x \mapsto 1$ and we get finally

$$v = \nabla u = 0\ .$$

0-forms). We observe indeed (see [HeSj93]), that, by doing the conjugation, $u = \exp \Phi/2 \cdot \tilde{u}$, we get the operator:

$$\tilde{A} = -\Delta + \frac{1}{4}|\nabla\Phi|^2 - \Delta\Phi \quad \text{on} \quad L^2(\mathbb{R}^m; dx).$$

This operator is essentially selfadjoint, with compact resolvent and clearly $\tilde{A}$ is positive and has zero as an eigenvalue with multiplicity 1. Coming back to our initial operator it is clear that we can invert A_0 on the orthogonal of the first eigenfunction which is the constant function 1. Then we denote by $(A^{\text{red}})^{-1}$ the inverse of A_0 reduced to the orthogonal to the zero eigenfunction, and vanishing on the constants. We observe that the property of positivity preservation of $\exp -tA_0$ is transmitted to $(A^{\text{red}})^{-1}$ considered as the limit as ϵ tends to 0^+ of $(A^{\text{red}} + \epsilon)^{-1}$. (Note that the projector on the first eigenvector of A has also the property). Then once we know that $(A^{\text{red}})^{-1}$ has this property, we get easily the property for the correlation in the following way. Let us study $\text{Cor}\,(f, g)$ with $f, g \in \mathcal{C}^+$. We can write:

$$g - \langle g \rangle = A_0((A^{\text{red}})^{-1} g)$$

which is true because $g - \langle g \rangle$ is orthogonal to the zero eigenvector. Then $u = (A^{\text{red}})^{-1} g \in \mathcal{C}^+$ and we have:

$$\text{Cor}\,(f, g) = \langle \nabla f \cdot \nabla u \rangle \geq 0\,. \tag{3.7.5}$$

In [HeSj93], we proved *only* that $(A^{\text{red}})^{-1}$ has the property in the case when Φ was strictly convex using the Maximum Principle.
Now J. Sjöstrand observes (cf. (3.6.7)) the following identity

$$\text{Cor}\,(f, g) = \left(\int (A_1^{-1} \nabla f) \cdot \nabla g \, \exp -\Phi \, dx \right) / \int \exp -\Phi \, dx\,. \tag{3.7.6}$$

Again we have to study positivity preserving property for the operator A_1. This is done (through the study of the heat equation) by J. Sjöstrand in [Sj94] (Proposition 1.3) *without any assumption of strict convexity.*

We finish this subsection by recalling for further use two lemmas.

The first one is due to Brascamp-Lieb (cf. Corollary 3.5 in [BraLi76] or Theorem 13.2 in [Si79]):

Lemma 3.7.2 *If $F(x, y)$ is strictly positive, logconcave (with respect to the variables (x, y)) (and satisfies suitable integrability assumptions (essentially F is of gaussian type)), then $G(x) = \int F(x, y) dy$ is logconcave.*

Similarly we shall get:

Lemma 3.7.3 *If* $f(x,y) = -\ln F(x,y)$ *satisfies condition* (3.7.3) (*with respect to the variables* (x,y))(*and satisfies suitable integrability assumptions*), *then* $g(x) = -\ln(G(x))$, *with* $G(x) = \int F(x,y)dy$, *satisfies the same condition with respect to* x.

Let us sketch the proof of this last lemma. We differentiate two times g with respect to the variables x_i, x_j. We get after some computations:

$$(\partial^2_{x_ix_j}g)(x) = \langle(\partial^2_{x_ix_j}f)(x,\cdot)\rangle_x - \langle((\partial_{x_i}f) - \langle(\partial_{x_i}f)\rangle_x)((\partial_{x_j}f) - \langle(\partial_{x_j}f)\rangle_x)\rangle_x \tag{3.7.7}$$

where $\langle u\rangle_x$ is the expectation of the function u with respect to the measure $d\mu_x = F(x,y)dy/G(x)$. We now observe that the first term of the r.h.s of (3.7.7) is negative because

$$(\partial^2_{x_ix_j}f) \leq 0 \text{ if } i \neq j.$$

Now in order to examine the second term

$$\langle((\partial_{x_i}f) - \langle(\partial_{x_i}f)\rangle_x)((\partial_{x_j}f) - \langle(\partial_{x_j}f)\rangle_x)\rangle_x$$

we observe that we can apply the Cartier's Theorem because the function $y \to (\partial_{x_i}f)(x,y)$ is monotone with respect of each variable y_k due to the inequality $(\partial^2_{y_kx_i}f)(x,y) \leq 0$ which is satisfied by assumption. This proves the lemma. ∎

Remark 3.7.4 Using the Brascamp-Lieb inequalities and formula (3.7.6), we can recover Theorem 4.2 of Brascamp-Lieb [BraLi76] which gives the following inequality between $m \times m$ symmetric matrices:

$$(\partial^2_{x_ix_j}g)(x) \geq \langle(\partial^2_{x_ix_j}f)(x,\cdot) - \partial^2_{x_iy_k}f \cdot [(f"_{yy})^{-1}]_{k\ell}\partial^2_{y_\ell x_j}f\rangle_x \,. \tag{3.7.8}$$

3.8 Laplace integrals in large dimensions: semiclassical aspects

The formal Laplace's method

Let us recall different statements obtained by J. Sjöstrand in [Sj93a], [Sj93b] and later extended in [He94a] in the spirit of [HeSj92a]. We shall also mention improvements inspired by [Sj92b], [Sj93c] and [Sj94]. Let us introduce a set $\mathcal{V}$ as the disjoint union over $\mathbb{N}$ of sets $\mathcal{V}_m$,

$$\mathcal{V} = \cup_{m\in\mathbb{N}}\mathcal{V}_m \,,$$

where $\mathcal{V}_m$ is a subset of the set of C^∞ potentials on $\mathbb{R}^m$. Let us make on $\mathcal{V}_m$ the following assumptions For all Φ in $\mathcal{V}_m$

$$\Phi \text{ is holomorphic in } B(0,1) \text{ with } |\nabla\Phi(x)|_\infty = \mathcal{O}(1) \tag{3.8.1}$$

uniformly in $\mathcal{V}$ and $B(0,1)$, (Here $B(0,1)$ is the open unit ball in $\mathbb{C}^m$ with respect to the norm $|x|_\infty = \sup_{j=1,\dots,m} |x_j|$)

$$\Phi(0) = 0 \;,\; \nabla\Phi(0) = 0 \;,\; \text{Hess}\,\Phi(0) = D + A, \tag{3.8.2}$$

where D is diagonal (positive definite) and
There exists r_1 and r_0 (independent of Φ in $\mathcal{V}$) such that

$$||A||_{\mathcal{L}(\ell^p)} \leq r_1 < r_0 \leq \lambda_{min}(D) \tag{3.8.3}$$

for all p s.t. $1 \leq p \leq \infty$.
We also assume

$$||\text{Hess}\,\Phi(x)||_{\mathcal{L}(\ell^p)} = \mathcal{O}(1) \tag{3.8.4}$$

uniformly in $\mathcal{V}$, $B(0,1)$ and p.
Here we write

$$|x|_p = (\sum |x_j|^p)^{\frac{1}{p}} \text{ for } 1 \leq p < \infty \;,$$

and

$$|x|_\infty = \sup_j |x_j| \;.$$

As proved in [Sj93b] (Lemma 1.1), we deduce from (3.8.2) and (3.8.3) that

$$(\text{Hess}\,\Phi(0))^{\frac{1}{2}} = \tilde{D} + \tilde{A} \tag{3.8.5}$$

with $\tilde{D}$ diagonal and

$$||\tilde{A}||_{\mathcal{L}(\ell^p)} \leq \tilde{r}_1 < \tilde{r}_0 \leq \lambda_{min}(\tilde{D}) \tag{3.8.6}$$

for all p s.t. $1 \leq p \leq \infty$ and uniformly in $\mathcal{V}$.
Then, under these conditions, J. Sjöstrand proves the following statement (see Proposition 3.1 and Section 5 in [Sj93a])

Theorem 3.8.1 *Under assumptions* (3.8.1)-(3.8.4), *then there exists a formal series*

$$f(x,h) \sim \sum_j f_j(x) \cdot h^j \tag{3.8.7}$$

and a series

$$E(h,m) \sim h \sum_{j \geq 0} E_j(m) \cdot h^j \tag{3.8.8}$$

s.t. in the sense of the formal series in h but in a fixed sufficiently small ℓ^∞ - neighborhood of 0 B , *the following equation is satisfied*

$$|\nabla f|^2 - \Phi - h \ln \det(\text{Hess}\, f) \sim E(h,m) \;. \tag{3.8.9}$$

Moreover, the functions f_j *satisfy:*

$$f_j(0) = 0, |\nabla f_j(x)| \leq C_j \text{ in } B , \tag{3.8.10}$$

and the $E_j(m)$ *(which of course depend on* m *through* $\Phi = \Phi^{(m)}$ *in* $\mathcal{V}_m$*) satisfy*

$$|E_j(m)| \leq C_j \cdot m . \tag{3.8.11}$$

Of course this statement and the origin of the equation could appear mysterious and it is probably better to give the following "formal" corollary:

Corollary 3.8.2 If

$$J(h,m) := (1/h\pi)^{\frac{m}{2}} \cdot \left(\int \exp(-\frac{1}{h}\Phi^{(m)}(x))\, dx \right) ,$$

then, at the level of the formal expansions [10] ,

$$(\ln J(h,m))/m \sim \sum_j (E_j(m)/m)\, h^j , \quad \text{as } h \to 0 . \tag{3.8.12}$$

Proof of the corollary. This is just an application of the formal Laplace's method with a uniform control with respect to m (i.e. $\Phi^{(m)}$ in $\mathcal{V}_m$). In the formal integral giving $J(h,m)$ we admit first that modulo controlled error terms, we can reduce the integral to a small ℓ^∞-path in $\mathbb{R}^m$. We are then looking for a change of variable $y = \nabla f(x,h)$. Then the integral becomes

$$(1/h\pi)^{\frac{m}{2}} \left(\int \exp\left[-\frac{1}{h}\Phi^{(m)}(x) - \ln \det \text{Hess}\, f(x,h) \right] dy \right)$$

and, at least formally, it is then clear that Theorem 3.8.1 gives that

$$(1/\pi h)^{\frac{m}{2}} \cdot \left(\int \exp\left(-\frac{1}{h}\sum y_j^2 + \frac{1}{h}E(h,m) \right) dy \right)$$

So finally

$$\ln J^f(h,m) \sim h^{-1}E(h,m) .$$

Of course everything is for the moment formal but the control of the coefficients is what is basic for the future.

---

[10] This means that for the moment we have not controlled any remainder term. $J(h,m)$ is defined only at the level of the formal expansions given by the formal stationary phase theorem. We shall return later on to a control of the remainder terms. We prefer to write $J^f(h,m)$ or $J^{form}(h,m)$ when we speak of the formal expansion.

Remark 3.8.3 One can add in Theorem 3.8.1 a "subprincipal" symbol $\Phi_1^{(m)}(x)$ which satisfies the following conditions (see condition (5.13) in [Sj93a]),

$$|\nabla\Phi_1^{(m)}(x)|_\infty = \mathcal{O}(1) \tag{3.8.13}$$

in a small complex ℓ^∞-ball around 0, and

$$\Phi_1^{(m)}(0) = 0 \,. \tag{3.8.14}$$

Remark 3.8.4 The reader can also look at the discussion given by J. Sjöstrand in the section "Formal treatment of the stationary phase method " in [Sj92b]. This is done using the notion of 0-standard function. We refer to Remark 5.6.3 for a discussion of possible improvements.

Proof of Theorem 3.8.1 **and of Remark** 3.8.3 (following [Sj93a] and [He94a])
We look for f in the form (3.8.6) and the first equation that we find in order to solve (3.8.9) is

$$|\nabla f_0|^2 = \Phi \tag{3.8.15}$$

called the eiconal equation, which can be solved according to [Sj93a], in some small ℓ^∞-ball centered at 0. In particular f_0 has the properties,

$$|\nabla f_0|_\infty \leq C_0 \,, \tag{3.8.16}$$

in this ball and,

$$(\mathrm{Hess}\, f_0)^{-1} \,,\ \mathrm{Hess}\, f_0 \in \mathcal{L}(\ell^p) \,,\quad \text{for } 1 \leq p \leq \infty \,. \tag{3.8.17}$$

f_1 should fulfill the so-called first transport equation

$$\nabla f_0 \cdot \nabla f_1 = \ln \det \mathrm{Hess}\, f_0 \ + \ \Phi_1 \ + \ E_0 \tag{3.8.18}$$

and we observe[11] that

$$\partial_{x_\ell} \ln \det \mathrm{Hess}\, f_0 = \mathrm{Tr}\ ((\mathrm{Hess}\, f_0)^{-1} \cdot (\partial_{x_\ell} \mathrm{Hess}\, f_0)) \tag{3.8.19}$$

If e_j is the jth unit vector, then

$$|\langle \partial_{x_j} \mathrm{Hess}\, f_0, t \otimes s\rangle| = |\langle \nabla^3 f_0, e_j \otimes t \otimes s\rangle| \leq C\ |t|_\infty \cdot |s|_\infty \cdot |e_j|_1 \,. \tag{3.8.20}$$

Inequality (3.8.20) is a consequence of (3.8.17) and of the Cauchy estimates proved in the

[11] We recall that $(d/dt) \ln \det A(t) = \mathrm{Tr}\ (A(t)^{-1} \cdot A'(t))$.

Proposition 3.8.5 ([Sj93a]) *Let $0 < r_1 < r_0$. Then there is a constant $C_0 > 0$ independent of m s.t. for every holomorphic function on $B_\infty(0, r_0)$ in $\mathbb{C}^m$ satisfying*

$$|\nabla u|_\infty \leq 1 \text{ in } B_\infty(0, r_0) ,$$

we have, for $|x|_\infty < r_1$,

$$|\langle \nabla^2 u, t \otimes s \rangle| \leq C_0 \; |t|_p \cdot |s|_{p^*} , \tag{3.8.21}$$

for t and s in $\mathbb{C}^m$ and with $\frac{1}{p} + \frac{1}{p^*} = 1$, $1 \leq p \leq \infty$.

$$||\langle \nabla^3 u, t_1 \otimes t_2 \otimes t_3 \rangle| \leq C_0 \; |t_1|_{p_1} \cdot |t_2|_{p_2} \cdot |t_3|_{p_3} \tag{3.8.22}$$

for t_j in $\mathbb{C}^m$ and $1 = p_1^{-1} + p_2^{-1} + p_3^{-1}$.

Remark 3.8.6 The estimates given in Proposition 3.8.5 are just at the origin of the notion of 0-standardness introduced by J. Sjöstrand [Sj92b], that we shall meet at many places in this survey (see for example Definition 5.2.2).

The next step is to get from (3.8.20)

$$|\langle \partial_{x_\ell} \text{Hess}\, f_0, t \otimes s \rangle| \leq C \; |t|_\infty \cdot |s|_\infty . \tag{3.8.23}$$

This means that, for each ℓ, the matrix $\partial_{x_\ell}\text{Hess}\, f_0$ belongs to $\mathcal{L}(\ell^\infty, \ell^1)$ with a norm independent of N. We can get the same property for $(\text{Hess}\, f_0)^{-1} \cdot (\partial_{x_\ell}\text{Hess}\, f_0)$ using (3.8.17), and, with help of the lemma

Lemma 3.8.7 ([Sj93a])[12]

$$|\text{ Tr } A| \leq ||A||_{\mathcal{L}(\ell^\infty, \ell^1)} . \tag{3.8.24}$$

We get finally the property

$$|\nabla \ln \det \text{Hess}\, f_0|_\infty \leq C , \tag{3.8.25}$$

for all x in a complex ℓ^∞-ball (possibly smaller that the initial one). If we observe that $\Phi_1^{(m)}$ satisfies the same inequality, we have finally that the r.h.s. in the transport equation (3.8.18) satisfies exactly what we need in order to solve this equation. We now choose the constant E_0 in (3.8.18) so that the r.h.s. vanishes for $x = 0$, and we get a unique solution of (3.8.18) if we impose $f_1(0) = 0$. As in Section 3 of [Sj93a], we get also that

$$|\nabla f_1|_\infty \leq C , \tag{3.8.26}$$

[12] This lemma can be extended [Sor93a] as, $|\text{ Tr } A| \leq ||A||_{\mathcal{L}(E,E')}$, where $E = (\mathbb{R}^m, N_p)$ is the normed space $\mathbb{R}^m$ with the norm $N_p(x) = m^{-\frac{1}{p}} (\sum_i |x_i|^p)^{\frac{1}{p}}$.

4 Universal analysis for the Schrödinger operator

This section is devoted to the study of the Schrödinger operator

$$S_V(x, hD_x) = -h^2\Delta + V(x)$$

where V is a potential which is assumed to be C^∞ and semibounded. We consequently study the unique selfadjoint extension in $L^2(\mathbb{R}^m)$ of this differential operator initially defined on $C_0^\infty(\mathbb{R}^m)$. The semi-classical study will be presented in the next section, so we are interested more in this section in "universal" results connected with statistical mechanics.

4.1 Universal estimate of the splitting

The following theorem is due to J. Sjöstrand [Sj93b],

Theorem 4.1.1 *If V is strictly convex (with bounded second derivatives) and satisfies for some $\sigma > 0$*

$$\inf_x \lambda_{min}((\text{Hess}\, V)(x)) = \sigma\ , \tag{4.1.1}$$

then we have the following lower bound for the splitting between the two first eigenvalues of the Schrödinger operator S_V

$$\lambda_2(m,h) - \lambda_1(m,h) \geq h\sqrt{2\sigma}\ . \tag{4.1.2}$$

The lower bound ([Sj93b]) can be obtained by using the Maximum Principle (see the work of Singer, Wong, Yau and Yau [SWYY85] which was the inspiration of J. Sjöstrand) or the Brascamp-Lieb inequalities [BraLi76] (as explained in [HeSj92a]) and the strict convexity of the potential is the decisive and unique assumption. The estimate on the second derivatives of V can probably be eliminated by regularization. Let us recall the second appproach. By the min-max Principle we get immediately

$$\lambda_2(m,h) = \inf_{\substack{u\in C_0^\infty(\mathbb{R}^m)\\ \int_{\mathbb{R}^m} uu_1\,dx=0}} (S_V u|u)_{L^2(\mathbb{R}^m)} \tag{4.1.3}$$

and expressing u as $u = v\cdot u_1$ we get

$$\lambda_2(m,h) - \lambda_1(m,h) = \inf_{\substack{v\in C_0^\infty(\mathbb{R}^m)\\ \langle v\rangle=0}} \int |\nabla v|^2 u_1^2 dx \Big/ \int |v|^2 u_1^2 dx\ , \tag{4.1.4}$$

where by definition

$$\langle v\rangle = \int v u_1^2 dx \Big/ \int u_1^2 dx\ ,$$

that is the mean value of v with respect to the measure $u_1^2 dx$ (we recall that we usually normalize by taking $u_1 > 0$ and $\int u_1^2 dx = 1$). We shall prove in the next subsection

that u_1 is logconcave (cf. Proposition 4.2.1) and we can write consequently $u_1^2 \cdot dx$ as a measure $\exp -2\Phi(x) \cdot dx$ with Φ convex and we get the estimate from the Brascamp-Lieb inequality which was given in the preceding section.
We can also implement the improvement of the Brascamp-Lieb inequality due to J. Sjöstrand ([Sj94], Remark 1.1), which was recalled in Subsection 3.6 (3.6.11), and obtain

$$\lambda_2(m,h) - \lambda_1(m,h) \geq \rho_1 \ , \tag{4.1.5}$$

where we recall that ρ_1 is the first eigenvalue (strictly positive) of the operator A_1 which was introduced in Subsection 3.6. We recall also that we do not need strict convexity in order to get this last estimate.
As a corollary of the Theorem 4.1.1 and using additional arguments (see for example [PPW56] (for the upper bound for the splitting), [HeSj92a] and references therein) we get

Corollary 4.1.2

- Model (K)
  If $\nu < 1/4$ then the splitting between the two first eigenvalues λ_2 and λ_1 is controlled by

  $$h(1-4\nu)^{1/2} \leq \lambda_2(m,h,\nu) - \lambda_1(m,h,\nu) \leq 4\lambda_1(m,h,\nu)/m \ .$$

- Model (DK)
  If v is convex with $v''(x) \geq \sigma > 0$

  $$h\sqrt{2\sigma} \leq \lambda_2(m,h) - \lambda_1(m,h) \leq 4\lambda_1(m,h)/m \ .$$

- Model (BH) $(a > 0)$

  $$h \cdot \sqrt{a} \leq \lambda_2(m,h) - \lambda_1(m,h)$$

Other universal controls of the splitting are given in the work of W. Kirsch and B. Simon [KiSi87].

4.2 Logconcavity of the first eigenfunction

We observe that for one normalized positive Gaussian $\Omega_0 = \alpha \exp -a|x-x_0|^2$, which is clearly logconcave,

$$\lim_{t\to\infty} \exp t\lambda_1 \exp -tS_V\Omega_0 = \langle u_1 | \Omega_0 \rangle \cdot u_1 \ . \tag{4.2.1}$$

We now observe that $\langle u_1 | \Omega_0 \rangle \neq 0$, due to the strict positivity of Ω_0 and of the first eigenfunction u_1. So it suffices to prove the logconcavity of $\exp -t\ S_V\Omega_0$ which is easy to deduce[18] from the logconcavity of $\left(\exp -\frac{t}{n}V \cdot \exp \frac{t}{n}\Delta\right)^n \ \Omega_0$ by applying a particular case of the Trotter formula[19]

$$\exp -tS_V = \lim_{n\to\infty} \left(\exp -\frac{t}{n}V \cdot \exp \frac{t}{n}\Delta\right)^n . \tag{4.2.2}$$

But, if V is strictly convex, $\left(\exp -\frac{t}{n}V \cdot \exp \frac{t}{n}\Delta\right)^n \ \Omega_0$ is strictly logconcave by direct inspection of the integral defining its kernel and use of Lemma 3.7.2.
We conclude then that $-\ln u_1$ is convex if V is convex.

Let us now sketch the alternative proof of [SWYY85], [Sj93b] which was actually at the origin of all the proofs using the Maximum Principle presented here. We observe that, if $\Phi(x) = -\ln u_1(x)$, then we have

$$|\nabla\Phi|^2 = V + \Delta\Phi - \lambda_1 . \tag{4.2.3}$$

Taking the Hessian, we get

$$2\nabla\Phi \cdot \nabla \text{Hess}\,\Phi - \Delta \text{Hess}\,\Phi + 2\text{Hess}\,\Phi \circ \text{Hess}\,\Phi = \text{Hess}\,V \tag{4.2.4}$$

and we get, using the Maximum Principle (and modulo some regularization argument),

$$2\phi^2 \geq \sigma, \tag{4.2.5}$$

where

$$\sigma = \inf_x \lambda_{\min}(\text{Hess}\,V(x)) > 0 ,$$

and

$$\phi = \inf_x \lambda_{\min}(\text{Hess}\,\Phi(x))$$

We have finally obtained the

Proposition 4.2.1 *If $\sigma > 0$ and V satisfies*

$$|D_x^\alpha V(x)| \leq C_\alpha , \ \forall \alpha \ s.t. \ |\alpha| \leq 2 ,$$

then u_1 is logconcave and

$$\phi \geq \sqrt{\frac{\sigma}{2}} \tag{4.2.6}$$

[18] We refer to the book [Si79](p 136-139) for a more detailed proof giving logconcavity.

[19] We met this formula already in the matrix case (see Subsection 2.4) and we shall consider it again in Section 6.

Remark 4.2.2 On the same way one can prove the FKG estimates for the measure $u_1^2(x) \cdot dx$. The proof is quite analogous to the proof of the logconcavity using the Lemma 3.7.3.

Remark 4.2.3 There is a strong connection between the properties of the potential V and the corresponding properties of the phase $\Phi = -\ln u_1$. This permits to apply many results of Section 3 to the measure $u_1^2\, dx$. This will be explained now on an example but appears also in the semiclassical context.

As illustration of our preceding remark, let us see how one can deduce the properties of Φ from the properties of V using the Maximum Principle and equations like Equation (4.2.3) or similar equations deduced from this one by differentiation. Let us first sketch one step (following [Sj92a] (Proposition 2.1)) which leads to the following statement:

Proposition 4.2.4 *Let B be the space $\mathbb{R}^m$ equipped with some norm $||\cdot||_B$, and assume that for some fixed $0 < \theta < 1$*

$$\sup_{x\in\mathbb{R}^m} ||\frac{1}{2}\mathrm{Hess}\, V(x) - I||_{\mathcal{L}(B)} \leq \theta\ , \tag{4.2.7}$$

then

$$\sup_{x\in\mathbb{R}^m} ||\mathrm{Hess}\, \Phi(x) - I||_{\mathcal{L}(B)} \leq \tilde{\theta}\ , \tag{4.2.8}$$

where $\tilde{\theta} = \theta/(1 + (1-\theta)^{\frac{1}{2}})$.

The starting point of the proof is (4.2.3) and (4.2.4). Writing $\mathrm{Hess}\,\Phi = I + \mathrm{Hess}\,\Psi$, $\mathrm{Hess}\, V = 2\,I + \mathrm{Hess}\, W$, we get

$$2\nabla\Phi \cdot \nabla \mathrm{Hess}\,\Psi - \Delta \mathrm{Hess}\,\Psi + 2\mathrm{Hess}\,\Psi \circ \mathrm{Hess}\,\Psi + 4\mathrm{Hess}\,\Psi = \mathrm{Hess}\, W\ . \tag{4.2.9}$$

Modulo a regularization argument we can assume that $||\mathrm{Hess}\,\Psi(x)||_{\mathcal{L}(B)}$ tends to 0 as $|x| \to \infty$. Let x_0 be a point such that

$$M = ||\mathrm{Hess}\,\Psi(x_0)||_{\mathcal{L}(B)} = \sup_{x\in\mathbb{R}^m} ||\mathrm{Hess}\,\Psi(x)||_{\mathcal{L}(B)}\ . \ .$$

Let $\nu \in B$ be a normalized vector such that

$$||\mathrm{Hess}\,\Psi(x_0)\nu||_B = M$$

and let $\mu \in B^*$ such that

$$\langle \mathrm{Hess}\,\Psi(x_0)\nu|\mu\rangle_{B\times B^*} = M\ .$$

Then the function $x \mapsto \langle \text{Hess}\, \Psi(x)\nu|\mu\rangle_{B\times B^*}$ takes its maximum value at x_0 and we deduce from (4.2.9)

$$-2M^2 + 4M \leq 2\theta$$

and hence

$$\text{either } M \leq \theta/(1+(1-\theta)^{\frac{1}{2}}) \text{ or } M \geq 1+(1-\theta)^{\frac{1}{2}}\ .$$

The last possibility can be excluded by a deformation argument.

Remark 4.2.5 The reader can compare with the results of Subsection 3.4 where the Maximum Principle was already applied.

4.3 Thermodynamic limit: existence and speed of convergence

Theorem 4.3.1 *If we assume that $V^{(m)}$ is one of the three models (K), (DK) or (BH), then the limit,*

$$\Lambda(h) = \lim_{m\to\infty} (\lambda_1(m;h)/m) \tag{4.3.1}$$

exists.

This is what we call in this context the thermodynamic limit.

Remark 4.3.2 The research for a more general statement like we propose for the Laplace integrals in Section 3 (Proposition 3.3.2) is left to the reader. We can for example consider the assumptions given by J. Sjöstrand in [Sj92b](Section 8). The author studies also in this article the thermodynamic limit

$$\frac{-h}{mt} \ln \text{Tr } \exp -\frac{t}{h} S^{(m)}(x, hD_x)$$

and its asymptotic semiclassical expansion.

Preliminary comments. This is not surprising if one first looks at the simple case of the harmonic oscillator. We shall prove it following the same ideas of statistical mechanics ([HeSj92a]) as we used in the corresponding study of the Laplace integrals (see Proposition 3.3.2) and (3.3.4). Another ingredient is some a priori estimate for the Schrödinger operator. Let us also observe that the proof giving an approximate additivity property (modulo $\mathcal{O}(h)$) gives simultaneously

$$|\Lambda(h) - (\lambda_1(m;h)/m)| = h\mathcal{O}(1/m) \tag{4.3.2}$$

in the convex case and in the general case

$$|\Lambda(h) - (\lambda_1(m;h)/m)| = \mathcal{O}(1/m) \tag{4.3.3}$$

by easy arguments (see at the end of this subsection). We have consequently the analog of (3.3.10) for the Laplace integrals.

It is also interesting to analyze if this rate of convergence can be improved. J. Sjöstrand ([Sj92a]) proved actually an exponentially rapid convergence to the thermodynamic limit for a class containing Example (K). More precisely he get the following theorem

Theorem 4.3.3 *We assume the following properties,*

$$V^{(m)}(0) = 0\ ,\ \nabla V^{(m)}(0) = 0\ , \tag{4.3.4}$$

and that, for some $C > 0$, $\kappa > 0$ and $0 < \delta < 1$, the following properties are fullfilled

$$V^{(m)}(x_1, x_2, \ldots, x_m) = V^{(m)}(x_2, x_3, \ldots, x_m, x_1)\ . \tag{4.3.5}$$

The family

$$V^{(m,n)}(x,s) = (1-s)V^{(m+n)}(x) + s(V^{(m)}(x') + V^{(n)}(x''))$$

satisfies, for any $s \in [0,1]$, m and n in $\mathbb{N}$ $(m \geq 1; n \geq 1)$, $x = (x', x'') \in \mathbb{R}^{m+n}$,

$$\text{Hess}\,_x V^{(m,n)})(x,s) \geq\ 1/C\ , \tag{4.3.6}$$

$$||\text{Hess}\, V^{(m,n)})(x,s)||_{\mathcal{L}(B_1)} \leq C\ , \tag{4.3.7}$$

with

$$B_1 = (\mathbb{R}^{(m+n)}, ||\ \ ||_{\ell^\infty}). \tag{4.3.8}$$

Let us also introduce the family of normed spaces B (with underlying vector space $\mathbb{R}^{(m+n)}$) such that we have

$$B = \ell^\infty_{\rho_{m,n}} \tag{4.3.9}$$

where $\rho_{m,n}$ is defined on $\mathbb{Z}/(m+n)\mathbb{Z}$ and satisfies

$$\rho_{m,n}(1) = \rho_{m,n}(m) = 1 \tag{4.3.10}$$

$$e^{-\kappa} \leq \frac{\rho_{m,n}(\nu+1)}{\rho_{m,n}(\nu)} \leq e^{\kappa}\ ,\ \nu \in \mathbb{Z}/(m+n)\mathbb{Z} \tag{4.3.11}$$

We shall denote by $\mathcal{B}_2$ the family of these B's associated to some κ and we assume that for any B in $\mathcal{B}_2$:

$$||\text{Hess}\,_x V^{(m,n)}(x,s)||_{\mathcal{L}(B)} \leq C\ , \tag{4.3.12}$$

and more precisely

$$||(I - \text{Hess}_x V^{(m,n)})(x,s)||_{\mathcal{L}(B)} \leq \delta \ , \tag{4.3.13}$$

$$||\nabla_x^3 V^{(m,n)})(x,s)||_{\mathcal{L}(B_1 \times B, B)} \leq C \ , \tag{4.3.14}$$

$$|\nabla_x (V^{(m+n)} - (V^{(m)} \oplus V^{(n)}))(x)|_B \leq C \ , \tag{4.3.15}$$

$$||\nabla_x^2 (V^{(m+n)} - (V^{(m)} \oplus V^{(n)}))(x)||_{\mathcal{L}(B_1, B)} \leq C \ , \tag{4.3.16}$$

Under these assumptions we have,

$$|\lambda_1(m)/m - \left(\lim_{p\to\infty} \lambda_1(p)/p\right)| \leq E \exp -\kappa m/4 \ , \tag{4.3.17}$$

for some constant E (independent of m).

This is of course the analog of Theorem 3.3.11 but was proved by J. Sjöstrand earlier. We shall not give the proof here and refer to Subsection 3.3 for an idea of the proof. The assumptions are actually here a little weaker. We do not need indeed the control of

$$||\nabla_x^3 (V^{(m+n)} - (V^{(m)} \oplus V^{(n)}))(x)||_{\mathcal{L}(B_1 \times B_1, B)}$$

and of

$$||\nabla_x^4 (V^{(m,n)})(x,s))||_{\mathcal{L}(B_1 \times B_1 \times B, B)} \ .$$

Using the new techniques introduced in [Sj92b] it is also possible to treat the model (DK).

About the proof of Theorem 4.3.1. We just sketch the proof of Theorem 4.3.1 in the case $h = 1$ and for the model (DK) (adapting [HeSj92a] where the proof for a class of potentials containing the Kac model is already given). The proof for the (BH) model is given in [Sor93b]. Let us prove the approximate additivity:

$$|\lambda_1(m+p) - \lambda_1(m) - \lambda_1(p)| \leq D \ .$$

We start from the following decomposition of $S^{(m+p)}$

$$S^{(m+p)} = S^{(m)} + \hat{S}_p^{(m+1)} - W(x_m, x_1) + W(x_m, x_{m+1}) - W(x_{m+p}, x_{m+1}) + W(x_{m+p}, x_1)$$

with

$$\hat{S}_p^{(m+1)} = -\hat{\Delta}_p^{(m+1)} + \sum_{k=m+1}^{m+p-1} W(x_k, x_{k+1}) + W(x_{m+p}, x_{m+1}),$$

$$W(x,y) = a|x-y|^2,$$

and

$$\hat{\Delta}_p^{(m+1)} = \sum_{k=m+1}^{m+p} \partial_{x_k}^2 .$$

It is then clear that the infimum of the spectrum of $S_p^{(m+1)}$ is the same as the infimum of $S^{(p)}$. For the lower bound for $\lambda_1(m+p)$, one writes

$$\begin{aligned}\lambda_1(m+p) &= \langle S^{(m+p)} u_1^{(m+p)} | u_1^{(m+p)} \rangle \\ &\geq \langle S^{(m)} u_1^{(m+p)} | u_1^{(m+p)} \rangle + \langle \hat{S}_p^{(m+1)} u_{m+p} | u_{m+p} \rangle \\ &- \| W^{\frac{1}{2}}(x_m, x_1) u_1^{(m+p)} \|^2 - \| W^{\frac{1}{2}}(x_{m+p}, x_{m+1}) u_1^{(m+p)} \|^2 .\end{aligned}$$

We now observe that, under the assumption on v

$$v(t) \geq \frac{t^2}{C} - C ,$$

and, using the invariance by permutation,

$$\begin{aligned}\lambda_1(m) &= \langle S^{(m)} u_1^{(m)} | u_1^{(m)} \rangle \\ &\geq \sum_{j=1}^m \int_{\mathbb{R}^m} v(x_j) |u_1^{(m)}|^2 dx \\ &= m \int_{\mathbb{R}^m} v(x_1) |u_1^{(m)}|^2 dx \\ &\geq \frac{m}{C} \int_{\mathbb{R}^m} x_1^2 |u_1^{(m)}|^2 dx - Cm .\end{aligned}$$

From this we deduce[20]

$$\int x_j^2 |u_1^{(m)}|^2 dx \leq C^2 + C \frac{\lambda_1(m)}{m} \leq D_1 .$$

It is then easy to get:

$$\lambda_1(m+p) \geq \lambda_1(m) + \lambda_1(p) - D$$

for some new D independent of m, p and well controlled with respect to the semiclassical parameter if necessary.

For the upper bound, we proceed similarly using the function,

$$\hat{u}_{p,m} = u_1^{(m)}(x_1, \ldots\ldots, x_m) \cdot u_1^{(p)}(x_{m+1}, \ldots\ldots, x_{m+p}) .$$

Remark 4.3.4 For potentials of the form $V^{(m)}(x) = \sum_j v(x_j) + \sum_j W(x_j, x_{j+1})$, we have the natural lower bound

$$\lambda_1(m,h)/m \geq \Lambda^0(h)$$

[20] In the convex case, if $v''(t) \geq \sigma > 0$ and $v(0) = v'(0) = 0$, this can be improved and we can get a good control of the constants with respect to h, for example: $D_1 = \mathcal{O}(h)$.

where $\Lambda^0(h)$ is the smallest eigenvalue of the operator

$$L^2(\mathbb{R}^2) \ni u \mapsto \frac{1}{2}\left(-h^2\partial_t^2 - h^2\partial_s^2 + v(t) + v(s)\right) u + W(t,s)u \,.$$

We can also get, by applying the Minimax Principle and minimizing our Schrödinger operator on functions of the type

$$\mathbb{R}^m \ni x \mapsto \prod_{j=1}^{m} \phi(x_j) \,,$$

the following natural estimate

$$\lambda_1(m,h)/m \leq \inf_{\phi\,,\,\phi\in C_0^\infty(\mathbb{R})} J(\phi;h)/||\phi||^2 \,,$$

where

$$J(\phi;h) = h^2 \int_{\mathbb{R}} \phi'(t)^2\,dt \;+\; \int_{\mathbb{R}} v(t)\phi(t)^2\,dt \;+\; \int_{\mathbb{R}^2} W(t,s)\phi(t)^2\phi(s)^2\,dt\,ds \,.$$

This approach is of course reminiscent of $N-body$ techniques but does not give an optimal result. This can be indeed verified explicitely in the case when $v(t) = t^2$ and $W(t,s) = \frac{a}{2}(t-s)^2$. At least for $a < \frac{1}{4}$, $h = 1$, it can be computed that the infimum of $J(\cdot;h)$ is realized for even ϕ's and is equal to $\sqrt{1+a}$. By explicit computations (see next section), the thermodynamic limit is computed as equal to $\int_0^1 \sqrt{(1+a(1-\cos 2\pi\, t))}dt$ which is strictly smaller if $a > 0$.

The model (BH). The exponential convergence is no more true for the model (BH) as can be seen for example in the computations of V. Sordoni [Sor93b]. One way to see this property is to reduce, observing that we are interested in the first eigenfunction which is radial, the question to the study of the first eigenvalue of the selfadjoint realization of an ordinary differential equation in the radial variable. After a dilation (compare with Remark 3.3.7), the operator is unitary equivalent to the Friedrichs extension of the following ordinary differential equation, defined on $C_0^\infty(\mathbb{R}^+)$ by

$$m\cdot\left(-\frac{1}{m^2}\frac{d^2}{d\rho^2} + \mu\rho^2 + g\rho^{2n} + \frac{(m-1)(m-3)}{4m^2}\frac{1}{\rho^2}\right)\,, \tag{4.3.18}$$

for which the semiclassical analysis with $\frac{1}{m}$ as semiclassical parameter is relevant. The semiclassical analysis at the bottom which will be developed in the Section 5 permits to get a complete expansion in powers of $\frac{1}{m}$.

Speed of convergence and semiclassical limit. We emphasize again that if a semiclassical parameter is present we can control all the constants with respect to this parameter (see (4.3.2)). This will be very important for controlling the semiclassical expansion.

5 Semiclassical analysis for the Schrödinger operator

5.1 Introduction

In this section we present the results obtained by J. Sjöstrand or B. Helffer-J. Sjöstrand [Sj93a], [Sj93b], [HeSj92a] and [HeSj92b] for the study of the Schrödinger operator in large dimension. We choose to explain here how these techniques work also for the special model proposed by Dobrokhotov, Kolokol'tsov in [DoKo93]. This is done by following carefully what was done for the Kac model in [HeSj92a], [HeSj92b]. At the end of the section we shall recall also what was done for other models by V. Sordoni [Sor93a] and [Sor93b]. Let us consider $S^{(m)}(x, hD_x) = -h^2\Delta^{(m)} + V^{(m)}(x)$ where $V^{(m)}$ is one of the examples given in the introduction. Let $\lambda_j(m,h)$ be the sequence of the eigenvalues of $S^{(m)}$; we are interested in the problems (III) and (V) presented in the introduction. Let us present some of the results which were obtained in this case by Sjöstrand, Helffer-Sjöstrand or Sordoni ([Sj93a], [Sj93b], [Sj92a], [Sj92b], [HeSj92a], [HeSj92b] and [Sor93a]).

5.2 The thermodynamic limit: semiclassical expansions

We have studied in Section 4 the existence of the thermodynamic limit and we can now naturally ask for the existence of a semiclassical expansion in powers of h as was done for the Kac model in [HeSj92a], [HeSj92b] applying the techniques of [Sj93a], [Sj93b]. The first theorem was the following:

Theorem 5.2.1 *Model* (K). *If* $\nu \neq \frac{1}{4}$, *the thermodynamic limit* $\Lambda(h,\nu) = \lim_{m\to\infty}(\lambda_1(m;h,\nu)/m)$ *admits a complete asymptotic expansion:*

$$\Lambda(h,\nu) \sim h\left(\sum_{j\geq 0}\alpha_j(\infty,\nu)\cdot h^j\right) \tag{5.2.1}$$

as h tends to 0. Moreover, if we denote the corresponding semiclassical expansions for $\lambda_1(m;h,\nu)/m$ by,

$$(\lambda_1(m;h,\nu)/m) \sim h\left(\sum_{j\geq 0}\alpha_j(m,\nu)\cdot h^j\right), \tag{5.2.2}$$

there exists $k_0(\nu) > 0$ s.t. for each j, there exists a constant $C_j(\nu)$, s.t.

$$|\alpha_j(\infty,\nu) - \alpha_j(m,\nu)| \leq C_j(\nu)\cdot exp(-k_0(\nu)m)\ . \tag{5.2.3}$$

$k_0(\nu)$ and $C_j(\nu)$ can be chosen independent of ν in any compact in $\mathbb{R}^+ \setminus \{1/4\}$.

In particular we get for the (DK) model (and some of its extensions) the following asymptotics for the splitting

$$\lambda_2(m;h) - \lambda_1(m;h) = h^{1/2} A_m(h) \exp -\frac{m}{h}\beta , \tag{5.3.16}$$

with $A_{m,0} > 0$.

An explicit formula for $A_{m,0}$ is given by Dobrokhotov and Kolokoltsov [DoKo94] who prove also the following property

Proposition 5.3.4

$$\lim_{m\to\infty} (\ln(A_{m,0})/m) = a_0 . \tag{5.3.17}$$

Remark 5.3.5 On the Kac model. The situation is not clear for the Kac model. The study of the asymptotic behavior of the Agmon distance as $m \to \infty$ between the two wells of the Kac model, when $\nu > \frac{1}{4}$, has not be done completely. The geometry is more difficult to analyze and the study of the form of the minimal geodesic between the two wells is also not clear.

5.4 Presentation of the rigorous results in the large dimension semiclassical case for the (DK) model

We shall see in this subsection that the results obtained before for the thermodynamic limit for the Kac model (cf. Theorem 5.2.1) are still valid for the model (DK) of Dobrokhotov-Kolokoltsov.

Theorem 5.4.1 (*the single well case*) *If v satisfies the conditions:*

$$v''(x) \geq \rho\,, \quad \text{with } \rho > 0 \quad , \quad v(0) = 0\,, \tag{5.4.1}$$

and for some N_0, some constants C_k, some $C > 0$

$$|v(x)| \leq C\langle x\rangle^{N_0} \;,\; |v^{(k)}(x)| \leq C_k(1 + |v(x)|)\langle x\rangle^{-k} \;, \tag{5.4.2}$$

then the thermodynamic limit

$$\Lambda(h) = \lim_{m\to\infty} (\lambda_1(m;h)/m) \tag{5.4.3}$$

admits a complete asymptotic expansion

$$\Lambda(h) \sim h \sum_{j\geq 0} \alpha_j \cdot h^j \tag{5.4.4}$$

as h tends to 0.
Moreover, if we denote the corresponding semiclassical expansion for $\lambda_1(m;h)/m$ by[26]

$$(\lambda_1(m;h)/m) \sim h \sum_{j\geq 0} \alpha_j(m) \cdot h^j \,, \tag{5.4.5}$$

there exists $k_0 > 0$ s.t. for each j, there exists a constant C_j, s.t.

$$|\alpha_j - \alpha_j(m)| \leq C_j \cdot \exp(-k_0 m) \,. \tag{5.4.6}$$

We recall that we have already obtained (Corollary 4.1.2) that if v satisfies the preceding conditions, then the splitting between the two first eigenvalues λ_2 and λ_1 is controlled by

$$h\sqrt{2\rho} \leq |\lambda_2(m,h) - \lambda_1(m,h)| \leq h\tilde{C} \,. \tag{5.4.7}$$

We now consider the double well case.

Theorem 5.4.2 (*the double well case*) *If v satisfies the conditions:*

$$\begin{aligned} &v(x) = v(-x) \; ; \; v \geq 0 \; ; \\ &v(x) = 0 \text{ iff } x = \pm s \; , \; v''(s) > 0 \; , \\ &v(x) \geq \tfrac{1}{C}x^2 - C \; , \end{aligned} \tag{5.4.8}$$

and (5.4.2), *then the thermodynamic limit $\Lambda(h) = \lim_{m\to\infty}(\lambda_1(m;h)/m)$ admits a complete asymptotic expansion as in Theorem* 5.4.1.

Remark 5.4.3 As already mentioned, Theorems 5.4.1 and 5.4.2 correspond to Theorem 5.2.1 for the Kac model.

We now rewrite explicitely Theorem 5.3.1 in the case of the (DK) model,

Theorem 5.4.4 *Under the same assumptions as in Theorem* 5.4.2, *let us consider the set $\mathcal{N}$ of the pairs (m,h) in $\mathbb{N} \times \mathbb{R}^+$ defined by*

$$m \leq C\, h^{-N_0} \,,$$

for some C and N_0 ; then there exists C_0 , h and $\epsilon_0 > 0$ such that, for all the (m,h) in $\mathcal{N}$ satisfying $0 < h \leq h_0$, the splitting between the two first eigenvalues λ_2 and λ_1 is controlled by,

$$\lambda_2(m,h) - \lambda_1(m,h) \leq C_0 \cdot \exp -(\epsilon_0 \cdot m/h). \tag{5.4.9}$$

[26] It can be proved that this $\sim$ is uniform with respect with m

In the proof of Theorems 5.4.1, 5.4.2 and 5.4.4, we simply adapt [HeSj92a] and [HeSj92b] (which follow the strategy initiated by J. Sjöstrand in [Sj93a] and [Sj93b]). Let us recall that the basic idea in order to analyze the thermodynamic limit was to compare a formal expansion of the eigenvalue divided by the dimension (deduced from the WKB approximation whose construction with control with respect to the dimension m was also given in [Sj93a] and [Sj93b]) of a one well problem and the first eigenvalue divided by the dimension of our problem. As in the corresponding problem for the Laplace integral (compare with Subsection 3.8), we can distinguish three parts.

- In the first one, one compares the WKB approximation of the one well problem and the first eigenvalue of the Dirichlet problem in a sufficiently small ℓ^∞ -ball around the point where the minimum of the potential was attained.

- In the second one we compare the first eigenvalue of the Dirichlet problem in this small ℓ^∞ -ball with the global problem.

  In these two steps we work modulo $m.O_N(h^N)$ (for any N) but the dimension is controlled by $m = O(h^{-N_0})$.

- The last step consists in eliminating the restriction on the dimension and is identical to the convex case (due to the control of the convergence in the thermodynamic limit).

Actually we need to prove a more precise result permitting to analyze the splitting between the two first eigenvalues. To understand what is needed, let us recall the following classical formula for the splitting which was already used for our lower bound of the splitting in the convex case (see (4.1.4) or, in the literature, [HeSj92b] Section 3 or [KiSi87]),

$$\begin{aligned} &\lambda_2 - \lambda_1 & (5.4.10)\\ = \;& h^2 \cdot \inf_{\substack{\phi\;;\;\phi\in C_0^\infty\;,\\ \int \phi(x)\, u_1^{(m)}(x)^2\, dx = 0}} \left[\left(\int |\nabla\phi|^2\, u_1^{(m)}(x)^2\, dx \right) \Big/ \left(\int |\phi|^2\, u_1^{(m)}(x)^2\, dx \right) \right] . \end{aligned}$$

Here $u_1^{(m)}$ denotes the first normalized eigenfunction. The estimates about the splitting are then deduced from a good choice of ϕ and from a precise information on the decay of $u_1^{(m)}$ in suitable domains. We observe that if v admits two symmetric non-degenerate minima at $\pm s$, then the potential admits two minima and that there exists δ,[27] s.t. the region $W(\delta)$ defined by,

$$W(\delta) := \left\{ x \in \mathbb{R}^m \;,\; -\delta \leq \sum_i x_i \leq \delta \right\} ,$$

[27] In our case we will search for a δ satisfying $0 < \delta < ms$.

does not contain these two wells.
Let us recall also that according to the symmetries of the problem we have

$$u_1^{(m)}(-x) = u_1^{(m)}(x) \, . \tag{5.4.11}$$

We then obtain easily, as in [HeSj92b]and by choosing,[28]

$$\phi(x) = \theta\left(\sum_j x_j/\delta\right) \, ,$$

with θ, a C^∞ function satisfying $\theta = -1$ near $-s$ and $\theta = 1$ near $+s$, the following inequality,

$$\lambda_2(m,h) - \lambda_1(m,h) \leq C\, m \cdot (h/\delta)^2 \cdot a(m,h,\delta)^2/(1 - a(m,h,\delta)^2) \, , \tag{5.4.12}$$

with

$$a(m,h,\delta) = ||u_1^{(m)}||_{L^2(W(\delta))} \, . \tag{5.4.13}$$

The control of the splitting is a consequence of

Theorem 5.4.5 *There exists $\tilde{C}$, h_0 and $\delta > 0$ s.t.*

$$a(m,h,\delta) \leq \tilde{C}\, \exp(-m/\tilde{C}h) \tag{5.4.14}$$

if $m = \mathcal{O}(h^{-N_0})$ and $0 < h < h_0$.

An instructive model. In order to understand what is going on, it is perhaps useful to analyze in detail an "almost" solvable model. Let us consider[29]

$$S^{(m)}(x, hD_x) = -h^2\Delta_x + \frac{1}{2}|x|^2 - \left|\sum_{i=1}^m x_i\right| \, , \tag{5.4.17}$$

[28] We forget a regularization argument: our ϕ is not in C_0^∞.

[29] This model was inspired by the limit $\nu \to \infty$ of the Kac model which gives after rescaling the model

$$-h^2\Delta_x + \frac{1}{2}|x|^2 - \frac{1}{2}\sum_{i=1}^m |x_i + x_{i+1}| \, , \tag{5.4.15}$$

which seems less easy to analyze.
Another model inspired by the (DK) model could be

$$-h^2\Delta_x + \sum_{i=1}^m \left(\frac{1}{2}x_i^2 - |x_i|\right) + \frac{a}{2} \cdot \sum_{i=1}^m (x_i - x_{i+1})^2 \, . \tag{5.4.16}$$

In all these models, the "one well" problems are (at least in small neighborhoods of these wells) attached to harmonic oscillators.

and look for an estimate of the splitting. We first perform an orthonormal change of variables s.t.

$$y_1 = m^{-\frac{1}{2}} \sum_{i=1}^{m} x_i \,.$$

We get in the new coordinates the following operator with separate variables

$$-h^2 \Delta_y + \frac{1}{2}|y|^2 - \sqrt{m}\,|y_1| \,.$$

We now perform a new dilation $y = m^{\frac{1}{2}} \tilde{y}$ and get

$$m \left(- \left(\frac{h}{m}\right)^2 \Delta_{\tilde{y}} + \frac{1}{2}|\tilde{y}|^2 - |\tilde{y}_1| \right) ,$$

and this permits to reduce the analysis of the splitting to a corresponding analysis of the one dimensional model

$$m \left(- \left(\frac{h}{m}\right)^2 \partial_t^2 + \frac{1}{2}t^2 - |t| \right) .$$

The new semiclassical parameter is $\tilde{h} = \frac{h}{m}$ and the splitting can be analyzed by standard classical techniques.
There is actually a small difficulty due to the fact that the potential is only Lipschitz, but this does not affect the optimal result. One can indeed observe, in order to eliminate the problem relative to the discontinuity of the derivative at 0, that this splitting is also the splitting between the first eigenvalue of the Neumann problem in $]0, +\infty[$ and the first eigenvalue for the Dirichlet problem in $]0, \infty[$ for the harmonic oscillator $-(\tilde{h})^2 \partial_t^2 + \frac{1}{2}t^2 - t$. But this is also the splitting between the first eigenvalue of the Neumann problem and the first eigenvalue for the Dirichlet problem in $]-1, \infty[$ for the harmonic oscillator $-(\tilde{h})^2 \partial_s^2 + \frac{1}{2}s^2$. A similar problem was analyzed in detail by C.Bolley and B. Helffer in [BoHe93] and the same type of problems appears also in the instanton approach (see for example [Sor94]).
One can finally obtain

$$\lambda_2(m,h) - \lambda_1(m,h) \sim (hm)^{\frac{1}{2}} a(\tilde{h}) \exp -mS/h \,, \tag{5.4.18}$$

where S is the Agmon distance between the two minima ± 1 of the potential $v(t) = \frac{1}{2}t^2 - |t|$ relative to the metric $(v(t) + \frac{1}{2}) \cdot dt^2$,

$$S = 2 \int_0^1 \sqrt{v(t) + \frac{1}{2}} dt = 1/\sqrt{2} \,.$$

One can also observe that $\lambda_1(m,h)/m$ is the smallest eigenvalue of

$$\mathcal{S}(t, \tilde{h} D_t) = -(\frac{h}{m})^2 \partial_t^2 + \frac{1}{2}t^2 - |t| \,.$$

Taking the limit $m \to \infty$ corresponds to taking the limit $\tilde{h} \to 0$ and we get

$$\lim_{m\to\infty} \lambda_1^{(m)}(h)/m = -\frac{1}{2} + \frac{h}{\sqrt{2}}. \tag{5.4.19}$$

A superficial analysis could lead to think that we have an error in $m\mathcal{O}(\tilde{h}^2) = \frac{1}{m}\mathcal{O}(h^2)$ but, near $t = 1$,
(and modulo an exponentially small error in $m \exp -(S-\epsilon)\frac{m}{h}$ for any $\epsilon > 0$),
the first eigenvalue of $\mathcal{S}(t, \tilde{h}D_t)$ is the same as the first eigenvalue of

$$\hat{\mathcal{S}}(t, \tilde{h}D_t) = -(\frac{h}{m})^2\partial_t^2 + \frac{1}{2}t^2 - t\,.$$

which is exactly $-\frac{1}{2} + \frac{h}{\sqrt{2}}$.
We recover what we were waiting for this model, that is a convergence to the thermodynamic limit which is exponentially rapid and directly related to the tunneling effect. The operator $\hat{\mathcal{S}}$ corresponds indeed to a "right" approximation[30] of the double well problem which appears to be an exact harmonic oscillator. The same is of course true for the "left" approximation. It is also interesting to consider more generally the family of operators

$$\nu \mapsto S^{(m)}(x, hD_x, \nu) = -h^2\Delta_x + \frac{1}{2}|x|^2 - \nu|\sum_{i=1}^{m} x_i|\,. \tag{5.4.20}$$

Playing with ν could permit to understand the transition procedure around $\nu = 0$ for this model. The case $\nu = 0$ gives a splitting of order $m \cdot \sqrt{2}\,\tilde{h} = \sqrt{2}\,h$. In the case $\nu < 0$, the splitting is obtained (for $\frac{h}{m}$ small enough) by the splitting of $m\left(-\frac{h}{m}^2\Delta_{\tilde{y}'} + \frac{1}{2}|\tilde{y}'|^2\right)$ on $L^2(\mathbb{R}^{m-1})$ that is $\frac{h}{\sqrt{2}}$.
We have finally obtained the following lemma,

Lemma 5.4.6 *If $\lambda_j(m, h, \nu)$ are the eigenvalues of $S^{(m)}(x, hD_x, \nu)$, then we have*

$$\lim_{m\to\infty} \lambda_1(m, h, \nu)/m = -\frac{\nu_+^2}{2} + \frac{h}{\sqrt{2}}\,, \tag{5.4.21}$$

where $\nu_+ = \sup(\nu, 0)$ and,

$$\lim_{m\to\infty} (\lambda_2(m, h, \nu) - \lambda_1(m, h, \nu)) = \begin{cases} 0 & \text{if } \nu > 0\,, \\ \frac{h}{\sqrt{2}} & \text{if } \nu < 0\,. \end{cases} \tag{5.4.22}$$

[30] For lack of space, we have not defined clearly what these "left" or "right" approximations are. Let us simply say that in the study of the tunneling effect the strategy chosen in [HeSj84-87], part I (see also [CDV93] or [He88] for a presentation) is to introduce suitable Dirichlet realizations of the Schrödinger operator in open sets containing only one well. By suitable, it was meant that they contain a ball (with respect to the Agmon metric) centered at the bottom of one well and whose radius is strictly larger than the half of the minimal distance between the wells.

model (cf. [He91]).
Let us remember that we consider the operator

$$S^{(m)}(h) = -h^2 \sum_{k=1}^{m} \partial_{x_k}^2 + \frac{a}{2} \sum_{k=1}^{m} (x_k - x_{k+1})^2 + \sum_{k=1}^{m} v(x_k) ,$$

(We recall the convention $x_{m+1} = x_1$).

The inequality we want to prove can be rewritten in the form

$$- C \cdot m \cdot h^{\frac{6}{5}} \leq \lambda_1(m,h) - h \cdot \lambda^{app.}(m) , \tag{5.5.13}$$

where $\lambda^{app.}(m)$ is the already computed first eigenvalue of

$$- \sum_{k=1}^{m} \partial_{x_k}^2 + \frac{a}{2} \sum_{k=1}^{m} (x_k - x_{k+1})^2 + \frac{v''(s)}{2} |x|^2 ,$$

which is directly obtained as Tr $Q^{\frac{1}{2}}$ where $Q^{\frac{1}{2}}$ is the positive square root of the matrix Q attached to the quadratic form

$$\langle x|Q|x\rangle = \frac{a}{2} \sum_{k=1}^{m} (x_k - x_{k+1})^2 + \frac{v''(s)}{2} |x|^2 .$$

This potential is sometimes called the harmonic spin-chain potential (see for example Section 2.8, p.85-86 in [Roe94], Section 4.4 p.258 in [Ga83]).
A partition of unity. Let θ be a function in $C_0^\infty(]-1,+1[)$ such that

$$0 \leq \theta \leq 1 \text{ with } \theta = 1 \text{ on }]-1/2, 1/2[.$$

We now introduce[32] for $j \in \{-1, 1\}$, $k \in \{1, \dots, m\}$

$$\phi_k^j(x,h) = \theta((x_k - js)/h^{\frac{2}{5}}) , \tag{5.5.14}$$

and we choose θ and ϕ_k^0 such that the following relation is satisfied

$$\sum_{j\in\{-1,0,1\}} \phi_k^j(x,h)^2 = 1 . \tag{5.5.15}$$

We now introduce a multiindex α with

$$\alpha = (\alpha_1, \alpha_2, \dots, \alpha_m) \text{ with } \alpha_k \in \{-1, 0, 1\} .$$

[32]There are two possible natural choices. The first is to take ℓ^∞-boxes in the x variables. The second one is to take the ℓ^∞-boxes in the y variables adapted to the harmonic approximation. We choose here the first approach.

We now introduce

$$\phi_\alpha(x,h) = \prod_{k=1}^{m} \phi_{\alpha_k}(x,h) .$$

We deduce from (5.5.14) and (5.5.15) the following properties,

$$\sum_\alpha \phi_\alpha(x,h)^2 = 1 , \tag{5.5.16}$$

and

$$\sum_\alpha ||\nabla\phi_\alpha(x,h)||^2 \leq C \cdot m \cdot h^{-\frac{4}{5}} , \tag{5.5.17}$$

where C is independent of h and m.
The first step is a very elementary and old formula called in some context "IMS localization formula"

$$S(h) = \sum_\alpha \phi_\alpha S(h)\phi_\alpha - h^2 \sum_\alpha ||\nabla\phi_\alpha||^2 . \tag{5.5.18}$$

According to (5.5.17), Formula (5.5.18) permits to reduce the proof of (5.5.13) to the proof of the existence of C s.t. for all α, m, we have,

$$- Cmh^{\frac{6}{5}} + h \cdot \lambda^{app.}(m) \leq \inf_{\substack{u \in C_0^\infty(E_\alpha) \\ ||u||=1}} \langle S(h)u \, , \, u\rangle_{L^2} , \tag{5.5.19}$$

where E_α is defined as

$$E_\alpha = \cap_{k=1}^m B_k^{\alpha_k} ,$$

where,

- for $j = \pm 1$,

$$B_k^j = \{x \, , \, |x_k - j \cdot s| \leq h^{\frac{2}{5}}\} ,$$

  and

- for $j = 0$, by

$$B_k^0 = \{x \, , \, |x_k - \ell \cdot s| \geq \frac{1}{2} h^{\frac{2}{5}} \text{ for } \ell = \pm 1\} .$$

We now observe that, if

$$|z - \ell \cdot s| \geq \frac{1}{2} h^{\frac{2}{5}} \text{ for } \ell = \pm 1 ,$$

there exists $\delta > 0$ s.t.

$$v(z) \geq \delta \cdot h^{\frac{4}{5}} , \tag{5.5.20}$$

interval induced by $\mathbb{R}/m \cdot \mathbb{Z}$. We now write $\{1, \ldots, m\} \setminus \mathcal{C}$ as a union of disjoint maximal intervals I_ℓ

$$\{1, \ldots, m\} \setminus \mathcal{C} = \cup I_\ell \ . \tag{5.5.32}$$

We shall estimate from below by a decoupled potential where there are no interaction between the intervals I_ℓ. We have,

$$\text{Tr } (Q_{\mathcal{C}}^{\frac{1}{2}}) \geq \sum_\ell \text{Tr } (\tilde{Q}_{I_\ell}^{\frac{1}{2}}) \tag{5.5.33}$$

where for all interval I we have defined

$$\tilde{Q}_I(x) = \sum_{k \in \tilde{I}} \left(\frac{a}{2}(x_k - x_{k+1})^2 + \frac{v''(s)}{2} x_k^2 \right) \tag{5.5.34}$$

with $\tilde{I} = \{k \in I \ , \ s.t \quad k+1 \in I\}$.
Let us admit for a moment the following lemma

Lemma 5.5.8 *There exists a constant C_1 such that for all I we have,*

$$|\text{ Tr } (\tilde{Q}_I^{\frac{1}{2}}) - \lambda^{app.}(n_I)| \leq C_1 \ . \tag{5.5.35}$$

We recall also the lemma

Lemma 5.5.9 *There exists $\lambda_0 > 0$ and λ_1 such that*

$$\lambda^{app.}(p) = p \cdot \lambda_0 + \lambda_1 + \mathcal{O}(1/p) \ . \tag{5.5.36}$$

This is just a version of the Euler-Mac Laurin Formula on Riemann sums (or of the result on the thermodynamic limit given in Theorem 4.3.3).
Note that, because of the invariance by circular permutation, we have actually

$$\lambda^{app.}(p) = p \cdot \lambda_0 + \mathcal{O}(\exp -\kappa p) \ , \tag{5.5.37}$$

for some $\kappa > 0$ (see for example [Sj92a] or (4.3.17)).

We now observe that,

$$\text{the number of intervals } \mathrm{I}_\ell \text{ is estimated by } n_{\mathcal{C}} \ . \tag{5.5.38}$$

We then deduce from (5.5.35) and (5.5.36) that

$$\text{Tr } ((\tilde{Q}_{I_\ell})^{\frac{1}{2}}) \geq n_{I_\ell} \cdot \lambda_0 - C \ . \tag{5.5.39}$$

Summing up over all intervals, we get

$$\mathrm{Tr}\,(Q_{\mathcal{C}}^{\frac{1}{2}}) \geq \left(\sum_{\ell} n_{I_\ell}\right) \cdot \lambda_0 - C \cdot n_{\mathcal{C}} \geq (m\lambda_0 + \lambda_1) - C_1 \cdot (n_{\mathcal{C}} + 1) \, .$$

We then obtain the existence of a constant C_2 such that

$$\mathrm{Tr}\,(Q_{\mathcal{C}}^{\frac{1}{2}}) \geq \lambda^{app.}(m) - C_2 \cdot (n_{\mathcal{C}} + 1) \, .$$

We have also equality for $\mathcal{C} = \emptyset$, and consequently

$$\mathrm{Tr}\,(Q_{\mathcal{C}}^{\frac{1}{2}}) \geq \lambda^{app.}(m) - 2\,C_3 \cdot n_{\mathcal{C}} \, . \tag{5.5.40}$$

It remains only to prove Lemma 5.5.8.

Proof of Lemma 5.5.8. We change a little the notation in order to come back to the case when $I = \{1, \ldots, m\}$. We have now to compare the first eigenvalues $\tilde{\lambda}_{1,m}$ and $\hat{\lambda}_{1,m}$ of the two following operators on $L^2(\mathbb{R}^m)$

$$-\Delta + \frac{a}{2} \sum_{k=1}^{m-1} (x_k - x_{k+1})^2 + \frac{v''(s)}{2} \sum_{k=1}^{m} x_k^2 \, ,$$

and

$$-\Delta + \sum_{k=1}^{m} \left[\frac{a}{2}(x_k - x_{k+1})^2 + \frac{v''(s)}{2} x_k^2 \right] \, .$$

We clearly have the inequality

$$\tilde{\lambda}_{1,m} \leq \hat{\lambda}_{1,m} \, .$$

We have consequently to find an upper bound for $\hat{\lambda}_{1,m}$. Let us consider the first normalized eigenfunction of the first operator $\tilde{\psi}_m$ and we have now to prove the existence of a constant $C_0 > 0$ such that, for all m, we have,

$$\langle (x_1 - x_m)^2 \tilde{\psi}_m \mid \tilde{\psi}_m \rangle \leq C_0 \, . \tag{5.5.41}$$

We deduce (5.5.41) from

$$||x_1 \cdot \tilde{\psi}_m||^2 \leq \; C_1 \, ,$$

and

$$||x_m \cdot \tilde{\psi}_m||^2 \leq \; C_1 \, .$$

We now diagonalize the quadratic form $\tilde{Q}_m(x)$ in the form

$$\tilde{Q}_m(x) = \sum_k \tilde{\omega}_{km} y_k^2 \, .$$

The research of a formal WKB solution corresponds to the simultaneous research of $\phi(x;h)$ and of $E(h) \sim \sum_{j=0}^{\infty} E_j\, h^j$ such that

$$(-h^2\Delta + V(x) - hE(h)) \exp -\phi(x;h)/h \sim 0 ,$$

that is,

$$V - |\nabla\phi|^2 + h^2\Delta\phi - hE(h) \sim 0 . \tag{5.6.2}$$

This leads to the following system of equations which we shall call "eiconal" equation

$$\begin{array}{lll}
(\mathrm{E}) & V(x) - |\nabla\phi_0(x)|^2 & = 0 , \\
(\mathrm{T}_1) & 2\nabla\phi_0(x)\cdot\nabla\phi_1 & = \Delta\phi_0(x) - E_0 , \\
(\mathrm{T}_2) & 2\nabla\phi_0(x)\cdot\nabla\phi_2 & = \Delta\phi_1(x) - E_1 - |\nabla\phi_1|^2 , \\
\cdot & & \\
\cdot & & \\
\cdot & & \\
(\mathrm{T}_k) & 2\nabla\phi_0(x)\cdot\nabla\phi_k & = \Delta\phi_{k-1}(x) - (\nabla\phi_1\cdot\nabla\phi_{k-1} \\
 & & \quad +\nabla\phi_2\cdot\nabla\phi_{k-2} + \cdots + \nabla\phi_{k-1}\cdot\nabla\phi_1) - E_{k-1} .
\end{array}$$

This is quite similar to the equation we introduced in Subsection 3.8 (see in particular (3.8.9)). Without the control with respect to the dimension, this construction is similar to the construction of the other WKB solution.

The rigorous treatment with control with respect to the dimension was introduced by J. Sjöstrand [Sj93a], [Sj93b] and further developments were introduced in [HeSj92a]. We follow here the new improvment by J. Sjöstrand in [Sj92b] to which we refer for details (see Section 2 in this paper). The following theorem is due to J. Sjöstrand ([Sj93a] + [Sj92b])

Theorem 5.6.2 *Let $d > 0$. Let $V^{(m)}$ be a C^∞ potential defined in a ball $B_\infty(x_c, d)$ satisfying*

- *strict convexity:*

$$\mathrm{Hess}\, V^{(m)}(x) \geq \sigma > 0 \;, \; \text{for} x \in B_{\mathbb{R}}(x_c, d) \text{for} \sigma > 0 \;. \tag{5.6.3}$$

-

$$V^{(m)} \text{ is } 0 - \textit{standard in } B_\infty(0,d) \textit{ relatively to the weights } \rho_\ell \textit{ satisfying} \tag{5.6.4}$$

$$\exp(-\kappa) \leq \rho_\ell(j+1)/\rho_\ell(j) \leq \exp(\kappa) \;. \tag{5.6.5}$$

•

$$\text{Hess}\, V^{(m)}(0) = D^{(m)} + A^{(m)}, \tag{5.6.6}$$

where $D^{(m)}$ is diagonal (positive definite) and

$$||A^{(m)}||_{\mathcal{L}(\ell^p_\rho, \ell^p_\rho)} \leq r_1 \; < \; r_0 \leq \lambda_{min}(D)$$

for all p s.t. $1 \leq p \leq \infty$ and for all ρ satisfying (5.6.5).

Then $V^{(m)}$ admits a unique minimum $x_c^{(m)}$ in $B_\infty(0,d)$ and there is a unique formal solution[34] *$(E(h), \phi(x,h))$ of the "eiconal" system such that the ϕ_j are 0-standard and the E_j are $\mathcal{O}_j(m)$.*

Sketch of the proof. We assume to simplify that $x_c = 0$. We sketch the construction of ϕ_0 and show why this solution is 0-standard. This is done by a careful analysis of the outgoing manifold Λ starting in $T^*\mathbb{R}^m$ of the point $(0,0)$ and associated with the hamiltonian flow H_q where $q(x,\xi) = \xi^2 - V(x)$. The phase ϕ_0 is then the generating function of this manifold, that is locally

$$\Lambda = \{(x,\xi)\; | \xi = \nabla\phi_0(x)\}\ .$$

Then we observe, using Lemma 3.8.7, that $\Delta\phi_0$ is 0-standard and the necessary and sufficient condition to solve (T_1) is then

$$E_0 = (\Delta\phi_0)(0)\ .$$

This condition determines E_0, gives an estimate of E_0 as a consequence of the 0-standard character of ϕ_0 and $\Delta\phi_0$.
We remark here that in the first articles by J. Sjöstrand this argument was obtained through an estimate of ϕ_0 in a complex ℓ^∞-ball (see the assumptions we give in our corresponding semiclassical study of the Laplace integrals) from which we deduce the estimates on the derivatives through Cauchy estimates.
Then the solution of (T_1) is given, up to a constant, by

$$\phi_1(x) = \int_{-\infty}^{0} (\Delta\phi_0 - E_0)[\exp(t\nabla\phi_0 \cdot \partial_x)(x)]\; dt$$

which can be studied modulo a good knowledge of the flow attached to $\nabla\phi_0 \cdot \partial_x$.

Remark 5.6.3 This approach is quite similar to what we meet in our semiclassical study of the Laplace integrals (see Subsection 3.8). As was already mentioned by J. Sjöstrand, one can surely eliminate the assumptions of analyticity in the statements concerning the Laplace integrals and replace it by the corresponding assumptions of 0-standardness which are given here. This is essentially treated in Section 3 of [Sj92b].

[34] everything is m-dependent.

We now implement additional assumptions in order to control the convergence of each $E_j(m)/m$, and the assumptions are then analogous to what appears in our study of the thermodynamic limit first introduced in the semiclassical context in [HeSj92a].

Theorem 5.6.4 *We keep the conditions given in Theorem* 5.6.2. *We assume that* $x_c = 0$ *as well as the following additional conditions.*

- *Invariance by circular permutation:*
  For every m, $V^{(m)}$ *satisfies,*

$$V^{(m)}(x_m, x_1, \dots, x_{m-1}) = V^{(m)}(x_1, \dots, x_m) . \tag{5.6.7}$$

- *Short range interaction:*
  For all m, n, *for all* ρ *defined on* $\{1, \dots, m+n\}$ *satisfying (5.6.5) and*

$$\rho(j) = 1 \text{ for } j \geq m+1 , \text{ and } \rho(1) = 1 ,$$

  the interaction $W^{(m,n)} = V^{(m+n)} - (V^{(m)} \tilde{\oplus} V^{(n)})$, *satisfies,*[35] (5.6.1) *and* (5.6.4) *uniformly but for* $1 = \rho\rho_1 \dots \rho_k$, *instead of* $1 = \rho_1 \dots \rho_k$.

- *Uniform* 0*-standardness:*
  The family $(1-t)(V^{(m)} \tilde{\oplus} V^{(n)}) + tV^{(m+n)}$ *is, for* $0 \leq t \leq 1$, *uniformly* 0*- standard for the* ρ*'s satisfying (5.6.5) and*

$$\begin{aligned} \exp -\kappa \leq \rho(n+m)/\rho(1) \leq \exp \kappa , \\ \exp -\kappa \leq \rho(m)/\rho(1) \leq \exp \kappa . \end{aligned} \tag{5.6.8}$$

Then the $E_j(m)/m$ *are convergent as* $m \to \infty$ *to some* α_j *and we have, for some* $\kappa_0 > 0$,

$$|E_j(m)/m - \alpha_j| \leq C_j \cdot \exp -(\kappa_0 \cdot m) \tag{5.6.9}$$

Verification of the assumptions for the (DK) model. The verification of (5.6.4) is easy. We first observe (always with the convention that $x_{m+1} = x_1$) that a function of one variable is 0-standard. In particular $x \mapsto v(x_\ell)$ and $x \mapsto x_l^2$ are 0-standard. We observe also that $x \mapsto x_\ell \cdot x_{\ell+1}$ is 0-standard. Finally this property remains true for a sum of terms depending on different variables and this gives the property for $V^{(m)}$.

For (5.6.6) we get immediately

$$D^{(m)} = (v''(s) + 2a) I_m \tag{5.6.10}$$

where I_m is the identity in $\mathbb{R}^m$, so we have,

$$r_0 = \lambda_{min}(D^{(m)}) = (v''(s) + 2a) . \tag{5.6.11}$$

[35] This corresponds to a more general notion of being 0-standard with respect to the parameter t (See (1.2) in [Sj92b] and (3.3.35)-(3.3.37)).

If we denote by τ the operator on $\mathbb{R}^m$ defined by,

$$(\tau x)_i = x_{i-1} \ ,$$

we can write

$$A^{(m)} = -(2a)(\tau + \tau^{-1})$$

The eigenvalues of A have been already computed as $-2a \cdot \cos(2\pi \frac{k}{m})$ for $k = 0, 1, \ldots, m-1$. It is then easy to verify that for ρ satisfying (5.6.5):

$$||A||_{\mathcal{L}(\ell^p_\rho)} \leq 2a \cdot \exp(\kappa)$$

We observe that one can choose κ such that

$$r_1 = 2a \cdot \exp(\kappa) \ < \ (v''(s) + 2a) \tag{5.6.12}$$

and we shall make this choice now.
Let us now verify the condition on $W^{(m,n)}$ which is given by

$$W^{(m,n)} = \frac{a}{2}\left((x_{m+n} - x_1)^2 + (x_m - x_{m+1})^2 - (x_m - x_1)^2 - (x_{m+n} - x_{m+1})^2\right) \ .$$

Here it is sufficient to observe that the only j for which $\partial_{x_j} W^{(m,n)}$ are not 0 are $j = 1, m, m+1, m+n$.
Let us verify (5.6.6). We first observe that

$$D_t^{(m+n)} = (1-t)(D^{(m)} \oplus D^{(n)}) + tD^{(m+n)}$$

and

$$A_t^{(m+n)} = (1-t)(A^{(m)} \oplus A^{(n)}) + tA^{(m+n)} .$$

All the properties we need are stable by arithmetical means, so it is sufficient to treat the case $(V^{(m)} \tilde{\oplus} V^{(n)})$ for ρ satisfying (5.6.5) and (5.6.8) which can be reduced by separation of variables to the study of $V = V^{(m)}$ for ρ satisfying (5.6.5).

Conclusion. If we take d small enough in Theorem 5.6.2, we can, using the WKB solution centered at one of the critical points x_c and multiplying by a suitable cut-off function, get a complete expansion of the first eigenvalue of the Dirichlet problem modulo $m\mathcal{O}(h^\infty)$.
The condition $m = \mathcal{O}(h^{-N_0})$ is introduced in Section 5 in [Sj93a]. We shall also have to use the same properties for $V^{(m)}$ replaced by

$$\tilde{V}^{(m,N)} := V^{(m)} - \sum_j (x_j - (x_c(m))_j)^{2N} \ .$$

By symmetry we also have the same properties near $-x_c(m)$. The introduction of this modified potential was introduced in [Sj93a] in order to absorbe errors due to a partition of unity. As is formally natural, it was proved there that the first eigenvalue of the corresponding Dirichlet operator attached with this modified potential changes only by $m \cdot \mathcal{O}(h^{N-1})$.

Similarly we write

$$I_j <_{\epsilon_0} 0 \text{ if } I_j \subset]-\infty, -\epsilon_0[, I_j \leq_{\epsilon_0} 0 \text{ if } I_j \subset]-\infty, 2\epsilon_0[\ .$$

Clearly $I_+ >_{\epsilon_0} 0$ and $I_- <_{\epsilon_0} 0$. We say that (j,k) with $j,k \in \mathbb{Z}/m\mathbb{Z}$ is a change of sign for a given Ω if $I_j >_{\epsilon_0} 0$, $I_k <_{\epsilon_0} 0$ or $I_j <_{\epsilon_0} 0$, $I_k >_{\epsilon_0}$ and if $I_\nu \ni_{\epsilon_0} 0$ (that is I_ν meets $[-\epsilon_0, \epsilon_0]$) for all $\nu \in]j,k[=_{def} \{j+1, j+2, ..., k-1\}$.
We define similarly $[j,k[,]j,k], [j,k]$ and write for instance $j < \nu \leq k$ if $\nu \in]j,k]$. Let us observe that the number of changes of sign is even because we are on $\mathbb{Z}/m\cdot\mathbb{Z}$. Let (j_1,k_1) and (j_2,k_2) be two consecutive changes of sign so that $j_1 < k_1 \leq j_2 < k_2 \leq j_1$, in $\mathbb{Z}/m\cdot\mathbb{Z}$.
In a first step we try to send a general box on a box where all the intervals I_j have essentially the same sign and try to measure the effect on the potential.
For a given Ω, we consider the map κ from $\mathbb{R}^m$ into $\mathbb{R}^m$

$$x \mapsto \tilde{x} = \kappa(x)$$

with the property that

$$\tilde{x}_\nu = -x_\nu \text{ for } k_1 \leq \nu \leq j_2\ ,\ \tilde{x}_\nu = x_\nu \quad \text{ for all other } \nu\ .$$

For $x \in \Omega$, we consider

$$\begin{aligned} V(x) - V(\kappa(x)) &= \tfrac{a}{2}[(x_{k_1-1} - x_{k_1})^2 - (x_{k_1-1} + x_{k_1})^2] \\ &+\tfrac{a}{2}[(x_{j_2} - x_{j_2+1})^2 - (x_{j_2} + x_{j_2+1})^2] \\ &= -2ax_{k_1-1}\cdot x_{k_1} - 2ax_{j_2}\cdot x_{j_2+1} \\ &= \mathcal{J}_1 + \mathcal{J}_2\ . \end{aligned}$$

If $j_1 = k_1 - 1$, then either $I_{k_1-1} <_{\epsilon_0} 0$, $I_{k_1} >_{\epsilon_0} 0$ or $I_{k_1-1} >_{\epsilon_0} 0$, $I_{k_1} <_{\epsilon_0} 0$, and the $\mathcal{J}_1$ can be bounded (using the definitions) from below by

$$\mathcal{J}_1 \geq 2a\epsilon_0^2.$$

If $j_1 \neq k_1 - 1$, then $I_{k_1-1} \in_{\epsilon_0} 0$ and we have only[37]

$$\mathcal{J}_1 \geq -(\epsilon_0 + |x^\Omega_{k_1}|)\epsilon_0\ ,$$

where x^Ω is the center of the box.
The same discussion holds for $\mathcal{J}_2$ where we distinguish between the cases $k_2 = j_2 + 1$ and $k_2 \neq j_2 + 1$.
Let us call the change of sign (j,k) strict if $k = j + 1$.

[37]We observe here that the situation is a little different from the situation in the Kac model for which the interaction was globally Lipschizian.

Applying the same procedure several times, we get a map

$$\kappa : \Omega = \prod_1^m I_j \mapsto \prod_1^m \tilde{I}_j = \tilde{\Omega}$$

$$x \mapsto \kappa(x) = \tilde{x} \text{ with } \tilde{x}_j = \pm x_j, \tilde{I}_j = \pm I_j \ ,$$

such that $\tilde{I}_j \geq_{\epsilon_0} 0$ and such that, for $x \in \Omega$,

$$V(x) - V(\kappa(x)) \geq 2a\epsilon_0^2 \alpha_+(\Omega) - 2a\epsilon_0 \sum{}' |x_k^\Omega| \ , \tag{5.7.7}$$

where $\sum'$ is a sum of at most $\alpha_0(\Omega)$ terms. Here $\alpha_+(\Omega)$ is the number of strict changes of sign and $\alpha_0(\Omega)$ is the number of I_j's meeting $[-\epsilon_0, \epsilon_0]$. Notice that $\alpha_0(\tilde{\Omega}) = \alpha_0(\Omega)$ and $|x_k^\Omega| = |(\kappa(x))_k^{\tilde{\Omega}}|$. Using these remarks, we get

$$V(x) - V(\kappa(x)) \geq 2a\epsilon_0^2 \alpha_+(\Omega) - 2as\epsilon_0 \alpha_0(\Omega) - \sum ||x_k^\Omega| - s| \ . \tag{5.7.8}$$

We can now assume that, after possibly a change of all the signs, $\Omega = \prod_1^m I_j$ (we forget the old tilde) with $I_j \geq_{\epsilon_0} 0$. We keep in mind what we have gained or lost in the first step.

The second step consists in the measure of the effect of a translation which will send the box Ω in a box contained in the fixed ball $\Omega_+^{(2)} =: \prod_{j=1}^m I_+^{(2)}$ where $I_+^{(2)} = [s - 2\delta_0, s + 2\delta_0]$ Let $\kappa : \Omega \mapsto \Omega_+ = (I_+)^m$ be the translation such that $\kappa(x^\Omega) = \tilde{x}^{\Omega_+}$. Here $\tilde{x}_j^{\Omega_+} = s$, if $I_j = I_+$ and in the other case $\tilde{x}_j^{\Omega_+} = s \pm \rho\delta_0$ for a suitable ρ, independent of Ω and δ_0, satisfying, $0 < \rho < 1$, where we take the sign + if I_j is on the right of s and the sign − if I_j is on the left of s.

The choice of ρ. In order to precise the choice of ρ, we observe that the main problems occur for the boxes which are in the vicinity of one well. Far from the wells we meet also a small problem of control at ∞ for which the assumption

$$v(x) \geq \frac{1}{C}|x| - C \ ,$$

is sufficient. We decompose the potential V in the form

$$V(x) = \sum_j W(x_j - s, x_{j+1} - s) \ ,$$

with

$$W(y, z) = \frac{a}{2}(y - z)^2 + \frac{1}{2}(v(y + s) + v(z + s)) \ .$$

We study the effect of the translation for each j. We consider a simplified function by taking the quadratic approximation of v at s and by approximating the function $(y, z) \mapsto W(y, z)$ by

$$(y, z) \mapsto W_0(y, s) = C(y - z)^2 + D(y + z)^2$$

with $C > 0$ and $D > 0$. We have actually,

$$C = \frac{1}{4}(2a + v''(s)) \ ; \ D = v''(s)\frac{1}{4} \ .$$

We now work for some times in these new coordinates. The new I_+ is $[-\delta_0, \delta_0]$. For all the pairs such that I_j and I_{j+1} are near s, and this corresponds to the most difficult situation, it is better to analyze what is going on for the quadratic approximation. We are now working with boxes in $\mathbb{R}^2$ and the more delicate case[38] is when this box is a product of an interval of type 2 with I_+. We consider the evolution of this potential restricted to a vertical interval[39] $L(\tau, \delta_0) = \{\tau\} \times [-\delta_0, \delta_0]$ centered at the point $(y = \tau, z = 0)$ and of length $2\delta_0$ and if we want to translate it at a suitable $\rho\delta_0$ with $0 < 1 < \rho$, with ρ fixed universally.
We consider consequently the function:

$$]0, +\infty[\ni \tau \mapsto W_0(\tau, z) \text{ for } z \in [-\delta_0, \delta_0] \ .$$

We have,

$$(\partial_y W_0)(\tau, z) \ = \ 2C(\tau - z) + 2D(\tau + z) \ = \ (a + v''(s))\tau - az \ .$$

Then $(\partial_y W_0)(\tau, z)$ as a function of z is minimal at $z = \delta_0$ and $(\partial_y W_0)(\tau, \delta_0)$ is positive for $\tau \geq a\delta_0/(a + v''(s))$.
We take ρ such that $2a/(2a + v''(s)) < \rho < 1$ and we finally take

$$\rho = (2a + v''(s))/2(a + v''(s)) \ .$$

With this choice we observe that we have

$$W_0(\tau, z) - W_0(\rho\delta_0, z) \geq \frac{1}{C}\delta_0^2$$

for a suitable C independent of δ_0 and for $\tau > (1 + \rho)\delta_0/2$. A similar argument is obtained for $\tau < 0$. Now if $0 < \tau \leq C_1\delta_0$, the inequality is still valid for W if one chooses δ_0 small enough

$$W(\tau, z) - W(\rho\delta_0, z) \geq \frac{1}{2C}\delta_0^2 \ .$$

If now

$$\tau \geq C_2\delta_0$$

[38] We have nothing to do if the two intervals are equal to I_+ and the situation is easier in the case when the two intervals are of type 2.

[39] Actually we need also to follow the argument for a box $[\tau - \epsilon_0, \tau + \epsilon_0] \times [-\delta_0, \delta_0]$ but the condition on ϵ_0 justifies the omission which follows in the argument. Let us observe also that the situation is symmetric in y and z. So we can treat similarly the horizontal interval.

(C_2 large but independent of δ_0) we now use that

$$W(\tau, z) \geq \frac{a}{2} \cdot (C_2 - 1)^2 \delta_0^2 ,$$

$$W_0(\rho\delta_0, z) \leq C_3 \delta_0^2 ,$$

where C_3 depends only on v and a. Finally we obtain the inequality

$$W(\tau, z) - W(\rho\delta_0, z) \geq \frac{1}{2C} \delta_0^2 ,$$

for a suitable universal C for any τ such that $|\tau| > (1+\rho)\delta_0/2$ and any $z \in [-\delta_0, \delta_0]$. Until now we have not used the behavior of v at ∞. Using this property we get finally for possibly a new C and for δ_0 small enough

$$W(\tau, z) - W(\rho\delta_0, z) \geq \frac{1}{2C} \delta_0 \cdot (\delta_0 + |\tau|) , \tag{5.7.9}$$

for all τ satisfying

$$\tau > -3s/2 , \; |\tau| > (1+\rho)\delta_0/2 ,$$

and any $z \in [-\delta_0, \delta_0]$.

In order to be complete, we now control the case when the box is the product of two intervals $[y_1 - \epsilon_0, y_1 + \epsilon_0] \times [z_1 - \epsilon_0, z_1 + \epsilon_0]$ with $y_1, z_1 \geq \delta_0 - \epsilon_0$ (after taking the origin at (s, s)). We have now to verify after dilation and modulo a perturbation argument the condition

$$W_0(y, z) > W_0(\rho, \rho) , \; \forall y \geq 1 , \; \forall z \geq 1 ,$$

which is immediate.
Similarly we observe

$$W_0(y, z) > W_0(\rho, -\rho) , \; \forall y \geq 1 , \; \forall z \leq -1 .$$

We now come back to our initial Ω and compose all the different κ's. Due to our choice of ρ , we have obtained for a suitable C that, choosing ϵ_0 and δ_0 satisfying $\epsilon_0 << \delta_0 << 1$, for $x \in \Omega$,

$$\begin{aligned} & V(x) - V(\kappa(x)) \geq \\ & \quad \tfrac{1}{C} \left[\delta_0^2 (\beta(\Omega) + 2a\alpha_+(\Omega) + \delta_0 \textstyle\sum_j ||x_j^\Omega| - s|\right] - 2a\, s\, \epsilon_0 \alpha_0(\Omega) - 2a\epsilon_0 \textstyle\sum ||x_j^\Omega| - s| , \end{aligned}$$

where $\beta(\Omega)$ is the number of intervals I_j which do not contain s (or $-s$). Notice that $\beta(\Omega)$ is unchanged by the first "κ" and that

$$\alpha_0(\Omega) \leq \beta(\Omega) .$$

So we obtain

$$\begin{aligned} V(x) - V(\kappa(x)) & \geq (\tfrac{1}{C}\delta_0^2 - 2a\, s\, \epsilon_0)(\beta(\Omega) + \tfrac{2a}{C}\alpha_+(\Omega) \\ & \quad +(\tfrac{1}{C}\delta_0 - 2a\epsilon_0)\textstyle\sum_j ||x_j^\Omega| - s| \\ & \geq \tfrac{1}{2C}\delta_0^2\beta(\Omega) + 2a\tfrac{1}{C}\alpha_+(\Omega)\,, \end{aligned}$$

by choosing ϵ_0 small in comparison with δ_0.
We notice also that the final κ (which is a composition of symmetries and translations) satisfies,

$$\kappa : \Omega \mapsto \tilde{\Omega}_+ \subset \Omega_+^{(2)}\,,$$

where Ω is the original box and $\Omega^{(2)}$ was introduced as $\Omega^{(2)} = \prod_1^m I_+^{(2)}$,

$$\kappa(x^\Omega) = \tilde{x}^{\Omega_+}\,,$$

and

$$V(x) - V(\kappa(x)) \geq \frac{1}{C}(\alpha_+(\Omega) + \beta(\Omega))\,,\ \forall x \in \Omega\,. \tag{5.7.10}$$

C is here a strictly positive constant, independent of Ω, once we have fixed ϵ_0 and δ_0 conveniently as explained before.
Let $S_{V,\Omega}$ denote the Dirichlet realization of $-h^2\Delta + V$ in Ω and let λ_+ denote the lowest eigenvalue of $S_{V,\Omega_+^{(2)}}$. Let λ_0 denote the lowest eigenvalue of $-h^2\Delta + V$ on $\mathbb{R}^m$.
By the min-max principle, we have,

$$\lambda_0 \leq \lambda_+ \tag{5.7.11}$$

and recall from Section 5.6 that, under the assumption $m = \mathcal{O}(h^{-N_0})$, we have a good knowledge of the asymptotics for λ_+ deduced from WKB constructions.
Formula (5.7.10) shows that, on $C_0^\infty(\Omega)$,

$$S_{V,\Omega} - \lambda_+ \geq \frac{1}{C}\left(\alpha_+(\Omega) + \beta(\Omega)\right)\,. \tag{5.7.12}$$

Using κ we will also get, *under the assumption* $m = \mathcal{O}(h^{-N_0})$,

$$S_{V,\Omega} - \sum_1^m (x_j - x_j^\Omega)^{2M} - \lambda_+^M(h) \geq (1/C)(\alpha_+(\Omega) + \beta(\Omega))\,, \tag{5.7.13}$$

if $\Omega \neq \Omega_\pm$, with a new constant $C > 0$, where,
λ_+^M is the first eigenvalue for the Dirichlet realization of

$$-h^2\Delta + V - \sum_j |x_j - s|^{2M}$$

in $\Omega_+^{(2)}$, and

$$\lambda_+^M(h) - \lambda_+(h) = m\mathcal{O}(h^{M-1}) . \tag{5.7.14}$$

We observe indeed that
$\sum_1^m (\tilde{x}_j - \tilde{x}_j^\Omega)^{2M} = \sum' (\tilde{x}_j - s)^{2M} + \sum'' (\tilde{x}_j - s \pm \rho\delta_0)^{2M}$,
where $\sum'$ corresponds with the sum over the j such that $s \in I_j$. We now obtain the estimate

$$\sum_1^m (\tilde{x}_j - \tilde{x}_j^\Omega)^{2M} \leq \sum_1^m (\tilde{x}_j - s)^{2M} + \beta(\Omega)\epsilon_0^{2M} \ , \ \forall \tilde{x} \in \kappa(\Omega) \ ,$$

and then get (5.7.14) as in [Sj93a] (Section 6), [Sj93b] or [HeSj92b] and, in this survey, at the end of Subsection 5.6.

From this point we imitate as in [HeSj92b] the argument of [Sj93a], [Sj93b] and we only present the main steps. We have a C^∞ positive function Ψ_Ω with

$$|\nabla\Psi_\Omega(x)|^2 \leq \frac{1}{2}\sum_1^m (x_j - x_j^\Omega)^{2M}, \Psi_\Omega(x^\Omega) = 0, \tag{5.7.15}$$

χ_Ω in $C_0^\infty(\Omega)$ and a measure $\mu(d\Omega)$ such that

$$\int \chi_\Omega(x)^2 \cdot \exp(-2\Psi_\Omega(x)/h)\mu(d\Omega) = 1 + \mathcal{O}(\exp(-1/Ch)), \tag{5.7.16}$$

where $\mathcal{O}$ is uniform with respect to x. We now want to prove that

$$\lambda_0 = \lambda_+ + \mathcal{O}(h^\infty) . \tag{5.7.17}$$

We first observe that

$$\lambda_0 \leq \lambda_+ \tag{5.7.18}$$

is clear by the min-max principle. We will now prove that

$$\lambda_+ \leq \lambda_0 + \mathcal{O}(h^\infty) , \tag{5.7.19}$$

and for this we go through the proof of

$$\lambda_+^M \leq \lambda_0 + \mathcal{O}(h^\infty) . \tag{5.7.20}$$

We are immediately reduced to consider the case when,

$$\lambda_+^M > \lambda_0 . \tag{5.7.21}$$

In order to analyze this case, let $\lambda < \lambda_+^M$; we have,

$$\begin{aligned}&\langle (S_V - \lambda)\chi_\Omega \exp(-\Psi_\Omega/h)u \mid \chi_\Omega \exp(-\Psi_\Omega/h)u\rangle \\ &\quad = \langle \chi_\Omega \exp(-\Psi_\Omega/h)(S_V - \lambda)u \mid \chi_\Omega \exp(-\Psi_\Omega/h)u\rangle \\ &\qquad + \langle |h\nabla(\chi_\Omega \exp(-\Psi_\Omega/h)|^2\, u \mid u\rangle\ ,\end{aligned} \tag{5.7.22}$$

which can be written as

$$\begin{aligned}&\langle (S_V - \lambda - |\nabla\Psi_\Omega|^2)\chi_\Omega \exp(-\Psi_\Omega/h)u \mid \chi_\Omega \exp(-\Psi_\Omega/h)u\rangle \\ &\quad = \langle \chi_\Omega \exp(-\Psi_\Omega/h)(S_V - \lambda)u \mid \chi_\Omega \exp(-\Psi_\Omega/h)u\rangle \\ &\quad + \langle (h^2|\nabla\chi_\Omega|^2 - 2h\chi_\Omega\nabla\chi_\Omega \cdot \nabla\Psi_\Omega)\exp(-\Psi_\Omega/h)u \mid \exp(-\Psi_\Omega/h)u\rangle .\end{aligned} \tag{5.7.23}$$

Combining with (5.7.14) and (5.7.15), we get:

$$\begin{aligned}&(\lambda_+^M - \lambda)\ ||\chi_\Omega \exp(-\Psi_\Omega/h)u||^2 \\ &\quad \le \langle \chi_\Omega \exp(-\Psi_\Omega/h)(S_V - \lambda)u \mid \chi_\Omega \exp(-\Psi_\Omega/h)u\rangle \\ &\qquad + \langle (h^2|\nabla\chi_\Omega|^2 - 2h\chi_\Omega\nabla\chi_\Omega \cdot \nabla\Psi_\Omega)\exp(-\Psi_\Omega/h)u \mid \exp(-\Psi_\Omega/h)u\rangle\end{aligned} \tag{5.7.24}$$

Integrating with respect to $\mu(d\Omega)$ gives

$$\begin{aligned}&(\lambda_+^M - \lambda)(1 + \mathcal{O}(\exp(-1/Ch)))||u||^2 \\ &\quad \le (1 + \mathcal{O}(\exp(-1/Ch)))\cdot(S_V - \lambda)u \mid u) + \mathcal{O}(\exp(-1/Ch)||u||^2\ .\end{aligned} \tag{5.7.25}$$

We then take $\lambda = \lambda_0$ and get (5.7.19). Finally (playing with arbitrary M and using (5.7.11) and (5.7.14)), we obtain

$$\lambda_0 = \lambda_+ + \mathcal{O}(h^\infty)\ ,$$

(under the condition $m = \mathcal{O}(h^{-N_0})$).

We next look at exponentially weighted estimates.
Let $\psi_+(t)$ be a positive function such that

$$\begin{array}{lll}\psi_+ & = 0 & \text{near } I_+\ ,\\ \psi_+ & = a_0 > 0 & \text{near } I_-\ ,\\ \lim_{|t|\to+\infty}\psi_+(t) & = +\infty & \end{array}$$

and ψ_+ has a small, strictly positive, constant slope (in absolute value) outside.
Let us define

$$\psi_-(t) = \psi_+(-t)\ ,$$

and

$$\phi(x) = \inf\left(\sum_1^m \psi_+(x_j), \sum_1^m \psi_-(x_j)\right)\ . \tag{5.7.26}$$

Then

$$\phi|_{\Omega_\pm} = 0$$

and

$$|\nabla\phi(x)|^2 \leq \frac{1}{3C} \cdot \beta(\Omega) \text{ if } x \in \Omega \,. \tag{5.7.27}$$

Let u be the positive normalized eigenfunction associated to λ_0. Then:

$$(S_V - \lambda_0)u = 0 \,.$$

Put

$$\phi_\Omega(x) = \phi(x) - \Psi_\Omega(x) \,. \tag{5.7.28}$$

Then

$$\langle (S_V - \lambda_0)\chi_\Omega \exp(\phi_\Omega/h)u \mid \chi_\Omega \exp(\phi_\Omega/h)u\rangle = \langle (|h\nabla(\chi_\Omega \exp(\phi_\Omega/h)|^2 u \mid u\rangle$$

which we write as

$$\begin{aligned} & \langle ((S_V - \lambda_0 - |\nabla\phi_\Omega|^2)\chi_\Omega \exp(\phi_\Omega/h)u \mid \chi_\Omega \exp(\phi_\Omega/h)u\rangle \\ = \; & \langle ((2h\chi_\Omega\nabla\chi_\Omega \cdot \nabla\phi_\Omega + h^2|\nabla\chi_\Omega|2) \exp(\phi_\Omega/h)\, u|\chi_\Omega \exp(\phi_\Omega/h)u\rangle. \end{aligned} \tag{5.7.29}$$

Here

$$|\nabla\phi_\Omega|^2 \leq 2|\nabla\phi|^2 + 2|\nabla\Psi_\Omega|^2$$

and combining (5.7.15), (5.7.16), (5.7.22), (5.7.27) and (5.7.29), we get, when $\Omega \neq \Omega_\pm$,

$$\begin{aligned} & \|\chi_\Omega \exp(\phi_\Omega/h)u\|^2 \\ \leq \; & C\langle (2h\chi_\Omega\nabla\chi_\Omega \cdot \nabla\phi_\Omega + h^2|\nabla\chi_\Omega|^2) \exp(\phi_\Omega/h)u|\chi_\Omega \exp(\phi_\Omega/h)u\rangle \end{aligned} \tag{5.7.30}$$

and integrating with respect to $\mu(d\Omega)$, yields

$$\begin{aligned} & \int (1 + \mathcal{O}(\exp(-1/Ch)) \exp(2\phi/h)|u|^2 \, dx \\ \leq \; & \mathcal{O}(\exp(-1/Ch))\|\exp(\phi/h)u\|^2 + \|u\|^2_{\Omega_+} + \|u\|^2_{\Omega_-} \end{aligned} \tag{5.7.31}$$

which implies:

$$\|\exp(\phi/h)u\|^2 = \mathcal{O}(1) \tag{5.7.32}$$

(under the condition $m = \mathcal{O}(h^{-N_0})$).
Let $x \in \mathbb{R}^m$. We notice that, for some $\tilde{\alpha} > 0$

$$\psi_+(t) \geq \tilde{\alpha}\text{dist}(t, I_+^{(2)}) \geq \tilde{\alpha}(|t - s| - 2\delta_0) \,.$$

Hence

$$\begin{aligned}\sum_1^m x_j &= \sum_1^m (x_j - s) + m\,s \\ &\geq -\sum_1^m |x_j - s| + m\,s \\ &\geq -\sum_1^m (|x_j - s| - 2\delta_0) + m\,s - 2m\,\delta_0 \\ &\geq -\tfrac{1}{\tilde{\alpha}}(\sum_1^m \psi_+(x_j)) + (s - 2\delta_0)\cdot m\,.\end{aligned}$$

So

$$\sum_1^m \psi_+(x_j) \geq \tilde{\alpha}(s-2\delta_0)\cdot m \; - \tilde{\alpha}\cdot \left|\sum_1^m x_j\right|\,.$$

The same estimate holds for $\sum_1^m \psi_-(x_j)$ so

$$\min\left(\sum_1^m \psi_+(x_j), \sum_1^m \psi_-(x_j)\right) \geq \tilde{\alpha}\,\cdot\,(s-2\delta_0)\cdot m \; - \tilde{\alpha}\left|\sum_1^m x_j\right|\,.$$

We have $s - 2\delta_0 > 0$ and, for m large enough, we have finally

$$\min\left(\sum_1^m \psi_+(x_j), \sum_1^m \psi_-(x_j)\right) \geq \frac{m}{C}\,, \tag{5.7.33}$$

if

$$\sum_1^m x_j \in [-1,+1]\,. \tag{5.7.34}$$

Then, according to (5.7.26) and (5.7.32), Theorem 5.4.5 is a immediate consequence of (5.7.33). ∎

6 From statistical mechanics to quantum mechanics

6.1 Introduction

In collaboration with J. Sjöstrand, we gave in [HeSj93] a rather general approach for the estimate of the decay of correlations attached to "Gaussian like" measures of the type

$$\exp -\Phi(X)dX \tag{6.1.1}$$

on $\mathbb{R}^n$ with Φ convex. We have recalled these results in Section 3. More general and refined estimates were given by J. Sjöstrand in [Sj93c] and this was applied in [He94b] in order to get the estimate of the splitting between the two largest eigenvalues of the Kac's operator. We shall recall these results in Section 7. Another proof of this estimate (more correctly a variant), based on the analysis of the transfer matrix method, originally due to Kramers-Wannier [KrWa41], that we have already seen in

Section 2 for the study of the Ising model (see for example [Tho72]), was proposed in [He94c]. But this method can be applied in a more general context (see Chapter 12 in the book by G.Parisi [Pa84]) and we give a systematic treatment in [He94c] and [He94d]. We present here briefly the technique trying to take account of these contributions and to analyze in the same spirit the last subsection of Chapter 12 in [Pa84] called: "The zero lattice spacing limit", which can be seen as an initiation to some problems in Quantum Field Theory. In order to present this approach (other cases are considered in Section 12.4 of [Pa84]), we consider a particular case when $n = mp$ and we then take a matrix notation for the variable X:

$$X = x_j^i$$

where the index $i = 1, \dots, m$ corresponds to the row and the index $j = 1, \dots, p$ corresponds to the column. We use alternatively the notation $X = (x_1, \dots, x_p)$ with $x_j \in \mathbb{R}^m$ or $X = (x^1, \dots, x^m)$ with $x^j \in \mathbb{R}^p$. We now consider the particular potential Φ,

$$\Phi^{(p)}(X) \equiv \Phi(X) \equiv \sum_{j=1}^{p} V(x_j) + \frac{|x_j - x_{j+1}|^2}{4h^2}\,, \tag{6.1.2}$$

where we take the convention that $x_{p+1} = x_1$ and where h is possibly a semiclassical parameter which is sometimes chosen equal to one if we are not interested in the "semiclassical" aspects. More generally, we could more generally consider examples of the form,

$$\Phi(X) = \sum_{j=1}^{p} \left(V(x_j) + I(x_j, x_{j+1})\right)\,, \tag{6.1.3}$$

where I is a symmetric "interaction" potential on $\mathbb{R}^m \times \mathbb{R}^m$. Let us mention however that the example (6.1.2) appears naturally in quantum field theory when the so called "lattice approximation" is introduced. For this special class of potentials, we shall demonstrate that the informations given by the transfer operator method is complementary to the other approachs we have explained before. We now recall the "dictionary" between the properties of the measure $\exp -\Phi(X)dX$ and the spectral properties of the transfer operator (which is also called Kac operator in other references)

$$K_V(x,y) = (2\pi h)^{-\frac{m}{2}} \exp -\frac{V(x)}{2} \cdot \exp -\frac{|x-y|^2}{4h^2} \cdot \exp -\frac{V(y)}{2}\,. \tag{6.1.4}$$

(We shall forget sometimes the coefficient $(2\pi h)^{-\frac{m}{2}}$ which is simply a normalization and appears only in the problem of the thermodynamic limit).
This dictionary permits to obtain interesting connections between estimates for the

quotient μ_2/μ_1 of the two first largest eigenvalues of the transfer operator and corresponding estimates controlling the speed convergence of thermodynamic quantities or the decay of the correlations. In particular, for fixed m and h, this speed of convergence can be exponentially rapid as $p \to \infty$. For this we do not need convexity of the potential. The phenomenon is here similar to what we have observed for the one-dimensional Ising model. In particular there is apparently no phase transition. But, as will be discussed in an example in Subsection 6.5, all these problems reappear in the semiclassical context $h \to 0$ or in the limit $m \to \infty$. This was actually our initial motivation, following M. Kac, for the study of these questions.

6.2 A link between statistical mechanics and quantum mechanics

We first recall some properties of the Kac operator (or transfer operator). Other properties will be presented in Section 7. This operator is associated with a potential V. We do not try to give the minimal assumptions which are needed for V and we assume in order to simplify the discussion the two following conditions

$$V(y) \geq \frac{1}{C}|y|^2 - C \tag{6.2.1}$$

for some strictly positive constant C, and

$$\forall \alpha \in \mathbb{Z}^m \;,\; \exists C_\alpha \text{ s.t. } |D_y^\alpha V(y)| \leq C_\alpha < y >^{(2-|\alpha|)_+} \;. \tag{6.2.2}$$

Under these assumptions, K_V is a compact strictly positive selfadjoint operator with strictly positive kernel. In particular it satisfies the assumptions of the extended Perron-Frobenius Theorem (see for example Theorem 3.3.2 in [GlJa87]) also called Krein-Rutman Theorem and K_V admits consequently a largest eigenvalue μ_1 equal to $||K_V||$ which is simple and corresponds to a unique strictly positive normalized eigenfunction which we denote by u_1. Let μ_j the sequence of eigenvalues that we order as a decreasing sequence tending to 0:

$$0 \leq \mu_{j+1} \leq \mu_j \leq .... \leq \mu_2 < \mu_1 .$$

Thermodynamic limit. Let us first look at the thermodynamic limit. We start from the decomposition:

$$\exp -\Phi(X) = K_V(x_1, x_2) \cdot K_V(x_2, x_3) \cdot \cdots \cdot K_V(x_{p-1}, x_p) \cdot K_V(x_p, x_1) \;, \tag{6.2.3}$$

and we observe that

$$\int_{\mathbb{R}^{mp}} \exp -\Phi(X)\, dX = \int_{\mathbb{R}^m} K_V^{(p)}(y, y)\, dy \tag{6.2.4}$$

where $K_V^{(p)}(x,y)$ is the distribution kernel of $(K_V)^p$. Our assumption on V permits to see that $(K_V)^p$ is trace class and we rewrite (6.2.4) in the form,

$$\int_{\mathbb{R}^{mp}} \exp -\Phi(X)\, dX = \mathrm{Tr}\ [(K_V)^p] = \sum_j \mu_j^p \,. \tag{6.2.5}$$

In particular we get,

$$\lim_{p\to\infty} \frac{\ln \int_{\mathbb{R}^{mp}} \exp -\Phi(X)\, dX}{p} = \ln \mu_1 \,, \tag{6.2.6}$$

and the speed of the convergence is easily estimated by,

$$\begin{aligned} & -\ln \left| \frac{\ln \int_{\mathbb{R}^{mp}} \exp -\Phi(X)\, dX}{p} - \ln \mu_1 \right| \\ = \ & -p \ln(\frac{\mu_2}{\mu_1}) - \ln k_2 + \ln p + \mathcal{O}(\exp -\delta_2 p) \,, \end{aligned} \tag{6.2.7}$$

where k_2 is the multiplicity of μ_2.

Mean value. Let us continue the dictionary. We consider now the mean value with respect to our measure of some cylindrical functions. More precisely, if f is a function of x_1 in $C_0^\infty(\mathbb{R}^m)$, let us introduce

$$\langle f^{(1)} \rangle_p = \frac{\int f(x_1) \exp -\Phi(X)\, dX}{\int \exp -\Phi(X) dX} \,. \tag{6.2.8}$$

An immediate computation gives

$$\langle f^{(1)} \rangle_p = \frac{\sum_j \mu_j^p (\int f(y)\, u_j(y)^2\, dy)}{\sum_j \mu_j^p} \,. \tag{6.2.9}$$

In particular, we get the following convergence

$$\lim_{p\to\infty} \langle f^{(1)} \rangle_p = \int f(y)\, u_1(y)^2\, dy \,, \tag{6.2.10}$$

and we have moreover the following control of the convergence

$$\langle f^{(1)} \rangle_p - \int f(y)\, u_j(y)^2\, dy = \mathcal{O}((\frac{\mu_2}{\mu_1})^p) \,. \tag{6.2.11}$$

Pair correlation. Let us consider the correlation functions. If f and g are two functions in $C_0^\infty(\mathbb{R}^m)$, we denote by $f^{(j)}$ the corresponding function on $\mathbb{R}^{mp}$:

$$X \to f(x_j) \,.$$

We shall now compute the pair correlation

$$\mathrm{Cor}^{(p)}(f^{(i)}, g^{(j)}) = \langle (f^{(i)} - \langle f^{(i)}\rangle_p)(g^{(j)} - \langle g^{(j)}\rangle_p)\rangle_p \,.$$

(If $f(x) = g(x) = x$, we write more simply $\mathrm{Cor}^{(p)}(i,j)$.) Because everything is invariant by circular permutation we can take $i = 1$. We first have

$$\langle f^{(1)} g^{(j)}\rangle_p \quad = \frac{\sum_{k,\ell} \mu_k^{j-1} \mu_\ell^{p-j+1} [\int f(y) u_k(y) u_\ell(y) dy][\int g(y) u_k(y) u_\ell(y) dy]}{\sum_i \mu_i^p} \,.$$

In particular, if $f = g$ (and this will be the interesting case for us), we get

$$\langle f^{(1)} f^{(j)}\rangle_p = \frac{\sum_{k,\ell} \mu_k^{j-1} \mu_\ell^{p-j+1} [\int f(y)\, u_k(y)\, u_\ell(y)\, dy]^2}{\sum_i \mu_i^p} \,,$$

and

$$\begin{aligned}
&\left(\sum_i \mu_i^p\right)^2 \mathrm{Cor}^{(p)}(f^{(1)}, f^{(j)}) \\
&= \left[\sum_k \mu_k^p \left(\int f(y)\, u_k(y)^2\, dy\right)^2\right] \left(\sum_i \mu_i^p\right) \\
&\quad - \left[\sum_k \mu_k^p \left(\int f\, u_k^2\, dx\right)\right]^2 \\
&\quad + \left[\sum_{k\neq\ell} \mu_k^{-1+j} \mu_\ell^{p-j+1} \left(\int f(y)\, u_k(y)\, u_\ell(y)\, dy\right)^2\right] (\sum_i \mu_i^p)
\end{aligned}$$

and, taking the limit as $p \to \infty$, we finally obtain the following identity

$$\lim_{p\to\infty} \mathrm{Cor}^{(p)}(f^{(1)}, f^{(j)}) = \sum_{k>1} \left(\frac{\mu_k}{\mu_1}\right)^{j-1} \left[\int f(y)\, u_k(y)\, u_1(y)\, dy\right]^2 \,, \tag{6.2.12}$$

(see (12.19) in [Pa84]).
From this we get the inequality

$$\left(\frac{\mu_2}{\mu_1}\right)^{j-1} \mathrm{var}(f) \geq \lim_{p\to\infty} \left[\mathrm{Cor}^{(p)}(f^{(1)}, f^{(j)})\right] \geq \left(\frac{\mu_2}{\mu_1}\right)^{j-1} \left[\int f(y)\, u_2(y)\, u_1(y)\, dy\right]^2 , \tag{6.2.13}$$

where

$$\mathrm{var}(f) = \int_{\mathbb{R}^m} u_1^2\, f^2\, dx - \left(\int_{\mathbb{R}^m} u_1^2\, f\, dx\right)^2 \tag{6.2.14}$$

is the variance of f with respect to the measure $u_1^2(x)\,dx$.
All these formulas are of course reminiscent of what we get in Subsections 2.2, 2.3, 2.4 and 2.5.

If we know by other means how to control the decay with respect to j uniformly with respect to p ($p \gg j$), and if we can choose f such that $\int f(y)\,u_2(y)\,u_1(y)\,dy \neq 0$, then we shall find a way to get a new estimate for μ_2/μ_1. Conversely, a good estimate for μ_2/μ_1 will give a very precise estimate for the corresponding correlation. Let us precise this argument which was already developed in [He94c]. Suppose that we prove as a first step that,

$$\begin{aligned} &\exists\,\kappa \text{ s.t. } \forall\epsilon > 0\ ,\ \exists\, j_0 \text{ and } C_\epsilon \text{ s.t } \forall j \geq j_0\ ,\ p > 2j\ , \\ &\left[\mathrm{Cor}^{(p)}(f^{(1)}, f^{(j)})\right] \leq C_\epsilon \exp -(\kappa - \epsilon)(j-1)\ , \end{aligned} \tag{6.2.15}$$

with $[\int f(y)\,u_2(y)\,u_1(y)\,dy] \neq 0$, then we get, taking successively the limit $p \to \infty$, the limit $j \to \infty$ for $\frac{1}{j}\ln\left(\mathrm{Cor}^{(p)}(f^{(1)}, f^{(j)})\right)$ and then the limit $\epsilon \to 0$ in the estimate obtained by the r.h.s inequality in (6.2.13) and by (6.2.15),

$$\frac{\mu_2}{\mu_1} \leq \exp -\kappa\ . \tag{6.2.16}$$

But using now the l.h.s in (6.2.13), we get the improved estimate replacing (6.2.15), that there exists κ and C such that, for all j,

$$\lim_{p\to\infty}\left[\mathrm{Cor}^{(p)}(f^{(1)}, f^{(j)})\right] \leq C \exp -\kappa(j-1)\ . \tag{6.2.17}$$

Let us now recall that estimates for the correlations were obtained in the case where f is C^∞ but with bounded gradient (See [HeSj93]), and a possible choice for f is $f = u_2$. Here we observe that if μ_2 is not of multiplicity one u_2 is not uniquely determined but corresponds to some arbitrary choice of a second eigenvector. Another natural choice was $f(y) = y^\ell$ for some ℓ, but it is not always true that,

$$\int y^\ell\,u_2(y)\,u_1(y)\,dy \neq 0\ , \tag{6.2.18}$$

is satisfied for some ℓ and for some choice of u_2 (remember the case when μ_2 has multiplicity larger than 1).
We note here that this condition appears for example in [Si79] and our possibility to choose another f is quite useful. On the other hand, one is usually interested in the correlation associated to $f(y) = y^\ell$ and we can not avoid in this case the question of the validity of (6.2.18). Note that in the case when V is a non degenerate quadratic form this condition (6.2.17) can be explicitely verified. Under this condition we get that the logarithm of the correlation is effectively asymptotic to $j\ln(\frac{\mu_2}{\mu_1})$ and the correlation length is $\ln(\frac{\mu_2}{\mu_1})$.

So our idea is now to use our recent study [HeSj93] of the correlation functions (and some extension based on an improvement of J. Sjöstrand [Sj93c] in the same spirit as in [He94b] or [He94c]) in order to prove an universal upper bound for μ_2/μ_1.
This was used in [He94c] in order to get

Theorem 6.2.1 *If, for some* $\sigma > 0$,

$$\text{Hess}\, V(x) \geq \sigma \; , \; \forall\, x \in \mathbb{R}^m \; , \tag{6.2.19}$$

then

$$\frac{\mu_2}{\mu_1} \leq \exp - \left[\cosh^{-1}[h^2\sigma + 1]\right] . \tag{6.2.20}$$

Remark 6.2.2 The result is optimal in the sense that the quadratic case when $V(y) = \sum_{j=1}^m \sigma_j y_j^2/2$ gives for $\sigma = \inf \sigma_j$ the equality:

$$\frac{\mu_2}{\mu_1} = \exp - \left[\cosh^{-1}[h^2\sigma + 1]\right] . \tag{6.2.21}$$

Remark 6.2.3 In the semiclassical limit, the semiclassical techniques permit to prove (see for example [BruHe91]) that:

$$\frac{\mu_2(h)}{\mu_1(h)} = \exp(\lambda_1(h) - \lambda_2(h)) + \mathcal{O}(h^2) . \tag{6.2.22}$$

An easy computation shows that the results on the splitting for the Kac model and the result on the splitting for Schrödinger are coherent. The link between the two statements is analyzed precisely in Section 4 in [He94c].

6.3 Study of the zero spacing limit

We follow the presentation by Parisi ([Pa84]) and analyze how the ideas coming from the preceding sections can be implemented. Our starting point is the potential:

$$\Phi(X) = \sum_{j=1}^{p} \left(aV(x_j) + \frac{|x_j - x_{j+1}|^2}{2a} \right) \tag{6.3.1}$$

We remark that a (which will tend to 0) does not play the same role as in the semiclassical case (because a appears also before V). Here $x_j \in \mathbb{R}^m$ but Parisi looks at the case $m = 1$. As in the preceding sections we can consider the transfer matrix (or more correctly transfer operator) whose kernel is

$$K_{V_a}(x,y) = (2\pi a)^{-\frac{m}{2}} \exp\left[-\frac{a}{2}(V(x) + V(y)) - \frac{|x-y|^2}{2a} \right] . \tag{6.3.2}$$

Here $(2\pi a)^{-\frac{m}{2}}$ is a normalization factor whose role will be clear in the limit $a \to 0$ and corresponds to the property that

$$(2\pi a)^{-\frac{m}{2}} \exp -\frac{|x-y|^2}{2a}$$

is the distribution kernel of $\exp a\Delta$. We now observe the trivial property for $t > 0$ and $a > 0$ such that $\frac{t}{a}$ is an integer. We have,

$$(\exp a\Delta)^{\frac{t}{a}} = \exp t\Delta \ ,$$

and we are interested in computing

$$T_t =: \lim_{a\to 0} (K_{V_a})^{\frac{t}{a}} \ ,$$

hoping that it has a limit in a rigorous sense. We can for example consider the question for the sequence $a = \frac{t}{N}$ and this gives the problem of the existence of the limit

$$T_t := \lim_{N\to\infty} (K_{V_{\frac{t}{N}}})^N \ . \tag{6.3.3}$$

We do not try to enter for the moment in the study of these mathematical problems but prefer to continue the heuristic exposition as it is presented in [Pa84] (we write in italic when we reproduce [Pa84] modulo some small change of notations) and try to precise the mathematical statements corresponding to what is said.
After the $a \to 0$ limit, the correlation function (we consider the case $m = 1$) in a box of size L can be written as

$$\langle x(t)x(0)\rangle_L = \frac{\mathrm{Tr}\,(T_t \cdot x \cdot T_{L-t} \cdot x)}{\mathrm{Tr}\,(T_L)} \ , \tag{6.3.4}$$

and Parisi means here implicitely that he is interested in the computation of the thermodynamic limit

$$\lim_{L\to\infty} \langle x(t)x(0)\rangle_L \ . \tag{6.3.5}$$

Now T_t can also be written as $\exp(-t\mathcal{H})$ for some selfadjoint operator and an heuristic argument, based on a formal application of the Campbell-Hausdorff formula

$$\exp -\frac{a}{2}A \cdot \exp -aB \cdot \exp -\frac{a}{2}A = 1 - a(A+B) + \mathcal{O}(a^2) \ , \tag{6.3.6}$$

shows that:

$$\mathcal{H} = -\frac{1}{2}\Delta + V \ . \tag{6.3.7}$$

If λ_1 is the smallest eigenvalue of $\mathcal{H}$ with corresponding first eigenvector v_1, we get:

$$\begin{aligned}\lim_{L\to\infty}\langle x(t)x(0)\rangle_L &= \langle v_1|x\ \exp t\lambda_1 \cdot T_t\ x|v_1\rangle \\ &= \langle v_1|x\ \exp -t(\mathcal{H}-\lambda_1)\ x|v_1\rangle\ .\end{aligned} \tag{6.3.8}$$

If we then take the limit $t\to\infty$ we get (if $\langle x\, v_1|v_1\rangle\ = 0$) the formula

$$\lim_{t\to+\infty}\exp t(\lambda_2-\lambda_1)\lim_{L\to\infty}\langle x(t)\,x(0)\rangle_L = \sum_{j\,;\lambda_j=\lambda_2}\langle x\,v_1|v_j\rangle^2\ . \tag{6.3.9}$$

We finally get

$$\begin{aligned}\textstyle\int d\mu[x]g_1(x(t))g_2(x(0) &\equiv \langle g_1(x(t)g_2(x(0))\rangle \\ &= \textstyle\sum_n\langle v_1|g_1|v_n\rangle\langle v_n|g_2|v_1\rangle\exp-(\lambda_n-\lambda_1)t\ ,\end{aligned} \tag{6.3.10}$$

with

$$\int d\mu[x] = 1\ ,$$

and

$$d\mu[x]\approx d[x]\exp\left[-\int(\frac{1}{2}(\frac{dx}{dt})^2+V(x))dt)\right]\ . \tag{6.3.11}$$

Equation (6.3.10) *gives a precise relation between the eigenvalues and eigenvector of the Schrödinger equation and a functional integral over a field; a knowledge of the correlation functions of statistical mechanics is sufficient to extract the spectrum of $\mathcal{H}$.* For a mathematical discussion around the aspect "functional integrals", the reader should consult other references and for example the book of B. Simon [Si79] or [Roe94]. We now try to analyze the other points more carefully.
Consider first the existence of the limit in (6.3.3) that we can rewrite:

$$T_t := \lim_{N\to\infty}(\exp-\frac{t}{2N}V\circ\exp\frac{t}{2N}\Delta\circ\exp-\frac{t}{2N}V)^N\ . \tag{6.3.12}$$

One recognizes here a variant of the Trotter formula (see [Tr59], [ReSi72] (Theorem VIII.31) or [Si79] (Theorem 1.1 p.4-6))

Theorem 6.3.1 *If A and B are selfadjoint semi-bounded operators with domains $D(A)$ and $D(B)$ and $A+B$ is essentially selfadjoint on $D(A)\cap D(B)$, then*

$$s-\lim_{n\to\infty}(\exp-\frac{t\,A}{N}\exp-\frac{t\,B}{N})^N = \exp-t(A+B)\ . \tag{6.3.13}$$

The variant is the following

Theorem 6.3.2 *If A and B are selfadjoint semi-bounded operators with domains $D(A)$ and $D(B)$ and $A+B$ is essentially selfadjoint on $D(A)\cap D(B)$, then*

$$s-\lim_{N\to\infty}\left(\exp-\frac{tB}{2N}\circ\exp-\frac{tA}{N}\circ\exp-\frac{tB}{2N}\right)^N=\exp-t(A+B)\,. \tag{6.3.14}$$

In our situation, we deduce that, for all $f\in L^2$, the following limit exists

$$\lim_{N\to\infty}T_{t,(N)}\,f\;=\;\exp-t(-\frac{1}{2}\Delta+V)f\,. \tag{6.3.15}$$

with

$$T_{t,(N)}=\left(\exp-\frac{t}{2N}V\cdot\exp\frac{t}{2N}\Delta\cdot\exp-\frac{t}{2N}V\right)^N\,, \tag{6.3.16}$$

and this explains rigorously why we get (6.3.6) and (6.3.7)

$$T_t=\exp-t\mathcal{H}\,. \tag{6.3.17}$$

We now analyze the discussion around (6.3.4) and (6.3.5). The starting point is probably the expression

$$\frac{\mathrm{Tr}\,(T_{t,(N)}\cdot x\cdot T_{L-t,(N)}\cdot x)}{\mathrm{Tr}\,(T_{t,(N)}\cdot T_{L-t,(N)})} \tag{6.3.18}$$

for a fixed choice of t,N,L.
This appears similar to the computation of some correlation pair of the type considered in the preceding sections at least if we take $L=kt$ with $k\in\mathbb{N}$ and if we modify (6.3.18) in

$$\frac{\mathrm{Tr}\,(T_{t,(N)}\cdot x\cdot (T_{t,(N)})^{k-1}\cdot x)}{\mathrm{Tr}\,((T_{t,(N)})^k)}\,. \tag{6.3.19}$$

With this modified definition, we find the correlation

$$\mathrm{Cor}^{\,(kN)}(1,N+1)$$

where k,N are related to the notations of the first section by considering

$$p=kN, N=i+1$$

and

$$\Phi^{(p)}(x_1,\dots,x_p)=\sum_j\frac{t}{2N}V(x_j)+\frac{N}{2t}|x_j-x_{j+1}|^2\,.$$

We can now first take the limit $k \to \infty$ and, applying the formula given in (6.2.12), we get

$$\lim_{k\to\infty} \mathrm{Cor}^{(kN)}(1, N+1) = \sum_{\ell>1} \left(\frac{\mu_\ell}{\mu_1}\right)^{N+1} \left[\int y u_\ell(y) u_1(y) dy\right]^2 , \tag{6.3.20}$$

where the μ_ℓ are the eigenvalues of the operator with distribution kernel

$$K_{(V,t,N)}(x,y) = \frac{t}{2N}(V(x) + V(y)) + \frac{N}{2t}|x_j - x_{j+1}|^2 .$$

If we now take the limit $N \to \infty$, we obtain the preceding limit. This is connected with the preceding formula and with the comparison between the Kac operator and the Schrödinger operator. We have proved indeed the

Lemma 6.3.3 *Let V be a C^∞ potential satisfying*

$$V(x) \geq 0 , \tag{6.3.21}$$

and

$$|D_x^\alpha V(x)| \leq C_\alpha , \forall \alpha \text{ s.t. } |\alpha| \geq 2 . \tag{6.3.22}$$

Let us denote by $\mu_1(s)$ and $\mu_2(s)$ the two largest eigenvalues of

$$K(s) = \exp -\frac{s}{2}V \cdot \exp s\Delta \cdot \exp -\frac{s}{2}V$$

for $s \in \mathbb{R}^+$. Then

$$\ln \mu_1(s) - \ln \mu_2(s) = s(\lambda_2 - \lambda_1) + \mathcal{O}(s^2) .$$

Let us analyze this more in detail. The proof is based on the following

Lemma 6.3.4 *Under the assumptions of the previous lemma, we have the following estimate,*

$$||K(s) - \exp -s(-\Delta + V)||_{\mathcal{L}(L^2)} = \mathcal{O}(s^2) \tag{6.3.23}$$

as $s \to 0$.

Using the min-max principle (cf. [ReSi75], Theorem XIII.1) for the two first eigenvalues gives

$$\ln \mu_j(s) = \exp -s \cdot \lambda_j + \mathcal{O}(s^2) . \tag{6.3.24}$$

The lemma will give a more precise result than the result deduced from the variant of the Trotter formula. We deduce from (6.3.23) that,

$$\left\| K\left(\frac{t}{N}\right) - \exp -\frac{t}{N}(-\Delta + V) \right\|_{\mathcal{L}(L^2)} = \mathcal{O}\left(\left(\frac{t}{N}\right)^2\right), \tag{6.3.25}$$

and we get, observing that, as a consequence of the assumption $V \geq 0$, the two operators $K(\frac{t}{N})$ and $\exp -\frac{t}{N}(-\Delta + V)$ are contractions,

$$\left\| K\left(\frac{t}{N}\right)^N - \exp -t(-\Delta + V) \right\|_{\mathcal{L}(L^2)} = N\mathcal{O}\left(\left(\frac{t}{N}\right)^2\right). \tag{6.3.26}$$

We can easily get the result by recursion. If A and B are two contractions such that $||A - B|| \leq \epsilon$ we can indeed write:

$$A^n - B^n = A^{n-1}(A - B) + (A^{n-1} - B^{n-1})B\ ,$$

which gives

$$||A^n - B^n|| \leq \epsilon + ||A^{n-1} - B^{n-1}||\ .$$

We get finally (6.3.26) and this proves (and improves) (6.3.6), (6.3.7) and (6.3.15). The next subsection will be devoted to detail the proof of Lemma 6.3.4.

6.4 Comparison between $\exp -t(-\Delta + V)$ and $\exp -\frac{t}{2}V \cdot \exp t\Delta \cdot \exp -\frac{t}{2}V$

We present here the results given in [He94c] and [He94d]. They are based on semi-classical analysis.

Some preliminaries

Let V be a C^∞ potential satisfying

$$V(x) \geq 0 \tag{6.4.1}$$

and

$$|D_x^\alpha V(x)| \leq C_\alpha\ ,\ \forall \alpha\ s.t.\ |\alpha| \geq 2\ . \tag{6.4.2}$$

It is a classical result, as a consequence of (6.4.2) and of the positivity of V, that the potential V satisfies for some constant C the following inequality,

$$|\nabla V(x)| \leq C\ V^{1/2}(x)\ ,\forall\ x \in \mathbb{R}^m\ . \tag{6.4.3}$$

Study of the Kac's operator

Let us now consider our operator,

$$K(t) = \exp -\frac{t}{2}V \circ \exp t\Delta \circ \exp -\frac{t}{2}V \ , \tag{6.4.4}$$

as a h-pseudodifferential operator $k^w(x, hD_x, t)$ (with $h = \sqrt{t}$), associated by the Weyl quantification, with the symbol $(x, \xi, t) \to k(x, \xi, t)$

$$\begin{aligned} &k^w(x, hD_x, t)u(x) = \\ &\quad (2\pi)^{-m} \int\int \exp -i\left((x-y)\cdot\xi\right) \cdot k((\tfrac{x+y}{2}), h\xi, t)u(y)dy\, d\xi \ . \end{aligned} \tag{6.4.5}$$

Conversely k is called the h-symbol of $K(t)$ (with $h = \sqrt{t}$). The parameter $\sqrt{t}$ plays here the role of the semi-classical parameter h. Sometimes, it will be interesting to consider differently t and h by considering $-t\Delta$ as $-h^2\Delta$ with Weyl symbol ξ^2 and by considering $V_t = tV$ as a potential depending on a small parameter t. Let us introduce the class

$$\begin{aligned} &\Sigma^s \\ = \ &\{p \in C^\infty(\mathbb{R}^m \times \mathbb{R}^m \times]0, h_0]) \ s.t.\ |\partial_x^\alpha \partial_\xi^\beta p(x, \xi, h)| \leq C_{\alpha\beta} h^s \ , \ \forall \alpha, \beta \in \mathbb{N}^m\} \end{aligned} \tag{6.4.6}$$

Here C^∞ is meant with respect with the (x, ξ) variables. For $s = 0$, this class Σ^0 is stable by composition, and the usual law of h-composition for the Weyl symbols can be used. We recall that the law of composition

$$r = p \sharp_h q$$

of two symbols p and q can be defined by

$$p_h^w(x, hD_x, h) \ o \ q_h^w(x, hD_x, h) = r_h^w(x, hD_x, h) \ . \tag{6.4.7}$$

Of course it is well known that if p and q are in Σ^0 then $p \sharp_h q - p\cdot q \in \Sigma^1$, but we shall need for our proof to do all the computations modulo $\mathcal{O}(h^4)$ and we use consequently the three first terms of the law of computation. Let us recall the formula (see [Ro87], p.80)

$$p \sharp_h q = \sum_{|\alpha|+|\beta|\leq 2} \frac{1}{(\alpha!\beta!)} \left(\frac{1}{2}\right)^{|\alpha|} \left(\frac{-1}{2}\right)^{|\beta|} \cdot h^{(|\alpha|+|\beta|)} \cdot (\partial_\xi^\alpha D_x^\beta p) \cdot (\partial_\xi^\beta D_x^\alpha q) + \mathcal{O}(h^3) \ . \tag{6.4.8}$$

Although we want to work modulo $\mathcal{O}(h^4)$, we shall not need to know explicitely the coefficient of h^3 in the expansion. We observe indeed the following important properties.

Property P1. *If we start from real symbols p and q, we observe that the coefficients of the expansion of $p \sharp_h q$ are alternatively real and imaginary. The same is true for $q \sharp_h p \sharp_h q$.*

The other classical point is that

Property P2. *The Weyl symbol of a selfadjoint operator is real.*

Using the two arguments, one get the property that the symbol

$$q \sharp_h p \sharp_h q \text{ has a natural expansion in powers of } h^2 . \tag{6.4.9}$$

This permits us to avoid in the computation some coefficients of h^3 which will automatically vanish.
We now consider $K(t)$ as the composition of three h-pseudodifferential operators. The h-symbol $p(x,\xi)$ of $\exp h^2\Delta$ is

$$p(x,\xi) = \exp -\xi^2 , \tag{6.4.10}$$

and the h-symbol of $\exp -\frac{t}{2}V$ is

$$q(x,\xi) = \exp -\frac{t}{2}V(x) . \tag{6.4.11}$$

For the moment $q(x,\xi)$ is considered as depending on a parameter $t \in]0,1]$ which may be independent of h, and we observe that q belongs to Σ^0 uniformly with respect to $t \in]0,1]$.
We want to compute the Weyl symbol of $K(t)$. We now establish the estimate for the symbol. We prove that it belongs to Σ^0 and is actually in a better class. Inequality (6.4.3) gives, observing the identity,

$$\nabla_x(\exp -\frac{t}{2}V(x)) = -\frac{t}{2}(\nabla V)(x) \cdot (\exp -\frac{t}{2}V(x)) ,$$

the following estimate

$$|\nabla_x(\exp -\frac{t}{2}V(x))| \leq C_0 t^{\frac{1}{2}} . \tag{6.4.12}$$

We now look at the second derivative. This is a sum of terms where the preceding argument is the same outside for one term

$$t(\nabla^2 V)(x)(\exp -\frac{t}{2}V(x))$$

which is clearly $\mathcal{O}(t)$. So we get

$$|\nabla_x^2(\exp -\frac{t}{2}V(x))| \leq C_2 t . \tag{6.4.13}$$

We now compute the Weyl symbol of $K(t)$ using the composition law of p.d.o. We shall first prove the,

Lemma 6.4.1

$$\sigma_h^w(K(t)) - \left[\exp -(\xi^2 + tV(x))\right] \in \Sigma^4 , \tag{6.4.14}$$

for $t = h^2$.

The proof uses then the following estimates saying that the symbol behaves a little better (for $t = h^2$) are a consequence of,

$$\nabla(\exp -\frac{t}{2}V(x)) \in \Sigma^1 , \tag{6.4.15}$$

and

$$\nabla^2(\exp -\frac{t}{2}V(x)) \in \Sigma^2 . \tag{6.4.16}$$

Study of $\exp -t(-\Delta + V)$

Let us now consider $\tilde{K}(t) = \exp -t(-\Delta + V)$, that we again prefer to consider as $\tilde{K}(t) = \exp -(-h^2\Delta + tV)$. Using the functional calculus of Helffer-Robert [Ro87], it is possible to prove similarly that

Lemma 6.4.2

$$\sigma_h^w(\tilde{K}(t)) - \left[\exp -(\xi^2 + tV(x))\right] \in \Sigma^4 . \tag{6.4.17}$$

Conclusion

From this we get that

$$K(t) - \tilde{K}(t) = O(t^2) \text{ in } \mathcal{L}(L^2), \tag{6.4.18}$$

where we have used the Calderón-Vaillancourt Theorem for the L^2-continuity of the pseudodifferential operators in order to control the remainder terms.
This permits to improve the result of the preceding section as previously announced.
As an extension F. Daumer ([Da94a]) has obtained the following extension,

Theorem 6.4.3 *Let* V *be a* C^∞ *potential satisfying* (6.4.1)

$$V(x) \geq 1 ,$$

and, for some $0 < \rho \leq 1$,

$$|D_x^\alpha V(x)| \leq C_\alpha V(x)^{(1-\rho|\alpha|)_+} , \forall \alpha \in \mathbb{N}^m , \tag{6.4.19}$$

then we get similarly,

$$K(t) - \tilde{K}(t) = O(t^{1+\inf(\frac{1}{2}, \rho)}) \text{ in } \mathcal{L}(L^2) . \tag{6.4.20}$$

This permits to extend our preceding results for more general V's (in particular potentials with quadratic terms like in the model (DK)). The condition (6.4.1) is a natural condition appearing in the construction of parametrixes for Schrödinger operators (see for example [He84] or [Ro87]). These results are connected with the work of Dzh.L. Rogava [Rog93] which announces an abstract version of the preceding result

Theorem 6.4.4 *Let A and B be positive definite selfadjoint operators in a Hilbert space H. Suppose that $D(B) \subset D(A)$. Suppose also that $A+B$ is selfadjoint on $D(B)$. Then for any $N \in \mathbb{N}$*

$$\left\| \left(\exp -\frac{tB}{2N} \circ \exp -\frac{tA}{N} \circ \exp -\frac{tB}{2N} \right)^N - \exp -t(A+B) \right\|_{\mathcal{L}(H)} = \mathcal{O}(\ln N/\sqrt{N}) .$$

It is however difficult to see from the sketch of the proof if the condition $D(B) \subset D(A)$ is important or not in the proof. Let us also mention the recent work of B.O. Dia and M. Schatzman [DiSc94]. Although the motivation was numerical analysis, the proof seems to contain the elements permitting to give an alternative proof (in an "abstract" style) of (6.4.18). We thank M. Schatzman for this remark.

6.5 Semiclassical computations around the transfer operator

We come back to the semiclassical study of the correlation using the technique of the transfer operator. We want in particular to compare with what was mentioned in Subsection 3.9 and more precisely (3.9.10). We start with a model generalizing the Kac model with near neighbors interaction,

$$\Phi(x) = \sum_{j \in \mathbb{Z}/m\cdot\mathbb{Z}} v(x_j) + \sum_{j \in \mathbb{Z}/m\cdot\mathbb{Z}} w(x_j, x_{j+1}) .$$

Following the general reduction we have to analyze the following transfer operator $K = K_V$ whose kernel is given by

$$K(x,y) = \exp -\frac{1}{2h} \cdot (v(x) + v(y)) \cdot \exp -h^{-1} w(x,y) .$$

This does not seem to give an easier problem in general but is interesting for the (DK) model when $w(x,y) = \frac{1}{4}(x-y)^2$. We recognize indeed the "Kac" operator

$$h^{\frac{1}{2}} \exp -\frac{v}{2h} \cdot \exp h\Delta \cdot \exp -\frac{v}{2h} ,$$

but the semiclassical parameter is not inserted as in the standard semiclassical analysis. In the case when the potential v is even and convex (with minimum at 0), we can apply the universal estimate for this operator (See Theorem 6.2.1 and its proof in Section 7) and obtain,

$$\mu_2/\mu_1 \leq \exp - \cosh^{-1}(\inf_x v''(x) + 1) , \tag{6.5.1}$$

and this gives (using (6.2.13)) for the correlation the following estimate,

$$\lim_{m\to\infty} \left[\mathrm{Cor}^{(m)}(x_0, x_j)\right] \leq (\frac{\mu_2}{\mu_1})^j (\int_{\mathbb{R}} u_1^2 x^2 \, dx) . \tag{6.5.2}$$

It remains to control $(\int_{\mathbb{R}} u_1^2 x^2 \, dx)$ with respect to h. One way is to use the proposition (see [Si74] and Proposition 6.4 in [He93b])

Proposition 6.5.1 *Let K, B bounded selfadjoint, with bounded inverse such that*

$$0 < K \leq B ,$$

then we have,

$$\ln K \leq \ln B \quad \text{ and } \quad B^{-1} \leq K^{-1} .$$

We apply it to the Kac operator $K = K_v$ and to $B = h^{\frac{1}{2}} \exp -v/h$ (we forget some regularization argument) and under the additional condition that

$$v(t) \geq \frac{1}{C} t^2 , \text{ for some } C > 0 ,$$

we get that

$$(\int_{\mathbb{R}} u_1^2 x^2 \, dx) \leq C \, h \, (-\ln(h^{-\frac{1}{2}}\mu_1) + 1) .$$

It remains to have a lower bound for μ_1 that we immediately obtain by using the minimax principle,

$$\mu_1 \geq \langle Ku \mid u\rangle // ||u||^2 , \forall u \in L^2(\mathbb{R}) .$$

If we assume that

$$0 \leq v(t) \leq C \, t^2 , \text{ for some } C > 0 ,$$

we can for example take $u = \exp -t^2/h$ and we get immediately a constant $\alpha > 0$ such that

$$\mu_1 \geq h^{\frac{1}{2}} \alpha , \forall h \in]0, h_0] .$$

One can also use the estimate (7.7.3). We finally get the probably optimal

$$\lim_{m\to\infty} \left[\mathrm{Cor}^{(m)}(x_0, x_j)\right] \leq C \, h \, (\exp -j \cdot \cosh^{-1}(\inf_x v''(x) + 1)) . \tag{6.5.3}$$

This is interesting to compare with (3.9.10) (which could be probably in the semiclassical regime be improved in $C_\epsilon \, h \, (\exp -j \cdot \cosh^{-1}((v''(0) - \epsilon) + 1))$, $\forall \epsilon > 0$.

We now want to compare with what we get in the double well case, for which we gave arguments for having (3.9.11). Using (6.2.13) for the correlation, we have,

$$\lim_{m\to\infty}\left[\mathrm{Cor}^{(m)}(x_0,x_j)\right] \leq (\frac{\mu_2}{\mu_1})^j(\int_{\mathbb{R}} u_1^2\, x^2\, dx)\ .$$

and this indicates that for fixed h we have still exponential decay of the correlation with respect to j. This shows that there is no easy commutation of the different limits $m\to\infty$, j large, and $h\to 0$. But we can control (think first to the analogy with the Schrödinger case) the splitting by,

$$1 > \frac{\mu_2}{\mu_1} = 1 - \mathcal{O}(\exp -\epsilon_0/h)\ , \tag{6.5.4}$$

for some $\epsilon_0 > 0$. As we have seen, the estimate on the splitting is directly related to the control of the decay of the first eigenfunction outside the wells. But we have,

$$u_1 = (h^{\frac{1}{2}}\cdot\mu_1^{-1})\cdot\exp -v/h\cdot(\exp h\Delta\,\exp -v/h\, u_1)\ ,$$

and this proves what we want once we have observed that we can control, as in the convex case, $(h^{\frac{1}{2}}\cdot\mu_1^{-1})$. By multiplying by suitable cutoff function we obtain two independents quasimodes relative to μ_1 modulo an exponentially small error. ▮

This is actually not enough to conclude (because we need an upper bound for μ_2/μ_1 similar to the lower bound) but this strongly suggests (and we hope to prove it in a near future) that the decay can be observable when j is larger than $\exp\epsilon_0/h$ and consequently m (which has to be in the discussion large with respect to $2j$). This is not perhaps completely surprising that we have met the condition $m=\mathcal{O}(h^{-N_0})$ in some of our semiclassical computations. It would be interesting to analyze all these formal arguments more carefully.

7 Around the Kac operator

7.1 Introduction

Our main object in this section is to complete the presentation of the properties of the Kac operator associated with a potential V by the formula

$$K_V = \exp -\frac{1}{2}V\cdot\exp h^2\Delta\cdot\exp -\frac{1}{2}V \tag{7.1.1}$$

where $V=V^{(m)}$ is a family of potentials defined on $\mathbb{R}^m$ by

$$V = \frac{1}{2}\sum_{j=1}^{m} x_j^2 + \sum_j W(x_j,x_{j+1})\ , \tag{7.1.2}$$

So finally we have essentially proved (compare with (4.2.5) for the Schrödinger case) the following

Proposition 7.3.2 *If V is strictly convex and satisfies*

$$\sigma = \inf_x \lambda_{\min}(\mathrm{Hess}\, V(x)) > 0 ,$$

then the first eigenfunction of the Kac operator u_1 is logconcave and more precisely $\Phi(x) = -\ln u_1$ satisfies,

$$\phi \geq (\sigma^2/4 + \sigma/2)^{1/2} , \tag{7.3.7}$$

with $\phi = \inf_x \lambda_{\min}(\mathrm{Hess}\, \Phi(x))$.

We observe that this is coherent with the Schrödinger case as σ is small.

Around FKG inequalities. We explain briefly how one can extend the result of Cartier to the case of the Kac operator and refer to [He93b] for details. We shall never try to give the most general assumptions on the potential (regularity...) but we are more interested in explaining the methods.

Theorem 7.3.3 *Let V be a C^∞ potential such that*

$$V \geq C|x|^2 - C \text{ with } C > 0 , \tag{7.3.8}$$

$$|D_x^\beta V| \leq C_\beta (1 + |x|)^{(2-|\beta|)_+} , \tag{7.3.9}$$

and such that the following condition is satisfied,

$$\partial_{ij}^2 V(x) \leq 0 \text{ for } i \neq j \ , \forall\, x \ \in \mathbb{R}^m \ . \tag{7.3.10}$$

Then, if u_1 is the first normalized eigenfunction of the Kac operator

$$K_V = \exp -V/2 \cdot \exp \Delta \cdot \exp -V/2, \tag{7.3.11}$$

the correlation attached to the measure $u_1^2\, dx$ satisfies,

$$\mathrm{Cor}\,(i,j) = \langle (x_i - \langle x_i \rangle)(x_j - \langle x_j \rangle) \geq 0 \ . \tag{7.3.12}$$

We want to start from the following expression for $u_1(x)^2$,

$$u_1(x)^2 = \lim_{p \to \infty} \mu_1^{-p} (K_V^{(p)})(x,x) , \tag{7.3.13}$$

where $K_V^{(p)}(x,y)$ is the distributional kernel of K_V^p. Under our strong assumptions, one can prove the

Lemma 7.3.4 *Suppose V is a C^∞ potential satisfying (7.3.8)-(7.3.9), then the function obtained by restriction to the diagonal of the distribution kernel of $\mu_1^{-2p} K_V^{2p}$,*

$$(\mu_1^{-2p} K_V^{2p})(x,x) \text{ converges exponentially rapidly as } p \to \infty \text{ to } u_1(x)^2 \tag{7.3.14}$$

in $\mathcal{S}(\mathbb{R}^m)$.

In particular, we can apply this result for the correlations and we get, at least for even p,

$$\int (\mu_1^{-p} K_V^{(p)})(x,x)\, dx \to \int u_1(x)^2\, dx \text{ as } p \to \infty, \tag{7.3.15}$$

$$\int x_i (\mu_1^{-p} K_V^{(p)})(x,x) dx \to \int x_i u_1(x)^2 dx \text{ as } p \to \infty\ , \tag{7.3.16}$$

and

$$\int x_i\, x_j\, (\mu_1^{-p} K_V^{(p)})(x,x)\, dx \to \int x_i x_j u_1(x)^2\, dx \text{ as } p \to \infty\ . \tag{7.3.17}$$

So the study of the correlation for the measure $u_1(x)^2 dx$ is naturally related to the study of particular correlation functions attached to the measure on $\mathbb{R}^{mp}$ of the form $\exp -\Phi \cdot dy_1 \cdot \cdots \cdot dy_p$ with

$$\Phi(y_1, \ldots., y_p) = \sum_{j=1}^{p} \left(V(y_j) + \frac{1}{4}(y_{j+1} - y_j)^2 \right)\ . \tag{7.3.18}$$

In order to apply Cartier's theorem to this measure, we then have only to analyze the condition (7.3.3) for Φ. So we have to look at the derivative, $X_j^i X_\ell^k \Phi$, where X_j^i corresponds to the derivative with respect to the variable y_{ji}. If $i = k$, then we have $j \neq \ell$, and the assumption on V gives the result. If $i \neq k$, then the only non vanishing terms correspond to $j = \ell$, and this derivative is equal to $-1/2 < 0$. We then obtain the theorem. ▮

Actually, we can do a little better and prove more directly that $-\ln u_1$ satisfies indeed condition (7.3.10) if V does. We emphasize the similarity with the way followed by Brascamp-Lieb in order to prove, if V is convex, the logconcavity of the first eigenfunction of the Schrödinger operator and we will treat in the case of the Kac operator the two problems similarly.

We first observe that it is sufficient to analyze the same property for the function $K^{(p)}(x,x)$ according to the Lemma (we can then take the limit $p \to \infty$). So we have to study, the function $-\ln K^{(p)}(x,x)$ and study if it satisfies condition (7.3.10), or if it is convex. This can be done by direct inspection of the kernel and use of Lemma 3.7.2.

7.4 The thermodynamic limit for the Kac operator

Let us consider our Kac operator,

$$K_{V^{(m)}} = K^{(m)} = \exp(-V(x)/2) \cdot \exp(\Delta^{(m)}) \cdot \exp(-V(x)/2) \,, \tag{7.4.1}$$

with

$$V(x) = \frac{1}{2}\sum_{k=1}^{m} x_k^2 + \sum_{k=1}^{m} W(x_k, x_{k+1}) \,, \tag{7.4.2}$$

(with the convention that $m + 1 = 1$) which is a compact operator on $L^2(\mathbb{R}^m)$. We just write here the following conditions on W

$$|\nabla W| \leq C \,. \tag{7.4.3}$$

We introduce the potential

$$\tilde{W}(s,t) = \frac{1}{4}(s^2 + t^2) + W(s,t) \,,$$

so we can also write,

$$V(x) = \sum_{k=1}^{m} \tilde{W}(x_k, x_{k+1}) \,,$$

and $\tilde{W}$ has (after possibly the addition of a constant) the following properties,

$$\tilde{W}(t,s) \geq 0 \,, \tag{7.4.4}$$

$$\tilde{W}(t,s) \geq C_1(t^2 + s^2) - D_1 \,, \tag{7.4.5}$$

$$\tilde{W}(t,s) \leq C_2(t^2 + s^2) + D_2 \,, \tag{7.4.6}$$

for some constants C_1, C_2, D_1, D_2 with $C_1 > 0$. Let us denote by $\mu_1(m)$ or more simply by $\mu(m)$ the first eigenvalue of $K^{(m)}$ and $u^{(m)} = u_1^{(m)}$ denotes the strictly positive normalized first eigenfunction. Then we have the

Theorem 7.4.1 *Under the assumptions* (7.4.3)-(7.4.6) *the sequence* $-\ln(\mu(m))/m$ *is convergent to a limit* E_∞ *when* m *tends to* ∞*. Moreover there exists a constant* C *such that*

$$|E_\infty - (-\ln(\mu(m))/m) \leq C/m \,. \tag{7.4.7}$$

7.5 Estimates on the correlations

We now come back to the correlations and we want to estimate the decay as $\mathrm{dist}(i,j)$ in $\mathbb{Z}/m\mathbb{Z}$ is large. In order to simplify the explanations we assume that $\Phi^{(p)}$ is even so that the mean value of $\langle x_i \rangle$ is equal to 0. We want to prove that, under the assumptions (7.2.2)-(7.2.5), the correlation function

$$\mathrm{Cor}^{(m)}(i,j) = \int x_i \cdot x_j u_1^{(m)}(x)^2 dx$$

can be estimated, for some $\delta > 0$ by

$$|\mathrm{Cor}^{(m)}(i,j)| \leq C \exp -\delta \ \mathrm{dist}_{\mathbb{Z}/m\cdot\mathbb{Z}}(i,j) \ . \tag{7.5.1}$$

The strategy is to come back to the study of Laplace integrals with more variables. It is then sufficient to prove the equivalent of (7.5.1) for the correlations attached to the measure $\exp -\Phi^{(p)}(x)dx$ but uniformly with respect to p. More precisely we have to control

$$\frac{\int_{\mathbb{R}^{(mp)}} x_{i1} x_{j1} \exp -\Phi^{(p)}(x) dx}{\int_{\mathbb{R}^{(mp)}} \exp -\Phi^{(p)}(x) dx} \ .$$

If we solve the basic equation

$$x_{i1} - \langle x_{i1} \rangle = (-\Delta + \nabla\Phi \cdot \nabla) u_i \ ,$$

an immediate integration by parts gives (this trick was introduced in [HeSj93] in the same context and presented in (3.5.3))

$$\mathrm{Cor}^{(m,p)}(i,j) = \langle \partial_{x_{j1}} u_i \rangle.$$

We are then reduced to a control of $\sup_{x \in \mathbb{R}^{(mp)}} |\partial_{x_{j1}} u_i(x)|$ which is a consequence of the new results we gave in Subsection 7.2.

As a particular case we can for example get the following theorem

Theorem 7.5.1 *Let W be a C^3 function on $\mathbb{R}^2$ such that $\nabla^p W$ is bounded (for $p = 1,2,3$). Then if*

$$\Psi^{(m)}(x,\beta) = \beta \sum_{i=1}^{m} W(x_i, x_{i+1})$$

and if β is small enough, then there exists $\kappa_0 > 0$ and C_0 s.t. the correlation,

$$\mathrm{Cor}^{(m)}(i,j) = \langle x_i x_j \rangle_m - \langle x_i \rangle_m \langle x_j \rangle_m \ ,$$

satisfies,

$$|\mathrm{Cor}^{(m)}(i,j)| \leq C_0 \exp -\kappa_0 \mathrm{dist}_{\mathbb{Z}/m\mathbb{Z}}(i,j) \ , \tag{7.5.2}$$

for all m and all pairs (i,j) .

This problem to extend results from the Schrödinger case to more general pseudodifferential operators was first met in the study by Helffer-Sjöstrand of the Harper's operator which appears as a pseudodifferential operator whose symbol is $\mathbb{R}^2 \ni (x,\xi) \mapsto \cos x + \cos \xi$ [HeSj88-90]. The observation is here that the decay is controlled via the Agmon distance attached to the same potential *as in the case of Schrödinger.* It is useful to remark here that the Kac operator can be inserted in a class of h-admissible operators whose Weyl symbol is holomorphic with respect to the variable ξ in a band in $\mathbb{C}^m$. In the case of Schrödinger, we used in order to control the decay of the eigenfunction attached to an eigenvalue $\lambda(h)$ near a fixed energy E by $\exp -\phi(x)/h$ to have an inequality of the type

$$|\nabla\phi(x)|^2 \leq (V(x) - E)_+$$

and this is why the Agmon distance appears. This can be interpreted as implying the ellipticity of the symbol

$$\mathbb{R}^m \times \mathbb{R}^m \ni (x,\xi) \mapsto (\xi + i\nabla\phi(x))^2 + V(x) - E\ .$$

This is this point of view which can be extended to more general symbols holomorphic with respect to the parameter ξ.

This result can be improved in the case of a nondegenerate minimum and one can also construct formal WKB solutions and control the splitting in the case of a double well situation under the same assumptions as in [HeSj84-87] (See also our work with B. Parisse [HePa94] where some related points are analyzed). Using these results one obtains for example (see Subsection 5.3 for corresponding results relative to the Schrödinger operator) the

Theorem 7.7.4 (Splitting for the double well case) *Under the assumptions of Theorem* 5.3.3, *we have*

$$-\ln(\mu_2/\mu_1) = h^{\frac{1}{2}} \cdot B_m(h) \cdot \exp -S(m)/h\ , \tag{7.7.8}$$

where,

$$B_m(h) = B_{m,0} + O_m(h)\ , \tag{7.7.9}$$

is the realization of a symbol.

But unfortunately all these results are not controlled with respect to m. A lot of things remain to be done in order to realize the program of studying the tunneling effect in large dimension. We recall that the program is not achieved even in the Schrödinger case ! However, an interesting step was done by J. Sjöstrand in [Sj92b] who introduces a formal class of h-pseudodifferential operators with control with the dimension. Let us briefly explain what was obtained in this context. J. Sjöstrand considers two 0-standard formal symbols a and b of the form

$$a \sim h^{-m_a} \sum_{j=0}^{\infty} a_j \cdot h^j\ ,$$

associated (See Definition 5.6.1) to weights of the form $\rho = (\rho_x, \rho_\xi)$ with $\rho_x = \rho_\xi$. Let us now look at the formal Weyl composition,

$$(a\sharp_h b)(x,\xi;h) \sim \sum_{k=0}^{\infty} \frac{h^k}{k!} \left(\frac{i}{2}\sigma(D_x, D_\xi; D_y, D_\eta)\right)^k (a(x,\xi;h)b(y,\eta;h))|_{y=x\,,\,\eta=\xi}\,, \tag{7.7.10}$$

where $\sigma(x,\xi;y,\eta) = \xi \cdot y - x \cdot \eta$. The first term $a(x,\xi)\,b(x,\xi)$ has no reason to be 0-standard but the Poisson bracket of two symbols is 0-standard. What can be observed at the semiclassical level is that the space of h-pseudodifferential operators is closed under the operation of taking commutators. Another interesting point is that if $P(x,\xi;h)$ is a 0-standard of order 0 (in h), and if $\phi(x)$ is a real 0-standard symbol of order 0, then at the formal level

$$Q = \exp -i\phi(x)/h \circ P^w(x, hD_x; h) \circ \exp i\phi(x)/h$$

can be express as a h-pseudodifferential operator with 0-standard symbol. These remarks could be the starting point for the construction of WKB solutions for the Kac operator.

8 Open problems

As we mentioned in the introduction this survey is in some sense a reactualization of the course given by M. Kac in 1966. Although many results are presented here, we must confess that the program by M. Kac is far to be completed. Moreover, when some of the questions are solved, like for the universal estimate for the Kac operator, the proof does not confirm the idea that the reduction of the problem through the technique of the transfer operator leads to simplification. But anyway the questions posed by this approach enter in more general problems and we hope to have convinced the reader that the "large dimension" techniques mainly introduced in the semi-classical context by J. Sjöstrand have to be developed.
We recall in conclusion some problems which remain unsolved and seem to us interesting.

Question 1

Coming back to our initial discussion, the problem of the thermodynamic limit, of the correlation and of the splitting is reasonably solved in the convex case. An unsolved problem seems to be the analyticity of the thermodynamic limit with respect to the temperature.
For the Kac model (see (1.3.1)), we have seen that the thermodynamic limit $\lim_{m\to\infty} \lambda_1(m)/m$ exists. What about the analyticity with respect to ν which plays the role of the inverse of the temperature. The convex case ($\nu < \frac{1}{4}$) which seems a priori the easiest is still unsolved. This is of course connected with similar questions in

statistical mechanics for spin systems or in quantum field theory where the Yang-Lee Theorem and the Vitali convergence theorem play an important role (see for example the book by Ruelle [Ru69] or Simon-Griffiths [SiGr73] (Section 6)).

Question 2

In the case $\nu > \frac{1}{4}$, we have proved with J. Sjöstrand an upper bound for the splitting between the two first eigenvalues $\lambda_2(m) - \lambda_1(m)$ of the order $\mathcal{O}(\exp -\epsilon_0 \frac{m}{h})$ for some $\epsilon_0 > 0$, but with the constraint $m = \mathcal{O}(h^{-N_0})$. It is basic to solve the problem of eliminating this condition which forbids to take the thermodynamic limit. Then, a second step will be to have a more realistic measure of the decay of the splitting and this seems easier (at least at the formal level) in the case considered by Dobrokhotov and Kolokolt'sov.
Of course the initial question concerned the Kac operator and M. Kac and C.J. Thomson have thought for some times that they have solved the problem (see the announcement in [KaTho66]) but publish an erratum a few months later explaining where the mistake was.

Question 3

The two first questions were in some sense preliminary discussions in order to understand the phase transition. This transition seems to appear, in the semiclassical regim, and for the case of the Kac model, near $\nu = \frac{1}{4}$ when the potential goes from the convex case ($\nu < \frac{1}{4}$) to the double well case ($\nu > \frac{1}{4}$). This was actually the title of the course of M. Kac and its main aim as he explains in the introduction:
" ... It is generally thought that the mechanisms underlying the phase transitions in the classical and Onsagarian theories are essentially different. As a result however, of some studies on a class of models, which will be presented here, a view of a mathematical mechanism of phase transitions begins to emerge which may well prove to be general. Although some of the material to be presented here will be in the form of a demonstration rather than a proof, a dimly lit bridge between the classical ideas of van der Waals and Curie and Weiss, and the new ones based on those of Onsager is beginning to emerge. "
We must recognize that all the results we are presenting are just turning around the crucial problem of the phase transition. The study of the critical value is already difficult for fixed m. The first try to transform "demonstration" into "proof" is done in [He92c] and will be pursued in [Da94b]. The simplest case is the following model

$$-h^2 \frac{d^2}{dx^2} + \frac{1}{4}((1-4\nu)x^2 + \frac{\nu^2}{3}x^4 \,.$$

We have also presented at the end of Subsection 5.4 an example which gives a good idea of the nature of the problem, that is the operator

$$-h^2\Delta + \frac{1}{2}|x|^2 - \nu|\sum x_i| .$$

References

[Ag] S.Agmon: *Lecture on exponential decay of solutions of second order elliptic equations.* Math. Notes, t.29, Princeton University Press (1982).

[AKR94] S.Albeverio, Yu.G.Kondratiev, and M.Röckner: Dirichlet operators via stochastic analysis. *Preprint SFB 256*, n^0 284, (1994).

[AvHerSi81] J.Avron, I.Herbst, and B.Simon: Schrödinger operators with magnetic fields, III Atoms in homogeneous magnetic fields. *Comm. Math. Phys.* **79** (1981), p.529-572.

[BeBru90] G.Ben Arous and M.Brunaud: Méthode de Laplace: Etude variationnelle des fluctuations de diffusions de type "champ moyen". *Stochastics and stochastics reports* **31** (1990), p.79-144.

[BoHe93] C.Bolley and B.Helffer: An application of semiclassical analysis to the asymptotic study of the supercooling field of a superconducting material. *Annales de l'institut Henri-Poincaré* (Section Physique Théorique) **58** (2) (1993), p.189-233.

[BraLi76] H.J.Brascamp and E.Lieb: On extensions of the Brunn-Minkovski and Prékopa-Leindler Theorems. including inequalities for Log concave functions. and with an application to diffusion equation. *Journal of Functional Analysis* **22** (1976), p.366-389.

[BreHi79] E.Brézin and S.Hikami: Large order behavior of $\frac{1}{N}$ expansion in zero and one dimensions. *J. Phys. A* **12** (6) (1979), p.759-770.

[BrFo82] J.Bricmont and J.R.Fontaine: Perturbation about the mean field critical point. *Comm. Math. Phys.* **86** (1982), p.337-362.

[BFLS81] J.Bricmont, J.R.Fontaine, J.L.Lebowitz, and T.Spencer: Lattice systems with continuous symmetry II. Decay of correlations. *Comm. Math. Phys.* **78** (1981), p.363-373.

[BruHe91] M.Brunaud and B.Helffer: Un problème de double puits provenant de la théorie statistico-mécanique des changements de phase (ou relecture d'un cours de M. Kac). *Preprint LMENS*, March 1991.

[BFS82] D.Brydges, J.Fröhlich, and T.Spencer: *Comm. Math. Phys.* **83** (1982), p.123.

[Mas84] V.P.Maslov: Global exponential asymptotics of the solutions of the tunnel equations and the large deviation problems (Russ.). *Tr. Mosk. Inst. Akad. Nauk* **163** (1984), p.150-180. (Engl. transl. in *Proc. Steklov Inst. Math* **4** (1985).)

[Pan84] T.F.Pankratova: Quasimodes and splitting of eigenvalues (Russ.). *Dokl. Akad. Nauk SSSR* **276**:4 (1984), p.795-798. (Engl. translation in *Sov. Math. Dok.* **29** (1984).)

[Pa84] G.Parisi: *Statistical field theory.* Frontiers in Physics, Addison-Wesley Publishing Company, Reading 1984.

[PPW56] L.E.Payne, G.Polya, and H.F.Weinberger: On the ratio of consecutive eigenvalues. *Journal of Math. and Physics* **35** (3), (Oct.1956).

[ReSi72] M.Reed, B.Simon: *Methods of modern mathematical Physics. I. Functional analysis.* Academic Press, New York 1972.

[ReSi75] M.Reed and B.Simon: *Methods of modern mathematical Physics. II. Fourier analysis, selfadjointness.* Academic Press, New York 1975.

[ReSi78] M.Reed and B.Simon: *Methods of modern mathematical Physics. IV. Analysis of operators.* Academic Press, New York 1978.

[Ro87] D.Robert: *Autour de l'approximation semi-classique.* Progress in Mathematics, Vol.68, Birkhäuser, Basel 1987.

[Roe94] G.Roepstorff: *Path integral approach to quantum Physics.* Texts and monographs in Physics, Springer, Berlin-Heidelberg-New York 1994.

[Rog93] Dzh.L.Rogava: Error bounds for Trotter-type formulas for self-adjoint operators. *Functional analysis and its applications* (1993), p.217-219.

[Ru69] D.Ruelle: *Statistical mechanics.* Math. Physics monograph series, W.A. Benjamin, Inc. New York 1969.

[Sh82] S.H.Shenker: Field theories and phase transitions. Les Houches, Session XXXIX (1982), Recent advances in field theory and statistical mechanics, Elsevier Science Publishers, Amsterdam-New York.

[Si74] B.Simon: *The* $P(\phi)_2$ *Euclidean quantum field theory.* Princeton Series in Physics (1974).

[Si79] B.Simon: *Functional integration and quantum physics.* Pure and Applied mathematics, vol.86, Academic Press, New York 1979.

[Si83] B.Simon: Instantons, double wells and large deviations. *Bull. AMS.* **8** (1983), p.323-326.

[Si84] B.Simon: Semi-classical Analysis of low lying eigenvalues.
I *Ann. Inst. H. Poincaré* **38**(1983), p.295-307.
II *Annals of Mathematics* **120** (1984), p.89-118.

[Si93] B.Simon: *The statistical mechanics of lattice gases.* Volume 1, Princeton Series in Physics, Princeton University Press, Princeton, New York 1993.

[SiGr73] B.Simon and R.B.Griffiths: The $(\phi^4)_2$ field theory as a classical Ising model. *Comm. Math. Phys.* **33** (1973), p.145-164.

[SWYY85] I.M.Singer, B.Wong, S.T.Yau, and S.S.T. Yau: An estimate of the gap of the first two eigenvalues of the Schrödinger operator. *Ann. Scuola Norm. Sup. Pisa* **4**(12) (1985), p.319-333.

[Sj82] J. Sjöstrand: Singularités analytiques microlocales. *Astérisque* **95** (1982).

[Sj91] J. Sjöstrand: Schrödinger in high dimensions. asymptotic constructions and estimates. Proceeding of the french-japanese symposium on algebraic analysis and singular perturbations, CIRM, Luminy, France, (October 1991).

[Sj92a] J. Sjöstrand: Exponential convergence of the first eigenvalue divided by the dimension for certain sequences of Schrödinger operator. *Astérisque* **210** (1992), p.303-326.

[Sj92b] J. Sjöstrand: Evolution equations in a large number of variables. Preprint, December 1992. *Mathematische Nachrichten.*

[Sj93a] J. Sjöstrand: Potential wells in high dimensions I. *Ann. Inst. Poincaré*, Section Physique théorique **58** (1), (1993), p.1-41.

[Sj93b] J. Sjöstrand: Potential wells in high dimensions II, more about the one well case. *Ann. Inst. H. Poincaré*, Section Physique théorique **58** (1), (1993), p.42-53.

[Sj93c] J. Sjöstrand: On ferromagnetic Laplace integrals and maximum principles. *Preprint de l'université d'Orsay*, June 1993, and *Ann. Inst. Fourier* **44**(2) (1992), p.601-628.

[Sj93d] J.Sjöstrand: Estimations sur les corrélations et principes du maximum. *Séminaire Equations aux dérivées partielles de l'Ecole Polytechnique*, October 1993.

[Sj94] J.Sjöstrand: Correlation asymptotics and Witten Laplacians. Preprint, June 1994.

[Sok82] A.D.Sokal: Mean field bounds and correlation inequalities. *J. Statist. Phys.* **28** (1982), p.431-439.

differential operators and corresponding equations. In this connection the problem of description of the domain of the generator, perturbation theory, smoothing properties of semigroups and the uniqueness problem become crucial.

One of the principal problems arising in perturbation theory of C_0-semigroups is the construction of the operator extension of the operator sum (difference) in such a way that the operator obtained will be the generator of a C_0-semigroup. It is well known that in general one cannot add an unbounded operator to the generator of a C_0-semigroup without destroying its property of being a generator even in the case when the perturbation is relatively small with respect to the unperturbed generator.

The problem of constructing the operator extension which generates a C_0-semigroup has a satisfactory solution in the case of Hilbert space. For this purpose the method of sesquilinear forms is used [K1, Ch. VI]. If the original operator is sectorial and the perturbation either sectorial or relatively form bounded with bound less than one, then the sum of the corresponding forms will be sectorial and according to the representation theorem defines the sectorial operator which generates a C_0-semigroup. The operator extension obtained is called form-sum. It enjoys several important properties such as approximation property (see [K1, Ch. VIII]) and Lie-Trotter-Kato formula [K2] and has clear physical meaning [Si].

In the special case of symmetric Markov semigroups it becomes possible to characterize the domain of the generator of the symmetric Markov semigroup on L^p by means of the domain of the quadratic form. Our purpose here is to give such a characterization. We extend previous results of the similar nature obtained by D.W.Stroock [S], N.Th.Varopoulos [V] and the authors [LSe1]. Using this characterization we derive some perturbation results for the generators of the symmetric Markov semigroups which can be considered as extensions of the well known KLMN-theorem (see [K1, Ch. VI], [RSi1]) to the L^p-spaces. We also investigate the analyticity and integral properties of the perturbed semigroups and obtain some useful inequalities for semigroups and their generators. Several classes of perturbations of Markov generators are considered, however the main attention is paid to the case of form-bounded perturbations. Some applications to second order elliptic operators are given, in particular, we develop L^p-theory of non-symmetric second order elliptic operators in terms of relative form-boundedness and reveal the fundamental role of coupling constants. We discuss the uniqueness problem for the case of generators of Markov semigroups perturbed by potentials. Our characterization of the domain of generators and inequalities obtained give a possibility to derive criteria of m-accretive closability which are shown to be sharp by the example of Schrödinger operators.

The paper is organized as follows:

1. Some preliminaries.
2. Fundamental properties of Markov generators.
3. Perturbation of symmetric Markov semigroups.I.
4. Non-symmetric Markov generators.
5. Perturbation of symmetric Markov semigroups.II.

6. Perturbations of symmetric Markov semigroups by potentials.
7. (L^p, L^q)-estimates for the resolvent and the semigroup, their integral properties and related problems.
8. Strong and Markov uniqueness.

In Section 1 we introduce Markov semigroups and Markov generators, the main objects of the paper, and discuss some simplest properties of these semigroups and their generators. Section 2 contains a description of the domain of the generator of a Markov semigroup in L^p and several important inequalities. In Section 3 we apply the results of the previous sections to construct the generators of C_0-semigroups in L^p corresponding to the formal sum of operators $A + B - C + iD$, where A, B, C, D are generators of Markov semigroups. Section 4 is devoted to studying the concrete non-symmetric second order elliptic operator. We obtain conditions on the first order perturbation expressed in terms of form boundedness with respect to symmetric second order part guaranteeing that the operator is a generator of a C_0-semigroup in L^p for p in a certain semiaxis which depends only on the form bound of the perturbation. In Section 5 the class of strongly relatively bounded perturbations is considered. Section 6 contains several remarkable facts on perturbation by potentials.

In Section 7 we require additional properties of the unperturbed Markov generator. It is assumed that the Markov generator has the property of Sobolev imbedding. Under this assumption we study (L^p, L^q) properties of the semigroup and the resolvent.

The last section contains several results and examples on Markov uniqueness and strong uniqueness, that is m-accretive closability.

1 Some preliminaries

In this section we introduce the notions of Markov semigroup and Markov generator and to do the paper self-contained collect some known facts which are used extensively later on.
Let $(M, \mathcal{M})$ be a measure space with σ-finite measure μ. $L^p = L^p(M, \mu)$, $1 \leqslant p \leqslant \infty$, be the Banach spaces of complex valued measurable functions with finite norm $\|f\|_p := (\int |f|^p d\mu)^{\frac{1}{p}}$ and $\|f\|_\infty := \inf\{\lambda; |\{x; |f(x)| > \lambda\}| = 0\}$, where $|\Gamma| \equiv \mu(\Gamma)$, $\Gamma \in \mathcal{M}$. By $\langle f, g\rangle$, $f \in L^p$, $g \in L^{p'}$, $p' = \frac{p}{p-1}$ we denote the duality, $\langle f, g\rangle \equiv \int \bar{f} g d\mu \equiv \langle \bar{f} g\rangle$.

Proposition 1.1 *Let $\mathcal{P}_f = \{p \in [1, \infty]; \|f\|_p < \infty\}$. Then* Clos $\mathcal{P}_f \ni p \longmapsto \|f\|_p$ *is a continuous function.*

Proposition 1.2 *Let $\{f_n\}_1^\infty$ be a sequence of measurable functions such that*
1. *$f_n \longrightarrow f$ pointwise μ-almost everywhere (a.e.).*
2. *$\lim_{n\to\infty} \|f_n\|_p = \|f\|_p$ for some $p \in (0, \infty)$.*

Then $f_n \longrightarrow f$ strongly in L^p. (We write sometimes $f_n \xrightarrow[L^p]{s} f$).

and

$$(f \in L^p_, \; |f| \leqslant 1) \Longrightarrow (|e^{-tA} f| \leqslant 1 \; \mu - \text{a.e.}) \tag{1.2}$$

We need the following two formulas from the theory of C_0-semigroups (see, e.g. [K1, Ch. IX])

$$e^{-tA} = s - L^p - \lim_{n \to \infty} (1 + \frac{t}{n} A)^{-n}, \; t \geqslant 0 \tag{1.3}$$

$$(\lambda + A)^{-1} = \int_0^\infty e^{-\lambda t} e^{-tA} dt, \; \lambda > 0, \tag{1.4}$$

($s - L^p - \lim$ means strong limit in L^p) from which we conclude that e^{-tA} preserves positivity for all $t \geqslant 0$ if and only if $(\lambda + A)^{-1}$ preserves positivity for all $\lambda > 0$.

Define $f^+ = f \vee 0$, $f^- = f^+ - f$, $f_\wedge := (|f| \wedge 1) \cdot \operatorname{sgn} f$, where as usual $\operatorname{sgn} z = \frac{z}{|z|}$ for $z \in \mathbb{C}$, $z \neq 0$ and $\operatorname{sgn} 0 = 0$. To characterize the semigroups satisfying (1.1), (1.2) by means of their generators let us introduce the following two conditions

$$[Af, f^+] \geqslant 0, \; \forall f \in \mathcal{D}(A) \cap \operatorname{Re} L^p \tag{1.5}$$

($[f, g] = \langle f, |g|^{p-1} \operatorname{sgn} g \rangle \|g\|_p^{2-p}$ is the semi-inner product in L^p),

$$\operatorname{Re} [Af, f - f_\wedge] \geqslant 0, \; \forall f \in \mathcal{D}(A). \tag{1.6}$$

In fact (see [Sta], [BrPa] and also [MaR], [Ou])

$$(1.1) + (1.2) \Longleftrightarrow (1.5) + (1.6).$$

We give the proof of this equivalence in the next two propositions.

Proposition 1.6 (*see* [AG]) *Let* $(e^{-tA}, t \geqslant 0)$ *be a* C_0*-semigroup of contractions on* $L^p(M, \mu)$. *Then it is positivity preserving if and only if it is real, i.e.* $e^{-tA} \operatorname{Re} L^p \subset \operatorname{Re} L^p$ *and the Phillips condition is satisfied*

$$[Af, f^+] \geqslant 0, \; \forall f \in \mathcal{D}(A) \cap \operatorname{Re} L^p$$

Proof If $e^{-tA}[L^p_+] \subset L^p_+$ then

$$[f, f^+] = \|f^+\|_p^2 \geqslant \|e^{-tA} f^+\|_p \|f^+\|_p \geqslant [e^{-tA} f^+, f^+] \geqslant [e^{-tA} f, f^+].$$

If $f \in \mathcal{D}(A)$ then the function $\varphi(t) = [e^{-tA} f, f^+]$ is differentiable and from the last inequality $d\varphi(t)/dt|_{t=+0} \leqslant 0$. Therefore $[Af, f^+] \geqslant 0$.

Conversely, let $g \in L^p_+$, $f = (1 + \varepsilon A)^{-1} g$, $\varepsilon > 0$. Then $f \in \mathcal{D}(A)$. We prove that $f \geqslant 0$. Indeed,

$$\|f^-\|_2^2 = [(-f), (-f)^+] \leqslant [(-f), (-f)^+] + \varepsilon [A(-f), (-f)^+] = -[g, f^-] \leqslant 0.$$

Therefore $f^- = 0$. □

Proposition 1.7 *Let $(e^{-tA}, t \geqslant 0)$ be a C_0-semigroup of contractions on $L^p(M,\mu)$. Then*

$$\|e^{-tA}\psi\|_\infty \leqslant \|\psi\|_\infty,\ \forall \psi \in L^p \cap L^\infty,\ t > 0$$

if and only if

$$\mathrm{Re}\,[Af, f - f_\wedge] \geqslant 0,\ \forall f \in \mathcal{D}(A).$$

Proof If $\|e^{-tA}\psi\|_\infty \leqslant \|\psi\|_\infty,\ \forall \psi \in L^p \cap L^\infty,\ t > 0$ then

$$\mathrm{Re}\,[Af, f - f_\wedge] = \lim_{t\downarrow 0} \frac{1}{t} \mathrm{Re}\,[(1 - e^{-tA})f, f - f_\wedge]$$
$$= \lim_{t\downarrow 0} \frac{1}{t} \left(\mathrm{Re}\,[(1 - e^{-tA})(f - f_\wedge), f - f_\wedge] + \mathrm{Re}\,[(1 - e^{-tA})f_\wedge, f - f_\wedge]\right).$$

The first term on the right-hand side is nonnegative because of contractiveness in L^p. The last term is equal to $\mathrm{Re}\,\langle |f| \wedge 1 - \overline{\mathrm{sgn}\, f} e^{-tA} f_\wedge, ((|f| - 1)^+)^{(p-1)}\rangle \|f - f_\wedge\|_p^{2-p}$. It is nonnegative since $\langle |f| \wedge 1, ((|f| - 1)^+)^{(p-1)}\rangle = \langle 1, ((|f| - 1)^+)^{(p-1)}\rangle$ and since $|e^{-tA} f_\wedge \mathrm{sgn} f| \leqslant 1$.

Conversely, let $g \in L^p$, $|g| \leqslant 1$ μ-a.e. and $f = (1 + \varepsilon A)^{-1} g$, $\varepsilon > 0$. Obviously $f \in \mathcal{D}(A)$ and $\varepsilon A f = g - f$. Using (1.6) we get

$$0 \leqslant \varepsilon \mathrm{Re}\,[Af, f - f_\wedge] = \mathrm{Re}\,[g - f, f - f_\wedge] = \langle(\mathrm{Re}\,\bar{g}\,\mathrm{sgn}\, f - |f|)((|f| - 1)^+)^{p-1}\rangle \|f - f_\wedge\|_p^{2-p}.$$

If $|f| \geqslant 1$ then $\mathrm{Re}\ \bar{g} \cdot \mathrm{sgn}\ f - |f| \leqslant 0$ since $|\bar{g} \cdot \mathrm{sgn}\ f| \leqslant 1$.
It follows that $(|f| - |f| \wedge 1)(\mathrm{Re}\ \bar{g} \cdot \mathrm{sgn}\ f - |f|) = 0$. Consequently $|f| \leqslant 1$. Using (1.3) yields $\|e^{-tA} g\|_\infty \leqslant 1$. It remains only to set $g = \psi/\|\psi\|_\infty$ for $\psi \in L^p \cap L^\infty$.□

In the particular case of symmetric Markov generators (1.5)+(1.6) coincide with the known Beurling–Deny criterion (see [Ou] where also the form-variant of the above propositions is proved).

Let $A \geqslant 0$ be a self-adjoint operator acting in $L^2(M,\mu)$ with domain $\mathcal{D}(A)$ and form-domain $Q(A) = \mathcal{D}(A^{\frac{1}{2}})$. For $f \in Q(A)$ define $t_A[f] := \|A^{\frac{1}{2}} f\|_2^2$.

We say that A satisfies the first Beurling–Deny condition if

$$u \in Q(A) \Longrightarrow |u| \in Q(A) \ \text{ and } t_A[|u|] \leqslant t_A[u], \tag{1.7}$$

and A satisfies the second Beurling–Deny condition if

$$0 \leqslant u \in Q(A) \Longrightarrow u \wedge 1 \in Q(A) \ \text{ and } t_A[u \wedge 1] \leqslant t_A[u]. \tag{1.8}$$

Theorem 1.1 (Beurling–Deny) *Let $A \geqslant 0$ be a self-adjoint operator acting in $L^2(M,\mu)$. Then* (1.7) $\Longleftrightarrow$ (1.1) *and* (1.7) + (1.8) $\Longleftrightarrow$ (1.1) + (1.2).

For the proof see, e.g. [Fu], [RSi2] or Remark 2.2.

Proposition 1.8 *Let A be a symmetric Markov generator. Define*

$$T_p^t := [e^{-tA} \restriction [L^1 \cap L^\infty]\,]^{\sim}_{L^p \to L^p},\ t > 0,\ p \in [1, \infty),$$

$$(\textit{closure as an operator from } L^p \textit{ to } L^p)$$

$$T_\infty^t := (T_1^t)^* \ (* \textit{ is the sign of adjoint operator}).$$

Then T_p^t is a C_0-contraction semigroup for any $p \in [1, \infty)$.

Proof Since the measure μ is σ-finite there exists the sequence $\{G_n\} \in \mathcal{M}$ such that $M = \bigcup_{n=1}^\infty G_n$, $G_n \subset G_{n+1}$ and $\mu(G_n) < \infty, n = 1, 2, \ldots$
Using first (1.1) and next the self-adjointness of T_2^t and (1.2) we have for arbitrary $f \in L^1 \cap L^\infty$

$$\|e^{-tA} f\|_1 = \lim_n \langle 1\!\!1_{G_n} | e^{-tA} f | \rangle \leqslant \lim_n \langle 1\!\!1_{G_n}, e^{-tA} |f| \rangle = \lim_n \langle e^{-tA} 1\!\!1_{G_n}, |f| \rangle \leqslant \|f\|_1.$$

(Here and below $1\!\!1_\Omega$ is the characteristic function of the set Ω.) Let $f \in L^1 \cap L^\infty$, $g = f/\|f\|_\infty$. Then using (1.1) and (1.2) we obtain

$$\|e^{-tA} g\|_\infty \leqslant \|e^{-tA} |g|\|_\infty \leqslant 1, \quad \text{therefore } \|e^{-tA} f\|_\infty \leqslant \|f\|_\infty.$$

Thus $e^{-tA} \restriction [L^1 \cap L^\infty]$ is a contraction in L^1 and in L^∞. By the Riesz–Thorin interpolation theorem

$$\|e^{-tA} f\|_p \leqslant \|f\|_p,\ t > 0,\ f \in L^1 \cap L^\infty,\ 1 < p < \infty.$$

Therefore the mapping T_p^t is correctly defined for all $p \geqslant 1$ and it is a semigroup. The strong continuity of T_p^t for $1 < p < \infty$ follows from the strong continuity of T_2^t and from Proposition 1.4, and for $p = 1$ it follows from the strong continuity of T_2^t and from Proposition 1.3 and Remark 1.1 (see also [BoHi] for the proof of strong continuity).□

Remark 1.2 According to Stein's theorem on Markov semigroups [St, p.65] T_p^t is a holomorphic contraction in the sector $S_p = \{t;\ |\arg t| < \frac{\pi}{2}(1 - |\frac{2}{p} - 1|)\}$, for all $p \in (1, \infty)$. We extend this result below.

Remark 1.3 A result analogous to Proposition 1.8 holds also for the non-symmetric case but for $p \in [2, \infty)$.

2 Fundamental properties of symmetric Markov generators

The properties of T^t and μ imply the existence of a positive additive set function $\mu_t(x, y)$ on $M \times M$, symmetric with respect to x and y on finite measure sets and such that

$$\langle f, T_\infty^t g \rangle = \int_{M \times M} \bar{f}(x) g(y) d\mu_t(x, y) \ \ \forall f, g \in L^1 \cap L^\infty.$$

To justify this claim let us introduce the Banach space $\mathcal{L}^\infty = \mathcal{L}^\infty(M,\mathcal{M})$ of all bounded measurable functions endowed with the norm $|||f||| = \sup_{x\in M}|f(x)|$. Let $\mathcal{N}^\infty = \mathcal{N}^\infty(M,\mathcal{M},\mu)$ be the set of all μ-negligible functions, so that $L^\infty = \mathcal{L}^\infty/\mathcal{N}^\infty$. Denote by $\pi : f \mapsto \tilde{f}$ the canonical mapping of $\mathcal{L}^\infty$ onto L^∞ (so one can identify L^∞ with $\pi(\mathcal{L}^\infty)$). Since μ is a σ-finite measure there exists a lifting $\rho : L^\infty \longrightarrow \mathcal{L}^\infty$ (see [IT],[M]). In particular, ρ is a linear multiplicative positivity preserving map such that $\rho(1\!\!1_G) = 1\!\!1_G$ for all $G \in \mathcal{M}$ with $\mu(G) < \infty$.

For any $t > 0$ define $T^t_\rho : \mathcal{L}^\infty \longrightarrow \mathcal{L}^\infty$ by $T^t_\rho f := \rho(T^t_\infty \tilde{f})$ so that T^t_ρ is a positivity preserving semigroup of contractions and

$$\langle \tilde{f}, T^t_\infty \tilde{g}\rangle = \langle f, T^t_\rho g\rangle, \quad \tilde{f},\tilde{g} \in L^1 \cap L^\infty.$$

The following set function is associated with the semigroup T^t_∞ which acts on L^∞:

$$P(t,x,G) := (T^t_\rho 1\!\!1_G)(x), \quad t>0,\ x\in M,\ G\in\mathcal{M}.$$

This function satisfies all the properties of a transition probability function except for the countable additivity property and the fact that $P(t,x,M) = 1$. Namely,

1° $P(t,x,\cdot)$ is finitely additive on $\mathcal{M}$.

2° $P(t,x,M) \leqslant 1$.

3° $\int P(t,\cdot,dy)f(y)$ exists and equals to $T^t_\rho f(\cdot)$ $\forall f \in \mathcal{L}^\infty$.

For the proof of 3° see, e.g. [Dy, Vol.1, p.51, Vol.II, p.203]. It should be pointed out that there exist measurable spaces $(M,\mathcal{M})$ such that any finitely additive measure on them is automatically also countable additive. In such a case $P(t,x,G)$ will be indeed a transition probability function. Note that the function $d\mu_t(x,y) = P(t,x,dy)d\mu(x)$ is symmetric with respect to x and y on sets of finite measure. This property is a direct consequence of the self-adjointness of T^t_2 and of the possibility to identify $T^t_\infty 1\!\!1_G$ and $T^t_2 1\!\!1_G$, $\forall G\in\mathcal{M}$ with $\mu(G)<\infty$.

By $-A_p$ we denote the generator of the semigroup T^t_p, so $T^t_p = e^{-tA_p}$, for all $t \geqslant 0$ and $p \in [1,\infty)$. Since T^t_p is a contraction the operator A_p is accretive with respect to the semi-inner product $[u,v] = \langle u, |v|^{p-1}\text{sgn } v\rangle \|v\|_p^{2-p}$ in L^p:

Re $[A_p u, u] \geqslant 0$ for all $u \in \mathcal{D}(A_p)$ and $[A_p u, u] \geqslant 0$ for all $u \in \mathcal{D}(A_p) \cap \text{Re } L^p$. The following theorem makes this property more precise for symmetric Markov generators.

We are going to use the quantity

$$\varkappa(p) = \sup\{(1+s^{\frac{1}{p}})(1+s^{\frac{1}{p'}})(1+s^{\frac{1}{2}})^{-2};\ s\in(0,1)\}.$$

Theorem 2.1 *Let $f \in \mathcal{D}(A_p)$ for some $p \in (1,\infty)$. Then $f_p := f|f|^{\frac{p-2}{2}} \in Q(A)$, and*

$$4\frac{p-1}{p^2} t_A[f_p] \leqslant \text{Re } \langle A_p f, |f|^{p-1}\text{sgn } f\rangle \leqslant \varkappa(p) t_A[f_p], \tag{2.1}$$

$$|\text{Im } \langle A_p f, |f|^{p-1}\text{sgn } f\rangle| \leqslant \frac{|p-2|}{2\sqrt{p-1}}\text{Re } \langle A_p f, |f|^{p-1}\text{sgn } f\rangle. \tag{2.2}$$

A_p generates a bounded holomorphic semigroup of angle

$$\theta_1 = \frac{\pi}{2} - \arctan\frac{|p-2|}{2\sqrt{p-1}}, \ \forall p \in (1,\infty).$$

If $f \in \mathcal{D}(A_p) \cap L^p_+$, then $f^{\frac{p}{2}} \in Q(A)$, and

$$4\frac{p-1}{p^2}t_A[f^{\frac{p}{2}}] \leqslant \langle A_p f, f^{p-1}\rangle \leqslant t_A[f^{\frac{p}{2}}]. \tag{2.3}$$

Remark 2.1 (2.3) is the Stroock–Varopoulos inequality [S], [V]. (2.2) is a generalization of the inequality from [MSo] (see [LP1], [LSe4]; see also [Ba] where (2.2) was proved for a particular case of diffusion semigroups and it was shown that the constant in (2.2) is the best possible). (2.1) with $f = \text{Re } f$ was proved in [LSe1].

For the proof of Theorem 2.1 we need the following lemma.

Lemma 2.1 *For all $0 < s, t < \infty$ and all $p \in [1,\infty)$ the following inequalities hold*

$$4\frac{p-1}{p^2}(s^{\frac{p}{2}} - t^{\frac{p}{2}})^2 \leqslant (s-t)(s^{p-1} - t^{p-1}) \leqslant (s^{\frac{p}{2}} - t^{\frac{p}{2}})^2 \tag{2.4}$$

$$(s^{\frac{p}{2}} + t^{\frac{p}{2}})^2 \leqslant (s+t)(s^{p-1} + t^{p-1}) \leqslant \varkappa(p)(s^{\frac{p}{2}} + t^{\frac{p}{2}})^2 \tag{2.5}$$

$$4\frac{p-1}{p^2}[s^p + t^p + 2b(st)^{\frac{p}{2}}] \leqslant s^p + t^p + b(st^{p-1} + ts^{p-1}), \ \forall b \in [-1,1] \tag{2.6}$$

$$|\sin\theta|\,|st^{p-1} - ts^{p-1}| \leqslant \frac{|p-2|}{2\sqrt{p-1}}(s^p + t^p - |\cos\theta|(t^{p-1}s + s^{p-1}t)), \ \forall p \in (1,\infty), \ \forall \theta \in \mathbb{R}^1 \tag{2.7}$$

$$s^p + t^p + b(st^{p-1} + ts^{p-1}) \leqslant \varkappa(p)[s^p + t^p + 2b(st)^{\frac{p}{2}}], \ \forall p \in (1,\infty), \ \forall b \in [-1,1] \tag{2.8}$$

Proof The right-hand side of (2.4) and the left-hand side of (2.5) are consequences of the inequality $2ab \leqslant a^2 + b^2$. The right-hand side of (2.5) follows from the definition of $\varkappa(p)$. The left-hand side of (2.4) follows from

$$\frac{4}{p^2}\left(s^{p/2} - t^{p/2}\right)^2 = \left(\int_t^s z^{\frac{p}{2}-1}dz\right)^2 \leq \left|\int_t^s dz\right| \cdot \left|\int_t^s z^{p-2}dz\right|.$$

(2.6) is a consequence of the left side of (2.4). (2.7) follows also from the left-hand side of (2.4) and from the implication $A^2 + B^2 \leqslant C^2 \Longrightarrow |A\sin\theta| + |Bcos\ \theta| \leqslant |C|$, where A, B, C are real numbers. (One needs only set

$$A = t^{p-1}s - s^{p-1}t, \quad B = \frac{|p-2|}{2\sqrt{p-1}}(t^{p-1}s + s^{p-1}t), \quad C = \frac{|p-2|}{2\sqrt{p-1}}(t^p + s^p)$$

and use the left-hand side of (2.4) to derive $A^2 + B^2 \leqslant C^2$.) And finally (2.8) follows from the right-hand side (2.5) and from the following elementary inequality:

$$\frac{A + b\,B}{A + b\,C} \leqslant \frac{A + B}{A + C}, \quad \forall b \in [-1, 1], \ \forall A \geqslant B \geqslant C > 0. \square$$

Proof of Theorem 2.1 We start from the proof of the following inequalities which are valid for all $f \in L^p$:

$$4\frac{p-1}{p^2}\langle(1 - T_2^t)f_p, f_p\rangle \leqslant \mathrm{Re}\ \langle(1 - T_p^t)f, |f|^{p-1}\mathrm{sgn}\ f\rangle \leqslant \varkappa(p)\langle(1 - T_2^t)f_p, f_p\rangle \quad (2.9)$$

and for all $f \in L_+^p$:

$$4\frac{p-1}{p^2}\langle(1 - T_2^t)f^{\frac{p}{2}}, f^{\frac{p}{2}}\rangle \leqslant \langle(1 - T_p^t)f, f^{p-1}\rangle \leqslant \langle(1 - T_2^t)f^{\frac{p}{2}}, f^{\frac{p}{2}}\rangle. \quad (2.10)$$

Since $Q(A) = \{g \in L^2 : \sup_{t>0} \frac{1}{t}\langle(1 - T_2^t)g, g\rangle < \infty\}$ and $\lim_{t\downarrow 0} \frac{1}{t}\langle(1 - T_2^t)g, g\rangle = t_A[g]$, $\forall g \in Q(A)$ the statements (2.1) and (2.3) follow from the spectral theorem for self-adjoint operators. (2.2) will follow from the inequality

$$|\mathrm{Im}\ \langle(1 - T_p^t)f, |f|^{p-1}\mathrm{sgn}\ f\rangle| \leqslant \frac{|p-2|}{2\sqrt{p-1}}\mathrm{Re}\ \langle(1 - T_p^t)f, |f|^{p-1}\mathrm{sgn}\ f\rangle. \quad (2.11)$$

First let $f \in L^1 \cap L^\infty$, supp $f \subset G$ with $\mu(G) < \infty$. Using the properties of $P(t, x, G)$ and denoting by $\int d\mu_t(x, y)f(x)h(y)$ the iterated integral $\int f(x)(\int h(y)P(t, x, dy))d\mu(x)$ we obtain

$$\begin{aligned}\langle T_p^t f, |f|^{p-1}\mathrm{sgn}\ f\rangle &= \frac{1}{2}\langle T_\infty^t f, |f|^{p-1}\mathrm{sgn}\ f\rangle + \frac{1}{2}\langle f, T_\infty^t(|f|^{p-1}\mathrm{sgn}\ f)\rangle \\ &= \frac{1}{2}\int d\mu_t(x, y)\left[\overline{f(x)}f(y)|f(y)|^{p-2} + \overline{f(y)}f(x)|f(x)|^{p-2}\right],\end{aligned}$$

$$\langle T_2^t f_p, f_p\rangle = \frac{1}{2}\int d\mu_t(x, y)\left[\overline{f_p(x)}f_p(y) + \overline{f_p(y)}f_p(x)\right],$$

$$\|f\|_p^p = \langle 1 - T_\infty^t 1, |f|^p\rangle + \langle T_\infty^t 1, |f|^p\rangle,$$

$$\begin{aligned}\langle T_\infty^t 1, |f|^p\rangle = \langle 1, T_\infty^t|f|^p\rangle &= \frac{1}{2}\langle P(t, \cdot, M), |f(\cdot)|^p\rangle + \frac{1}{2}\left\langle 1, \int P(t, \cdot, dy)|f(y)|^p\right\rangle \\ &= \frac{1}{2}\int d\mu_t(x, y)\left[|f(x)|^p + |f(y)|^p\right],\end{aligned}$$

Consequently setting $f(x) = |f(x)|e^{i\theta_x}$, $|f(x)| = s$, $f(y) = le^{i\theta_y}$, $\theta = \theta_x - \theta_y$, we get

$$\begin{aligned}\langle(1 - T_p^t)f, |f|^{p-1}\mathrm{sgn}\ f\rangle =& \langle 1 - T_\infty^t 1, |f|^p\rangle + \frac{1}{2}\int d\mu_t(x, y)[|f(x)|^p + |f(y)|^p \\ & - \overline{f(x)}f(y)|f(y)|^{p-2} - \overline{f(y)}f(x)|f(x)|^{p-2}] \qquad (2.12) \\ =& \langle 1 - T_\infty^t 1, |f|^p\rangle + \frac{1}{2}\int d\mu_t(x, y)[s^p + l^p - sl^{p-1}e^{-i\theta} - ls^{p-1}e^{i\theta}],\end{aligned}$$

$$\begin{aligned}\langle(1-T_2^t)f_p, f_p\rangle =&\langle 1-T_\infty^t 1, |f|^p\rangle + \frac{1}{2}\int d\mu_t(x,y)[|f(x)|^p + |f(y)|^p \\ &- \overline{f_p(x)}f_p(y) - \overline{f_p(y)}f_p(x)] \\ =&\langle 1-T_\infty^t 1\!\!1_G, |f|^p\rangle + \frac{1}{2}\int d\mu_t(x,y)[s^p + l^p - 2(sl)^{\frac{p}{2}}\cos\theta],\end{aligned} \tag{2.13}$$

$$\begin{aligned}&\mathrm{Re}\ \langle(1-T_p^t)f, |f|^{p-1}\mathrm{sgn}\ f\rangle \\ &\quad = \langle 1-T_\infty^t 1, |f|^p\rangle + \frac{1}{2}\int d\mu_t(x,y)[s^p + l^p - (sl^{p-1} + ls^{p-1})\cos\theta], \\ &\mathrm{Im}\ \langle(1-T_p^t)f, |f|^{p-1}\mathrm{sgn}\ f\rangle = \frac{1}{2}\int d\mu_t(x,y)(sl^{p-1} - ls^{p-1})\sin\theta.\end{aligned} \tag{2.14}$$

Since $\langle 1 - T_\infty^t 1, |f|^p\rangle \geqslant 0$, then (2.11) follows from (2.7) and (2.14). Using the inequality $4\frac{p-1}{p^2} \leqslant 1$ by virtue of (2.6) and (2.8) (2.9) follows from (2.12) and (2.13). (2.10) directly follows from (2.4). Thus (2.9), (2.10) and (2.11) are established for $f \in L^1 \cap L^\infty$ or $f \in L^1 \cap L_+^\infty$ with supp $f \subset G$, $\mu(G) < \infty$. To finish the proof note that these sets of functions are dense in L^p and in L_+^p respectively.
Analyticity of e^{-tA_p} in the sector $S_p = \{z \in \mathbb{C};|\arg z| < \theta_1\}$ follows from (2.2) and general theory (see, e.g. [G, p.37] or [K1, p.492]). $\square$

Remark 2.2 From the above proof it follows immediately that (1.7) implies (1.1). It is enough for this to consider the case $p = 2$ and to note that

$$\langle(1-T_2^t)|f|, |f|\rangle \leqslant \langle(1-T_2^t)f, f\rangle.$$

$(1.7) + (1.8) \Longrightarrow (1.2)$ follows from $(s\wedge 1 - t\wedge 1)^2 \leqslant (s-t)^2$. These observations give a proof to the Beurling–Deny theorem. Notice also that (2.12) for $p = 2$ gives

$$\langle(1-T_2^t)f, f\rangle = \langle 1-T_\infty^t 1, |f|^2\rangle + \frac{1}{2}\int d\mu_t(x,y)|f(x) - f(y)|^2.$$

From this it follows immediately that if $f \in Q(A)$ and g is such that $|g(x)| \leqslant |f(x)|$ and $|g(x) - g(y)| \leqslant |f(x) - f(y)|$ then $g \in Q(A)$ and $t_A[g] \leqslant t_A[f]$ (see [Fu, p.5]).

Remark 2.3 One can prove Theorem 2.1 without using the notion of lifting using simple functions of the form $f = \Sigma_{i=1}^k c_i 1\!\!1_{G_i}$, where $\{G_i\}$ are disjoint sets of finite measure, $c_i \in \mathbb{C}$, instead of functions from $L^1 \cap L^\infty$. In order to do this one should use directly the representations

$$T^t f(\cdot) = \sum_{i=1}^k c_i T^t 1\!\!1_{G_i}(\cdot) := \int_M P(t,\cdot,dy)f(y),$$

$$T^t f|f|^{p-2} = \sum_{i=1}^k c_i|c_i|^{p-2}T^t 1\!\!1_{G_i},\ T^t f_p = \sum_{i=1}^k c_i|c_i|^{\frac{p-2}{2}}T^t 1\!\!1_{G_i},$$

instead of property 3° of $P(t,x,G)$ (see [LSe1]).

Remark 2.4 Let us denote the sharp constants in inequality (2.1) by $\varkappa_L(A_p)$ and $\varkappa_U(A_p)$:

$$\varkappa_L(A_p)t_A[f_p] \leqslant \mathrm{Re}\,\langle A_p f, |f|^{p-1}\mathrm{sgn}\ f\rangle \leqslant \varkappa_U(A_p)t_A[f_p],$$
$$f \in \mathcal{D}(A_p),\ f_p \in Q(A).$$

For the whole class of symmetric Markov generators $\inf \varkappa_L(A_p) = 4\frac{p-1}{p^2}$ (for example, $(M,\mu) = (\mathbb{R}^d, dx)$, $A_p = -\Delta_p$) and there exist symmetric Markov generators with $\varkappa_U(A_p) > 1$. However we do not know examples of operators with $\varkappa_U(A_p) = \varkappa(p)$. The dependence $A \to \varkappa_L(A_p)$ is also unknown even for the simplest cases like

$$0 < \alpha < 1,\ A^\alpha = (-\Delta + m^{\frac{1}{\alpha}})^\alpha - m,\ (M,\mu) = (\mathbb{R}^d, dx).$$

Now we give one more result in the same direction. It reflects the fact that there exist test functions different from the power and still preserving ellipticity in the case when A is an elliptic second order differential operator. This has important applications in the uniqueness problem for the equation $A_p u - V_p^- u = 0$ (see section 8).

Lemma 2.2 *Let $\varphi : \mathbb{R}^1_+ \to \mathbb{R}^1_+$ be a function such that*

(i) $\varphi(t) = 0\ \forall\ t \in [0,b]$ *for some* $b \geq 0$*;*

(ii) $\varphi'(t) > 0\ \forall\ t > b$*;*

(iii) *the function* $g_\varphi(t)$ *is differentiable for* $t > b$ *,where*

$$g_\varphi(t) = \begin{cases} t\phi(t) - \kappa & \text{if } t \geqslant b, \\ 0 & \text{if } t < b, \end{cases}$$

$$\phi(t) = \sqrt{\varphi'(t)}\ ,\ \kappa \equiv t\phi(t)|_{t+0=b};$$

(iv) $\sup_{t>b}\left(1 + \frac{t\phi'(t)}{\phi(t)}\right)^2 = c_\varphi^{-1} < \infty$*;*

or

(i') $\varphi(0) = 0,\ \varphi(t) = \varphi(b)\ \forall\ t \geqslant b$ *for some* $b > 0$*;*

(ii') $\varphi'(t) > 0\ \forall\ t \in (0,b)$*;*

(iii') *the function* $g_\varphi(t)$ *is differentiable for* $t \in [0,b)$ *,where*

$$g_\varphi(t) = \begin{cases} t\phi(t) - \kappa & if\ 0 \leqslant t \leqslant b, \\ 0 & if\ t \geqslant b, \end{cases}$$

$$\phi(t) = \sqrt{\varphi'(t)}\ ,\ \kappa \equiv t\phi(t)|_{t-0=b};$$

(iv') $\sup_{0<t<b}\left(1+\frac{t\phi'(t)}{\phi(t)}\right)^2 = c_\varphi^{-1} < \infty.$

Then for all $t, s \in \mathbb{R}^1_+$

$$c_\varphi^{-1}\cdot(t-s)(\varphi(t)-\varphi(s)) \geqslant (G_\varphi(t)-G_\varphi(s))^2,$$

$$c_\varphi^{-1}\cdot(t+s)(\varphi(t)+\varphi(s)) \geqslant (G_\varphi(t)+G_\varphi(s))^2,$$

$$c_\varphi^{-1}\cdot s\cdot\varphi(s) \geqslant G_\varphi^2(s),$$

where $G_\varphi(t) := g_\varphi(t) - g_\varphi(0)$.

Proof For (i)-(iv) the last inequality follows from the first one setting $t = 0$. The second inequality is a consequence of the last one:

$$c_\varphi(g_\varphi(t)+g_\varphi(s))^2 \leqslant t\varphi(t)+s\varphi(s)+2\sqrt{s\varphi(s)\cdot t\varphi(t)} \leqslant (t+s)(\varphi(t)+\varphi(s)).$$

So we have to prove only the first inequality.

$$\begin{aligned}(g_\varphi(t)-g_\varphi(s))^2 &= \left(\int_s^t dg_\varphi(z)\right)^2\\ &= \left(\int_s^t (z\phi'(z)+\phi(z))dz\right)^2\\ &\leq \left(\int_s^t (1+\frac{z\phi'(z)}{\phi(z)})^2 dz\right)\left(\int_s^t \phi^2(z)dz\right)\\ &\leq c_\varphi^{-1}(t-s)(\varphi(t)-\varphi(s)) .\end{aligned}$$

The same arguments apply to the case (i')-(iv').□

Theorem 2.2 *Let A be a symmetric Markov generator and let $\varphi : \mathbb{R}^1_+ \to \mathbb{R}^1_+$ satisfy the conditions of Lemma 2.2. Let $\dot{G}_\varphi(z) := G_\varphi(|z|)\mathrm{sgn}\ z$ ($z \in \mathbb{C}$). If $f = \mathrm{Re}\ f \in \mathcal{D}(A_p)$ and $\varphi(|f|)A_pf \in L^1$ for some $p \in [1,\infty)$, then $\dot{G}_\varphi(f) \in Q(A)$ and*

$$c_\varphi\|A^{\frac{1}{2}}G_\varphi(|f|)\|_2^2 \leqslant c_\varphi\|A^{\frac{1}{2}}\dot{G}_\varphi(f)\|_2^2 \leqslant \langle A_pf, \varphi(|f|)\mathrm{sgn}\ f\rangle. \tag{2.15}$$

Proof Let $f = \mathrm{Re}\ f \in L^1\cap L^\infty$, supp $f \subset G$, $\mu(G) < \infty$. Then $\varphi(|f|) \in L^1 \cap L^\infty$, supp $\varphi(|f|) \subset \bar{G}$. Thus the arguments used in the proof of Theorem 2.1 and the property of the lifting $\rho(\varphi(|f|)) = \varphi(|\rho(f)|)$ give

$$\langle(1-T_2^t)\dot{G}_\varphi(f), \dot{G}_\varphi(f)\rangle = \langle 1-T_\infty^t 1, G_\varphi^2(|f|)\rangle + \frac{1}{2}\int d\mu_t(x,y)(\dot{G}_\varphi(f(x))-\dot{G}_\varphi(f(y)))^2,$$

$$\langle(1-T_p^t)f, \varphi(|f|)\mathrm{sgn}\ f\rangle = \langle 1-T_\infty^t 1, |f|\varphi(|f|)\rangle +$$

$$+\frac{1}{2}\int d\mu_t(x,y)(f(x)-f(y))(\varphi(|f(x)|)\text{sgn } f(x)-\varphi(|f(y)|)\text{sgn } f(y))$$

The statement of the theorem for $\varphi(|f|) \in L^{p'}$ follows now from Lemma 2.2. To treat the case $\varphi(|f|)A_p f \in L^1$ note that the basic inequality

$$c_\varphi \frac{1}{t}\langle (1-T_2^t)\dot{G}_\varphi(f), \dot{G}_\varphi(f)\rangle \leqslant \frac{1}{t}\langle (1-T_p^t)f, \varphi(|f|) \text{ sgn } f\rangle$$

holds if one replaces $\dot{G}_\varphi(f)$ by its truncation $\dot{G}^n_\varphi(f)$ and $\varphi(|f|)$ by $\varphi_n(|f|)$, where $\dot{G}^n_\varphi(f) = \dot{G}_\varphi((|f| \wedge n)\text{sgn } f)$, $\varphi_n(|f|) = \varphi(|f| \wedge n)$. Then

$$c_\varphi \|A^{\frac{1}{2}}\dot{G}^n_\varphi(f)\|_2^2 \leqslant \langle A_p f, \varphi_n(|f|) \text{ sgn } f\rangle \leqslant \|\varphi(|f|)A_p f\|_1 \text{ and } \dot{G}^n_\varphi(f) \underset{L^2}{\overset{s}{\longrightarrow}} \dot{G}_\varphi(f)\ (n \to \infty).$$

The assertion follows now from the closedness of $A^{\frac{1}{2}}$.□

Remark 2.5 In contrast to Theorem 2.1 there is no two-sided inequality for the whole class of functions φ. This can be readily seen from the example $\varphi(t) = \ln_+ t := \ln(t \vee 1)$.

3 Perturbations of symmetric Markov semigroups. I

Let A and B be symmetric Markov generators. Let $A_p + B_p$ be the algebraic sum of the operators A_p and B_p with $\mathcal{D}(A_p+B_p) = \mathcal{D}(A_p)\cap\mathcal{D}(B_p)$. Generally it is a difficult problem to construct the extension of A_p+B_p which is a generator of a C_0-semigroup. This problem has a relatively simple solution if $Q(A)\cap Q(B)$ is dense in L^2. In this case one can define the form-sum $A\dot{+}B$ which is the non-negative self-adjoint operator in L^2 associated with the sesquilinear form

$$t_{A+B}[u,v] = t_A[u,v] + t_B[u,v],\ u,v \in Q(A)\cap Q(B) := Q(A\dot{+}B).$$

The following theorem is valid.

Theorem 3.1 *If A and B are symmetric Markov generators such that $Q(A)\cap Q(B)$ is dense in L^2 then $A\dot{+}B$ is a symmetric Markov generator and*

$$(A\dot{+}B)_p \supset A_p + B_p,\ \forall p \in (1,\infty).$$

The proof of this theorem is given below. Before doing that we discuss some properties of the Yosida approximation for symmetric Markov generators.

Proposition 3.1 *Let B be a symmetric Markov generator. Let $B_{n,2} = nB(n+B)^{-1}$ be the Yosida approximation of B, $n = 1,2,\ldots$ Then $B_{n,2}$ is a symmetric Markov generator, $B_{n,p}$ is bounded for all $p \geqslant 1$ and $B_{n,2} \leqslant B_{m,2}$ if $n \leqslant m$.*

Proof $B_{n,p} = nB_p(n+B_p)^{-1} = n - n^2(n+B_p)^{-1}$ is a bounded operator in L^p, therefore

$$e^{-tB_{n,p}} = e^{-nt}\sum_{k=0}^{\infty}\frac{(n^2t)^k}{k!}(n+B_p)^{-k}.$$

It follows that

$$1.\ \|e^{-tB_{n,p}}\|_{p\to p} \leqslant 1 \text{ and } 2.\ e^{-tB_{n,p}}[L^p_+] \subset L^p_+.$$

Thus $B_{n,2}$ is a symmetric Markov generator.
Finally, the function $(0,\infty) \ni \lambda \longmapsto \lambda\nu(\lambda+\nu)^{-1}$, $\nu \in \mathbb{R}^1_+$, is monotone increasing and therefore by the spectral theorem $B_{n,2} \leqslant B_{m,2}$ if $n \leqslant m$.□

Proof of Theorem 3.1 It is an easy matter to check (1.7) and (1.8) for the form-sum or (1.1) and (1.2) using Lie-Trotter-Kato formula [K2], so that $A\dot{+}B$ is a Markov generator. By the monotone convergence theorem for quadratic forms [K1, Ch. VIII, Theorem 3.13a] for all $k \geqslant 1$

$$(A+B_{n,2})(1+\frac{1}{k}(A+B_{n,2}))^{-1} \xrightarrow[L^2]{s} (A\dot{+}B)(1+\frac{1}{k}(A\dot{+}B))^{-1}\ (n \longrightarrow \infty).$$

Now using Proposition 1.4 we obtain

$$e^{-t(A+B_{n,p})} \xrightarrow[L^p]{s} e^{-t(A\dot{+}B)_p},\ (1+A_p+B_{n,p})^{-1} \xrightarrow[L^p]{s} (1+(A\dot{+}B)_p)^{-1}.$$

Let $f \in \mathcal{D}(A_p)\cap\mathcal{D}(B_p)$. Then

$$(1+A_p+B_{n,p})f \xrightarrow[L^p]{s} (1+A_p+B_p)f,$$

$$f = (1+A_p+B_{n,p})^{-1}(1+A_p+B_{n,p})f \xrightarrow[L^p]{s} (1+(A\dot{+}B)_p)^{-1}(1+A_p+B_p)f.$$

Therefore $f \in \mathcal{D}((A\dot{+}B)_p)$ and $(A\dot{+}B)_pf = A_pf + B_pf$.
Analyticity of $e^{-t(A\dot{+}B)_p}$ in the sector $S_p = \{z \in \mathbb{C};|\arg z| < \theta_1\}$ follows from Theorem 2.1 and general theory (see, e.g. [G, p.37] or [K1, p.492]).□

The construction of the operator extension generating a C_0-semigroup for the difference $A_p - B_p$ is much more complicated than the same problem for the sum $A_p + B_p$. In the Hilbert space it is solved by the KLMN–theorem. Let $A \geqslant 0$ be a self-adjoint operator in a Hilbert space $\mathcal{H}$. By $PK_\beta(A)$ we denote the class of self-adjoint operators which are form-bounded with respect to A with form-bound β, i.e.

$$\mathcal{D}(|B|^{\frac{1}{2}}) \supset Q(A) \text{ and}$$

$$t_{|B|}[\varphi] \leqslant \beta t_A[\varphi] + c(\beta)\|\varphi\|_2^2,\ \varphi \in Q(A)$$

for some numbers $\beta > 0$ and $c(\beta) \in \mathbb{R}^1$.

The self-adjoint variant of KLMN–theorem asserts that if $0 \leqslant B \in PK_\beta(A)$ with $\beta < 1$ then the quadratic form $t_{A\dot{-}B}[\varphi] := t_A[\varphi] - t_B[\varphi]$ defined on $Q(A)$ is closed and semibounded.The self-adjoint operator $A\dot{-}B$ associated with this form (form-difference of the operators A and B) is semibounded and extends $A - B$ with $\mathcal{D}(A\dot{-}B)$ dense in $Q(A)$.

Although this theorem has various applications to PDE's, it has a serious disadvantage. It is not sensitive to the "regularity" of initial data of the equation $\frac{d}{dt}u(t) + (A\dot{-}B)u(t) = 0$, $u(0) = f$ and to the value of the form-bound β. It is clear at least intuitively that the solution $u(t)$ should become more "regular" with decreasing β as well as with choosing more "regular" f. This is exactly the case for symmetric Markov generators. We succeed in extending the theory to the case of Banach spaces. The results given below reveal the fundamental role of the value of β (and not only the fact that $\beta < 1$). In applications it leads to important consequences in problems of existence and smoothness of solutions as well as in the uniqueness problem ([KoSe1–4],[LSe2], [Se 4-5]).

Let $\varkappa(p)$ be the constant from Section 2. For fixed $\beta \in (0,1)$ the equation $\beta\varkappa(q) = 4\frac{q-1}{q^2}$, $1 < q < \infty$, has exactly two solutions $t_1 \in (1,2)$ and $t_2 = t_1' \in (2,\infty)$, where $t_1' = t_1/t_1 - 1$. This is a direct consequence of the definition of $\varkappa(p)$: $\varkappa(q) = \varkappa(q')$ and $4\frac{q-1}{q^2} = 4\frac{q'-1}{q'^2}$. Note also that $\varkappa(q)$ is a convex function which attains its minimum for $q = 2$.

Theorem 3.2 *Let A and B be symmetric Markov generators. Suppose that $B \in PK_\beta(A)$for some $\beta \in (0,1)$. Let $t(\beta) = t_1$ and $t'(\beta) = t_2$ be the corresponding roots of the equation $\beta\varkappa(q) = 4\frac{q-1}{q^2}$, $1 < q < \infty$. Then for every $p \in [t(\beta), t'(\beta)]$ and $t \geqslant 0$*

(i) $$\|e^{-t(A\dot{-}B)}f\|_p \leqslant e^{\varkappa(p)c(\beta)t}\|f\|_p,\ f \in L^2 \cap L^p$$

and consequently this defines a family of operators $\{(A\dot{-}B)_p\}_{t(\beta)\leqslant p\leqslant t'(\beta)}$ which generate quasi-contraction C_0-semigroups.

(ii) $$(A\dot{-}B)_p \supset A_p - B_p$$

(iii)
$\forall p \in (t_1, t_2)$ the semigroup $e^{-t(A\dot{-}B)_p}$ is holomorphic with respect to t of angle

$$\theta_1 = \frac{\pi}{2} - \arctan\frac{|p-2|}{2\sqrt{p-1}}\frac{4(p-1)/p^2 + \beta\varkappa(p)}{4(p-1)/p^2 - \beta\varkappa(p)},$$

(iv) $$\forall p \in (t_1, t_2)\ \forall f \in \mathcal{D}((A\dot{-}B)_p)\ :\ f_p := f|f|^{\frac{p-2}{2}} \in Q(A)\ \textit{and}$$

$$\begin{aligned}\operatorname{Re}\,\langle (A\dot{-}B)_p f, \operatorname{sgn} f\cdot|f|^{p-1}\rangle &\geqslant 4\frac{p-1}{p^2}\|A^{\frac12}f_p\|_2^2 - \varkappa(p)\|B^{\frac12}f_p\|_2^2\\ &\geqslant \left(4\frac{p-1}{p^2} - \beta\varkappa(p)\right)\|A^{\frac12}f_p\|_2^2 - \varkappa(p)c(\beta)\|f\|_p^p.\end{aligned}$$

Proof Let $B_{n,p}$ be the Yosida approximation of the operator B_p. Since $B_{n,p} \in \mathcal{L}(L^p)$, the operators $C_{p,n} = A_p - B_{n,p}$ with $\mathcal{D}(C_{p,n}) = \mathcal{D}(A_p)$ generate quasicontractive semigroups in L^p, for every $1 \le p < \infty$, $\forall n = 1, 2, \dots$.

Let $u_n(t) := e^{-tC_{2,n}} f$, $f \in L^2 \cap L^p$, $1 < p < \infty$. Then for any $t > 0$ $u_n(t) \in \mathcal{D}(C_{p,n})$ since 1) the semigroup $e^{-tC_{p,n}}$ is holomorphic according to Theorem 3.1 or due to E.Stein's interpolation theorem, 2) bounded perturbations preserve holomorphy of the original semigroup, 3) $e^{-tC_{2,n}} \restriction [L^2 \cap L^p] = e^{-tC_{p,n}} \restriction [L^2 \cap L^p]$ and $\frac{d}{dt} u_n(t) = -C_{p,n} u_n(t)$, $t > 0$.

Note that $e^{-tC_{2,n}}[\text{Re } L^2] \subset \text{Re } L^2$ (it is a consequence of $B_n[\text{Re } L^2] \subset \text{Re} L^2$ and the Trotter product formula). Now without loss of generality, we can assume $f = \text{Re } f$ in the definition of $u_n(t)$, so that $u_n = \text{Re } u_n$ (we write u_n instead of $u_n(t)$).

Multiplying both sides of the equation $-\frac{d}{dt} u_n = C_{p,n} u_n$ by $u_n |u_n|^{p-2}$, integrating over M we obtain

$$
\begin{aligned}
-\frac{1}{p}\frac{d}{dt}\|u_n\|_p^p &= \langle C_{p,n} u_n, u_n|u_n|^{p-2}\rangle = \langle A_p u_n, u_n|u_n|^{p-2}\rangle - \langle B_{n,p} u_n, u_n|u_n|^{p-2}\rangle \\
&\ge 4\frac{p-1}{p^2}\|A^{\frac{1}{2}}(u_n|u_n|^{\frac{p}{2}-1})\|_2^2 - \varkappa(p)\|B_n^{\frac{1}{2}}(u_n|u_n|^{\frac{p}{2}-1})\|_2^2 \\
&\ge \left(4\frac{p-1}{p^2} - \beta\varkappa(p)\right)\|A^{\frac{1}{2}}(u_n|u_n|^{\frac{p}{2}-1})\|_2^2 - c(\beta)\varkappa(p)\|u_n\|_p^p.
\end{aligned}
$$

(We used consecutively (2.1) and the condition $B_{n,2} \in PK_\beta(A)$ $(B_{n,2} \leqslant B)$). Consequently, for any $p \in [t(\beta), t'(\beta)]$

$$
\frac{d}{dt}\|u_n\|_p^p \le pc(\beta)\varkappa(p)\|u_n\|_p^p.
$$

Thus $\|u_n(t)\|_p \le e^{c(\beta)\varkappa(p)t}\|u_n(0)\|_p$ or

$$
\|e^{-tC_{p,n}} f\|_p \le e^{c(\beta)\varkappa(p)t}\|f\|_p, \ \forall f \in L^2 \cap L^p. \tag{3.1}
$$

For $n \geqslant m$ one has $t_{A\dot{-}B}[g] \leqslant t_{C_{2,n}}[g] \leqslant t_{C_{2,m}}[g]$ for $g \in Q(A)$. Thus according to a monotone convergence theorem [K1, Ch. VIII, Theorem 3.11]

$$
\exp(-t(A - B_n)) \xrightarrow[L^2]{s} \exp(-t(A\dot{-}B)). \tag{3.1'}
$$

From this and from (3.1) it follows that we can pass to the limit in (3.1) with respect to n and obtain (i).

The proof of (ii) is carried out similarly to that in Theorem 3.1 with using (3.1) and (3.1′). We prefer to postpone the proofs of (iii) and (iv) to the proof of the next theorem. □

Remark 3.1 The semigroup $(e^{-t(A\dot{-}B)_p}, \ t > 0)$ need not be positivity preserving. However, if the semigroups $e^{-tC_{2,n}}$ are positivity preserving for all sufficiently large

n and for all $t > 0$, i.e. $\exp(-tC_{2,n})[L^2_+] \subset L^2_+$ then setting $u_n(t) = e^{-tC_{2,n}}|f|$, $f \in L^2 \cap L^p$ we can repeat the proof of (i) from Theorem 3.2 using inequality (2.3) from Theorem 2.1. Instead of (3.1) we obtain

$$\|e^{-tC_{p,n}}|f|\|_p \leq e^{c(\beta)t}\|f\|_p, \; f \in L^2 \cap L^p.$$

Thus taking into account the inequality $|e^{-tC_{p,n}}f| \leq e^{-tC_{p,n}}|f|$, we get

$$\|e^{-t(A\dot{-}B)_p}\|_{p\to p} \leq e^{c(\beta)t}, \; \forall p \in [t(\beta), t'(\beta)],$$

where $t(\beta) = \frac{2}{1+\sqrt{1-\beta}}$, $t'(\beta) = \frac{2}{1-\sqrt{1-\beta}}$. At least in the last case the interval of quasi-contraction $[t(\beta), t'(\beta)]$ cannot be extended. To justify this claim one can consider the example $-\Delta - c|x|^{-2}$: for arbitrary $p \in (p(\beta), t(\beta))$, where $p(\beta) = \frac{2d}{(d+2)-(d-2)\sqrt{1-\beta}}$, there exists $c = c(p,\beta)$ such that $-\Delta_p - c|x|^{-2}$ generates a bounded C_0-semigroup (see Theorem 5.2 below) and at the same time it can be easily checked that this operator is not quasi-accretive in L^p.

Remark 3.2 Let A be a symmetric Markov generator. Let $\{\tau_n\}$ be a sequence of bounded symmetric forms such that the operator B_n associated with the form τ_n is the difference of two bounded symmetric Markov generators $B_n^{(1)}$, $B_n^{(2)}$: $B_n = B_n^{(1)} - B_n^{(2)}$, moreover $\forall n = 1, 2, \ldots$ $B_n^{(1)} \in PK_\beta(A + \frac{1}{\beta}B_n^{(2)})$ for some $\beta \in (0,1)$, β does not depend on n. Suppose that $A - B_n \longrightarrow C$ in strong resolvent sense, where C is some self-adjoint operator (for example, if $\tau_n[u,v] \longrightarrow \tau[u,v]$ $\forall u,v \in Q(A)$ and $B_n^{(2)} \in PK_\gamma(A)$ for some $\gamma < \infty$, γ does not depend on n [StoVo],[HSl]). Then all the assertions of Theorem 3.2 are valid for the operator C instead of $A\dot{-}B$.

Theorem 3.3 *Let A, B, C or $-C$ be symmetric Markov generators. Suppose that $B \in PK_{\beta_1}(A)$ for some $\beta_1 \in (0,1)$ and $C \in PK_{\beta_2}(A)$ for some $\beta_2 > 0$. Let t_1 and t_2 be the roots of the equation* $(\beta_1 + \frac{|q-2|}{2\sqrt{q-1}}\beta_2)\varkappa(q) = 4\frac{q-1}{q^2}$, $1 < q < \infty$. [1] *Let $H_2 \equiv H = A\dot{-}B\dot{+}iC$ be the form-sum, $i := \sqrt{-1}$. Then for every $p \in [t_1, t_2]$ and $t \geqslant 0$*

(i) $$\|e^{-tH}f\|_p \leqslant e^{\varkappa(p)\{c_B(\beta_1) + \frac{|p-2|}{2\sqrt{p-1}}c_C(\beta_2)\}t}\|f\|_p, \; f \in L^2 \cap L^p$$

and consequently this defines a family of operators $\{H_p\}_{t_1 \leqslant p \leqslant t_2}$ which generate quasi-contraction C_0-semigroups.

(ii) $$H_p \supset A_p - B_p + iC_p$$

(iii) *the semigroup e^{-tH_p} is holomorphic with respect to t of angle*

[1] The roots always exist and if $t_1 \in (1,2)$ then $t_2 = t_1' \in (2,\infty)$. This follows from the abovementioned properties of $\varkappa(p)$ and from $\frac{|q-2|}{\sqrt{q-1}} = \frac{|q'-2|}{\sqrt{q'-1}}$.

$$\theta_1 = \frac{\pi}{2} - \arctan \mathcal{K},\ 0 < \theta_1 < \frac{\pi}{2},\ p \in (t_1, t_2),\ \textit{where}$$

$$\mathcal{K} = \frac{\frac{|p-2|}{2\sqrt{p-1}}(4\frac{p-1}{p^2} + \varkappa(p)\beta_1) + \varkappa(p)\beta_2}{4\frac{p-1}{p^2} - \varkappa(p)\beta_1 - \frac{|p-2|}{2\sqrt{p-1}}\varkappa(p)\beta_2}$$

(iv) $$\forall p \in (t_1, t_2),\ \forall f \in \mathcal{D}(H_p)\ f_p = f|f|^{\frac{p-2}{2}} \in Q(A)\ \textit{ and}$$

$$\mathrm{Re}\ \langle H_p f, \mathrm{sgn}\ f \cdot |f|^{p-1}\rangle \geqslant (4\frac{p-1}{p^2} - (\beta_1 + \frac{|p-2|}{2\sqrt{p-1}}\beta_2)\varkappa(p))\|A^{\frac{1}{2}} f_p\|_2^2$$
$$-\varkappa(p)\{c_B(\beta_1) + \frac{|p-2|}{2\sqrt{p-1}}c_C(\beta_2)\}\|f\|_p^p.$$

Remark 3.3 For the operator C instead of the condition $C \in PK_{\beta_2}(A)$ obviously one can assume only $Q(C) \supset Q(A)$. We wanted to emphacise the dependence on β_2.

Proof It is enough to prove i) with B_n instead of B and with C_m instead of C, i.e. without loss we can assume that $B, C \in \mathcal{L}(L^2) \cap \mathcal{L}(L^p)$. Since Re $\langle \frac{d}{dt}u, u|u|^{p-2}\rangle = \frac{1}{p}\frac{d}{dt}\|u\|_p^p$, then

$$-\frac{1}{p}\frac{d}{dt}\|u\|_p^p = \mathrm{Re}\ \langle (A_p - B_p + iC_p)u, u|u|^{p-2}\rangle$$
$$= \mathrm{Re}\ \langle A_p u, u|u|^{p-2}\rangle - \mathrm{Re}\ \langle B_p u, u|u|^{p-2}\rangle - \mathrm{Im}\ \langle C_p u, u|u|^{p-2}\rangle$$
$$\geqslant 4\frac{p-1}{p^2}\|A^{\frac{1}{2}}u_p\|_2^2 - \varkappa(p)\|B^{\frac{1}{2}}u_p\|_2^2 - \frac{|p-2|}{2\sqrt{p-1}}\varkappa(p)\|C^{\frac{1}{2}}u_p\|_2^2$$

$u \in \mathcal{D}(A_p)$. (We used consecutively (2.1) and (2.2)). To finish the proof of i) it remains only to use the conditions $B \in PK_{\beta_1}$ and $C \in PK_{\beta_2}$.
The proof of ii) is standard.
In order to prove iii) let us estimate $|\mathrm{Im}\ \langle \bar{H}_p u, |u|^{p-1}\mathrm{sgn}\ u\rangle|$ by Re $\langle \bar{H}_p u, |u|^{p-1}\mathrm{sgn}\ u\rangle$, where $\bar{H}_p = H_p + \lambda$ for an appropriate λ which will be chosen below. As before we assume B and C to be bounded operators in L^2 and in L^p. Using (2.1),(2.2) and PK_β-conditions for B and C we have

$$|\mathrm{Im}\ \langle \bar{H}_p u, |u|^{p-1}\mathrm{sgn}\ u\rangle| \leqslant \mathcal{K}_+ \mathrm{Re}\ \langle A_p u, |u|^{p-1}\ \mathrm{sgn}\ u\rangle + \Lambda_+\|u\|_p^p,$$

where $\mathcal{K}_+ = \frac{|p-2|}{2\sqrt{p-1}}(1 + \frac{p^2}{4(p-1)}\varkappa(p)\beta_1) + \frac{p^2}{4(p-1)}\varkappa(p)\beta_2$, $\Lambda_+ = \varkappa(p)(\frac{|p-2|}{2\sqrt{p-1}}c_B(\beta_1)$ $+c_C(\beta_2))$,

$$\mathrm{Re}\ \langle \bar{H}_p u, |u|^{p-1}\mathrm{sgn}\ u\rangle \geqslant \mathcal{K}_- \mathrm{Re}\ \langle A_p u, |u|^{p-1}\ \mathrm{sgn}\ u\rangle + \Lambda\|u\|_p^p,$$

where $\mathcal{K}_- = 1 - \frac{p^2}{4(p-1)}(\beta_1 + \frac{|p-2|}{2\sqrt{p-1}}\beta_2)\varkappa(p)$, $\Lambda = \lambda - \varkappa(p)(c_B(\beta_1) + \frac{|p-2|}{2\sqrt{p-1}}c_C(\beta_2))$ $\equiv \lambda - \Lambda_-$.

Setting $\lambda = \Lambda_- + \frac{\mathcal{K}_-}{\mathcal{K}_+}\Lambda_+$, we obtain

$$|\mathrm{Im}\ \langle \bar{H}_p u, |u|^{p-1}\mathrm{sgn}\ u\rangle| \leqslant \frac{\mathcal{K}_+}{\mathcal{K}_-}\ \mathrm{Re}\ \langle \bar{H}_p u, |u|^{p-1}\mathrm{sgn}\ u\rangle. \tag{3.2}$$

In order to complete the proof of iii) it remains only to pass to the limit in (3.2). To do this we use the arguments from [Se1, Prop.2.3] and [BeSe, Remark 2 to Theorem 3]. Let $H_p(n,m) := A_p - B_{p,n} + iC_{p,m}$ and $u_{n,m} = \exp(-tH_p(n,m))f$, $u = \exp(-tH_p)f$, $f \in L^2 \cap L^p$. Using the general approximation theorem [K1,Ch.VI, Th.3.2] it is not difficult to check that $u_{n,m} \xrightarrow[L^2]{s} u$ and according to the proof of (i) and by Proposition 1.4

$$u_{n,m} \xrightarrow[L^p]{s} u \quad (n,m \longrightarrow \infty). \tag{3.3}$$

In (3.3) f can be considered as an arbitrary fixed element from L^p.
Furthermore by virtue of (3.2) the semigroups $\exp(-t\bar{H}_p(n,m))$ are holomorphic and uniformly bounded with respect to m,n in the sector $S_p = \{t; |\arg t| \leqslant \theta,\ \forall\theta \in (0,\theta_1)\}$. By (3.3) $u_{n,m}$ converges pointwise on the positive semiaxis. Consequently by the Vitali convergence theorem u is holomorphic in S_p and $u(z) = s - L^p - \lim_{n,m\to\infty} u_{n,m}(z)$, $z \in S_p$. In particular $u'(z) = s - L^p - \lim_{n,m\to\infty} u'_{n,m}(z)$, $z \in S_p$, and $\bar{H}_p(n,m)u_{n,m} \xrightarrow[L^p]{s} \bar{H}_p u$. Moreover, passing to a subsequence if required we have $g_{n,m} := |u_{n,m}|^{p-1}\mathrm{sgn}\ u_{n,m} \longrightarrow g := |u|^{p-1}\mathrm{sgn}\ u$ pointwise μ-a.e. and $\|g_{n,m}\|_{p'} = \|u_{n,m}\|_p^{p-1} \longrightarrow \|g\|_{p'} = \|u\|_p^{p-1}$ by virtue of (3.3). Hence according to Proposition 1.2 $g_{n,m} \xrightarrow[L^{p'}]{s} g$. Thus, in (3.2) we can pass to the limit $(n,m) \longrightarrow \infty$. This completes the proof of (iii) of the theorem and (iii) of Theorem 3.2.
The similar arguments can be carried out when proving iv) : one should pass to the limit $(n,m) \longrightarrow \infty$ in the following inequality

$$\begin{aligned}\mathrm{Re}\ \langle (A_p - B_{p,n} + iC_{p,m})u_{n,m}, u_{n,m}|u_{n,m}|^{p-2}\rangle &\geqslant 4\frac{p-1}{p^2}\|A^{\frac{1}{2}}(u_{n,m})_p\|_2^2 \\ &-\varkappa(p)\|B^{\frac{1}{2}}(u_{n,m})_p\|_2^2 - \frac{|p-2|}{2\sqrt{p-1}}\|C^{\frac{1}{2}}(u_{n,m})_p\|_2^2\end{aligned}$$

(see the proof of (i)). Thus statement iv) is proved for both Theorems 3.3 and 3.2.□

4 Non-symmetric Markov generators

Here we give some applications to differential operators of the form

$$-d\cdot a\cdot df(x) + b\cdot df(x) := -\sum_{i,j=1}^{d}\frac{\partial}{\partial x_i}(a_{ij}(x)\frac{\partial f(x)}{\partial x_j}) + \sum_{i=1}^{d} b_i(x)\frac{\partial f(x)}{\partial x_i} \tag{4.1}$$

acting in $L^p \equiv L^p(\Omega) \equiv L^p(\Omega, dx)$, where Ω is an open set in $\mathbb{R}^d$, dx is Lebesgue measure, $a : \Omega \longrightarrow \mathbb{R}^d \otimes \mathbb{R}^d$ is a symmetric matrix-valued measurable function, $b : \Omega \longrightarrow \mathbb{R}^d$ is a vector-valued measurable function.

Our purpose in this section is to construct the generator of a C_0-semigroup on L^p corresponding to the formal expression (4.1) under minimal restrictions on the coefficients and to investigate the main features of the semigroup and the generator. It should be pointed out that condition (1.6) can be checked for a wide class of differential operators. It also enables us not to use the auxiliary material concerning operators with smooth coefficients.

Suppose that

$$a(\cdot) \geqslant \nu I \text{ a.e. in the matrix sense for some } \nu > 0,$$

$$a_{ij},\ b_i \text{ and } b \cdot a^{-1} \cdot b \in L^1_{loc}(\Omega)$$

$$(b \cdot a^{-1} \cdot b := \sum_{i,j=1}^{d} b_i b_j a^{-1}_{ij},\ a^{-1}_{ij} \text{ are the elements of the matrix inverse to } a).$$

Let us define the following sesquilinear form in $L^2(\Omega)$

$$a[f,g] := \langle d\bar{f} \cdot a \cdot dg \rangle \equiv \int_\Omega \left(\sum_{i,j} a_{ij}(x) \frac{\partial \bar{f}(x)}{\partial x_i} \frac{\partial g(x)}{\partial x_j} \right) dx,$$

$$\mathcal{D}(a) = \{ f \in H^1(\Omega) : \langle d\bar{f} \cdot a \cdot df \rangle < \infty \}.$$

It is well-known that this form is closed. Let us define the Dirichlet form $t_A = (a \upharpoonright C_0^\infty(\Omega))^\sim$. Obviously t_A is a closed densely defined non-negative form with $\mathcal{D}(t_A) \equiv Q(A) \subset H^1_0(\Omega)$. Let A be the operator associated with the form t_A. Then $A = A^* \geqslant 0$. The semigroup $(e^{-tA}, t \geqslant 0)$ is positivity preserving $(e^{-tA}[L^2_+] \subset L^2_+ := \{ f \in L^2 : f \geqslant 0 \text{ a.e.}\})$, L^∞-contractive $(\|e^{-tA} f\|_\infty \leqslant \|f\|_\infty,\ \forall f \in L^2 \cap L^\infty)$. Thus Theorem 2.1 holds for the operator A.

If $b \cdot a^{-1} \cdot b \in PK_\beta(A)$ for some $\beta < 1$ (i.e. $b \cdot a^{-1} \cdot b \leqslant \beta A + c(\beta)$, $c(\beta) \in \mathbb{R}^1$, in form sense) then according to the general theory (see [K1,Ch. VI]) the sesquilinear form

$$t[f,g] = t_A[f,g] + \tau[f,g],\ \tau[f,g] := \langle d\bar{f} \cdot bg \rangle,\ \mathcal{D}(t) = Q(A),$$

is sectorial. Let T be the operator associated with the form t. The operator T is m-sectorial, $\mathcal{D}(T)$ is dense in $(Q(A), \|\cdot\| = \|(1+A)^{\frac{1}{2}} \cdot \|_2)$. The following theorem shows that T inherits some properties of A.

Theorem 4.1 *The semigroup $(e^{-tT}, t \geqslant 0)$ is a positivity preserving L^∞-contractive C_0-semigroup. By means of $e^{-tT_p} := [e^{-tT} \upharpoonright [L^2 \cap L^p]]^\sim_{L^p \to L^p}$ a family of C_0-semigroups of quasicontractions in L^p is defined for all $p \in [2, \infty)$,*

$$\|e^{-tT_p}\|_{p \to p} \leqslant e^{c(\beta)t/2(p-1)},\ 2 \leqslant p < \infty.$$

$$f \in \mathcal{D}(T_p) \Longrightarrow f_p \in Q(A) \text{ and}$$

$$\frac{2}{p}(2\frac{p-1}{p} - \sqrt{\beta})t_A[|f_p|] \leqslant \mathrm{Re}\ \langle (T_p + \frac{c(\beta)}{p\sqrt{\beta}})f, |f|^{p-1}\mathrm{sgn}\ f\rangle, \tag{4.2}$$

$$|\mathrm{Im}\ \langle T_p f, |f|^{p-1}\mathrm{sgn}\ f\rangle| \leqslant l_{\beta,p}\mathrm{Re}\ \langle (T_p + \Lambda_{\beta,p})f, |f|^{p-1}\mathrm{sgn}\ f\rangle, \tag{4.3}$$

$$\text{where } l_{\beta,p} = \frac{|p-2| + \frac{p\sqrt{\beta}}{2}}{2\sqrt{p-1-\frac{p\sqrt{\beta}}{2}}}, \quad \Lambda_{\beta,p} = \frac{c(\beta)}{p\sqrt{\beta}}\left(1 + \frac{1}{l_{\beta,p}}\sqrt{p-1-\frac{p\sqrt{\beta}}{2}}\right),$$

$$\|e^{-tT_p}\|_{p\to q} \leqslant e^{c(\beta)t/p\sqrt{\beta}}\left(c_{\nu,d}(2\frac{p-1}{p} - \sqrt{\beta})\right)^{-\frac{d}{2}(\frac{1}{p}-\frac{1}{q})} t^{-\frac{d}{2}(\frac{1}{p}-\frac{1}{q})},\ 2 \leqslant p < q \leqslant \infty.$$

To prove this theorem we need the following lemma.

Lemma 4.1 *Let $b \cdot a^{-1} \cdot b \in PK_\beta(A)$ for some $\beta < 1$ and T be the m-sectorial operator defined above. Then*

$$\|e^{-tT}f\|_\infty \leqslant e^{r(\beta)t}\|f\|_\infty,\ f \in L^2 \cap L^\infty,\ t > 0,$$

where $r(\beta) = \frac{c(\beta)t(\beta)}{4}$, $t(\beta) = \frac{2}{1+\sqrt{1-\beta}}$.

Proof Let us verify condition (1.6) for $e^{-t\Gamma}$ with $\Gamma = T + r(\beta)$. From $f \in \mathcal{D}(\Gamma) = \mathcal{D}(T) \subset Q(A)$ then $f_\wedge \in Q(A)$, $f - f_\wedge \in Q(A)$ and since $Q(A) \subset H^1(\Omega)$ one obtains $f_\wedge \in H^1(\Omega)$ and $d(f - f_\wedge) = \mathbb{1}_{|f|>1}[(1 - \frac{1}{|f|})df + \frac{f}{|f|^2}d|f|]$. Therefore

$$\begin{aligned}\langle \Gamma f, f - f_\wedge\rangle = &\langle \mathbb{1}_{|f|>1}(1 - \frac{1}{|f|})d\bar{f}\cdot a\cdot df\rangle + \langle \mathbb{1}_{|f|>1}\frac{f}{|f|}d\bar{f}\cdot a\cdot\frac{1}{|f|}d|f|\rangle \\ &+ \langle \mathbb{1}_{|f|>1}\frac{f}{|f|}d\bar{f}\cdot b(|f|-1)\rangle + \langle \mathbb{1}_{|f|>1}(|f|-1)|f|\rangle r(\beta)\end{aligned}$$

and

$$\mathrm{Re}\ \langle \Gamma f, f - f_\wedge\rangle \geqslant \langle \mathbb{1}_{|f|>1}d|f|\cdot a\cdot d|f|\rangle + \langle \mathbb{1}_{|f|>1}d|f|\cdot b(|f|-1)\rangle + \langle \mathbb{1}_{|f|>1}(|f|-1)|f|\rangle r(\beta).$$

Since a is a positive symmetric matrix then

$$\begin{aligned}|\langle \mathbb{1}_{|f|>1}d|f|\cdot b(|f|-1)\rangle| &\leqslant \langle \mathbb{1}_{|f|>1}\sqrt{d|f|\cdot a\cdot d|f|}\sqrt{(|f|-1)^2 b\cdot a^{-1}\cdot b}\rangle \\ &\leqslant \frac{\varepsilon}{2}\langle \mathbb{1}_{|f|>1}d|f|\cdot a\cdot d|f|\rangle + \frac{1}{2\varepsilon}\langle \mathbb{1}_{|f|>1}(|f|-1)^2 b\cdot a^{-1}\cdot b\rangle \\ &= \frac{\varepsilon}{2}\langle d\psi\cdot a\cdot d\psi\rangle + \frac{1}{2\varepsilon}\langle \psi b\cdot a^{-1}\cdot b\psi\rangle,\ \varepsilon > 0\end{aligned}$$

where $\psi := \mathbb{1}_{|f|>1}(|f|-1) \in Q(A)$. Using the condition $b\cdot a^{-1}\cdot b \in PK_\beta(A)$ we obtain

$$|\langle \mathbb{1}_{|f|>1}d|f|\cdot b(|f|-1)\rangle| = |\langle d\psi\cdot b\psi\rangle| \leqslant \left(\frac{\varepsilon}{2} + \frac{\beta}{2\varepsilon}\right)\langle d\psi\cdot a\cdot d\psi\rangle + \frac{c(\beta)}{2\varepsilon}\|\psi\|_2^2.$$

Hence we get

$$\text{Re}\,\langle \Gamma f, f - f_\wedge\rangle \geqslant (1 - \frac{1}{2}(\varepsilon + \frac{\beta}{\varepsilon}))t_A[\psi] + \frac{c(\beta)}{2}(\frac{t(\beta)}{2} - \frac{1}{\varepsilon})\|\psi\|_2^2.$$

Setting $\varepsilon = \frac{2}{t(\beta)} = 1 + \sqrt{1-\beta}$ we arrive at Re $\langle \Gamma f, f - f_\wedge\rangle \geqslant 0$.□

Proof of Theorem 4.1 Let $f \in \mathcal{D}(T)$, Then $\bar{f} \in \mathcal{D}(T)$ and $\overline{Tf} = T\bar{f}$. Therefore $e^{-tT}[\text{Re}\, L^2] \subset \text{Re}\, L^2$. Moreover if $f = \text{Re}\, f \in \mathcal{D}(T)$ then using arguments similar to that from the proof of Lemma 4.1 we obtain the dispersivity condition of $\Gamma = T + r(\beta)$: $\langle 1\!\!1_{f\geqslant 0}\Gamma f, f\rangle \geqslant 0$. Let us prove the inequality $\|e^{-tT}f\|_\infty \leqslant \|f\|_\infty$, $f \in L^2 \cap L^\infty$, $t > 0$. Let $f_t = e^{-tT}f$, $f \in L^2 \cap L^\infty$. It is clear that $f_t \in \mathcal{D}(T) \subset Q(A)$ and by virtue of Lemma 4.1 $f_t \in Q(A) \cap L^\infty$. Therefore using the characterization of Dirichlet forms [Fu, p.5] we conclude that $f_t|f_t|^{p-2}$ and $f_t|f_t|^{\frac{p-2}{2}}$ are from $Q(A) \cap L^\infty$ for $p \geqslant 2$. Multiplying both sides of the equation $-\frac{df_t}{dt} = Tf_t$ by $f_t|f_t|^{p-2}$ and integrating over Ω we get

$$\begin{aligned} &-\langle \frac{df_t}{dt}, f_t|f_t|^{p-2}\rangle = \langle Tf_t, f_t|f_t|^{p-2}\rangle \\ &= \langle |f_t|^{p-2}d\bar{f}_t \cdot a \cdot df_t\rangle + \langle d\bar{f}_t \cdot bf_t|f_t|^{p-2}\rangle + (p-2)\langle |f_t|^{p-3}f_t d\bar{f}_t \cdot a \cdot d|f_t|\rangle. \end{aligned}$$

Taking the real parts and using change of variables $v = f_t|f_t|^{\frac{p-2}{2}}$ we obtain

$$-\frac{1}{p}\frac{d}{dt}\|v\|_2^2 \geqslant 4\frac{p-1}{p^2}\langle d|v| \cdot a \cdot d|v|\rangle + \frac{2}{p}\langle d|v| \cdot b|v|\rangle.$$

Using the condition $PK_\beta(A)$ for the term $\langle d|v| \cdot b|v|\rangle$ yields

$$-\frac{d}{dt}\|v\|_2^2 \geqslant (4\frac{p-1}{p} - \varepsilon - \frac{\beta}{\varepsilon})\langle d|v| \cdot a \cdot d|v|\rangle - \frac{c(\beta)}{\varepsilon}\|v\|_2^2 \geqslant -\frac{p'c(\beta)}{2}\|v\|_2^2.$$

$$(\varepsilon = \frac{2}{p'} + \sqrt{(\frac{2}{p'})^2 - \beta})$$

Consequently $\|f_t\|_p \leqslant e^{\frac{c(\beta)t}{2(p-1)}}\|f\|_p$. From the last inequality and from the continuity of the mapping $p \longrightarrow \|g\|_p$ ($g \in L^{p_0} \cap L^\infty$) (see Proposition 1.1) it follows that $\|f_t\|_\infty \leqslant \|f\|_\infty$.

Furthermore, if $p > 2$ then $L^2 \cap L^\infty \subset L^p$ and $\|g\|_p \leqslant \|g\|_2^\theta \cdot \|g\|_\infty^{1-\theta}$, $g \in L^2 \cap L^\infty$, $\theta = \frac{2}{p}$. Hence $\|(e^{-tT} - 1)f\|_p \leqslant 2^{1-\theta}\|f\|_\infty^{1-\theta} \cdot \|(e^{-tT} - 1)f\|_2^\theta$ so that $(e^{-tT} \upharpoonright [L^2 \cap L^\infty])^{\sim}_{L^p \to L^p}$ is a C_0-semigroup of contractions, $\forall p \in (2, \infty)$.

Statements (4.2),(4.3) for $f \in (\lambda + T)^{-1}[L^2 \cap L^\infty]$, $\lambda > \frac{c(\beta)}{p\sqrt{\beta}}$ can be obtain by direct calculations. The passage to arbitrary f from $\mathcal{D}(T_p)$ is obvious. The (L^p, L^q)-estimate can be proven by the usual procedure based upon the Nash inequality (see [FS]) and by interpolation afterwards.□

Theorem 4.2 *Let the conditions of Theorem* 4.1 *be satisfied.*
1. *Suppose in addition that the Lebesgue measure* $|\Omega|$ *of the set* Ω *is finite. Then for* $\forall p \in [\frac{2}{2-\sqrt{\beta}}, 2)$ e^{-tT_p} *is* C_0*-semigroup and*

$$\|e^{-tT}f\|_p \leqslant e^{c(\beta)t/2(p-1)}\|f\|_p, \ \forall f \in L^2 \cap L^p, \ t > 0.$$

2. *Suppose in addition that* $b \cdot a^{-1} \cdot b \in L^1 + L^\infty$ ($|\Omega| < \infty$ *or* $|\Omega| = \infty$). *Then for* $\forall p \in [\frac{2}{2-\sqrt{\beta}}, 2)$ e^{-tT_p} *is* C_0*-semigroup and*

$$\|e^{-tT}f\|_p \leqslant e^{c(\beta)t/2(p-1)}\|f\|_p, \ \forall f \in L^2 \cap L^p, \ t > 0.$$

Proof Let $f \in L^1 \cap L^\infty$, $f_t = e^{-tT}f$, $\varphi_\varepsilon = f_t(\varepsilon + |f_t|)^{p-2}$, $v_\varepsilon = f_t(\varepsilon + |f_t|)^{\frac{p-2}{2}}$, $\nu_\varepsilon = (2-p)|f_t|/(\varepsilon + |f_t|)$, $\varepsilon > 0$. It can be seen from Lemma 4.1 and from the properties of Dirichlet forms that $\varphi_\varepsilon, v_\varepsilon \in Q(A) \cap L^\infty$. Since $|\Omega| < \infty$ and $p < 2$ then $\frac{df_t}{dt} \in L^p$, $\varphi_\varepsilon \in L^{p'}$ and

$$\frac{1}{p}\frac{d}{dt}\|v\|_2^2 = \lim_{\varepsilon \downarrow 0} \mathrm{Re}\ \langle \frac{df_t}{dt}, \varphi_\varepsilon\rangle.$$

Moreover,

$$\mathrm{Re}\ \langle A^{\frac{1}{2}}f_t, A^{\frac{1}{2}}\varphi_\varepsilon\rangle = \langle d|v_\varepsilon| \cdot \frac{1-\nu_\varepsilon}{(1-\frac{\nu_\varepsilon}{2})^2} a \cdot d|v_\varepsilon|\rangle \geqslant 4\frac{p-1}{p^2}\|A^{\frac{1}{2}}|v_\varepsilon|\|_2^2,$$

$$\mathrm{Re}\ \langle d\bar{f}_t \cdot b\varphi_\varepsilon\rangle = \langle d|v_\varepsilon| \cdot (1 - \frac{\nu_\varepsilon}{2})^{-1} b|v_\varepsilon|\rangle \leqslant \frac{2}{p}\langle d|v_\varepsilon| \cdot a \cdot d|v_\varepsilon|\rangle^{\frac{1}{2}} \langle |v_\varepsilon| b \cdot a^{-1} \cdot b|v_\varepsilon|\rangle^{\frac{1}{2}}.$$

Therefore ($\delta = \frac{2}{p'} + \sqrt{(\frac{2}{p'})^2 - \beta}$)

$$-\frac{d}{dt}\|v\|_2^2 \geqslant (4\frac{p-1}{p} - \delta - \frac{\beta}{\delta}) \liminf_{\varepsilon \downarrow 0} \langle d|v_\varepsilon| \cdot a \cdot d|v_\varepsilon|\rangle - \frac{c(\beta)}{\delta}\|v\|_2^2 \geqslant -\frac{p'c(\beta)}{2}\|v\|_2^2.$$

It follows that $\|e^{-tT}f\|_p \leqslant e^{c(\beta)t/2(p-1)}\|f\|_p$, $f \in L^1 \cap L^\infty$. Since $L^2 \subset L^p$ ($p < 2$ and $|\Omega| < \infty$), then $[e^{-tT} \upharpoonright [L^2 \cap L^p]]^\sim_{L^p \to L^p}$ is C_0-semigroup for $\forall p \in [\frac{2}{2-\sqrt{\beta}}, 2)$. This finishes the proof of 1).

Now we prove 2). Let first $b \in [C_0^\infty(\Omega)]^d$. Then $T^* = T + div\ b$ and according to Theorem 4.1

$$\|e^{-tT^*}f\|_\infty \leqslant e^{\|div\ b\|_\infty t}\|f\|_\infty, \ f \in L^2 \cap L^\infty.$$

From the last bound we deduce by duality

$$\|e^{-tT}f\|_p \leqslant e^{\|div\ b\|_\infty t}\|f\|_p, \ f \in L^2 \cap L^p, \ p \in [1, 2),$$

so that $e^{-tT_p} := [e^{-tT} \upharpoonright [L^2 \cap L^p]]^\sim_{L^p \to L^p}$ is C_0-semigroup of quasicontractions on L^p for all $p \in [1, \infty)$.

$$\frac{2}{2-\sqrt{\beta}} < p < q \leqslant \infty,$$

3) *if in addition* $|\Omega| < \infty$, *then for any* $q \in (\frac{2}{2-\sqrt{\beta}}, \infty)$ *and* $\lambda \in \rho(-\Gamma_q)$

$$(\lambda + \Gamma_q)^{-1}[L^\infty] \subset Q(A),$$

where Γ_q *is the generator of* P_q^t, *and the following equality holds*

$$\langle \Gamma_q f, g\rangle = \langle d\bar{f} \cdot a \cdot dg\rangle + \langle d\bar{f} \cdot bg\rangle, \ \forall f \in (\lambda + \Gamma_q)^{-1}[L^\infty], \ \forall g \in Q(A).$$

The last equality uniquely defines P_q^t. P_q^t *is holomorphic.*

Proof The proof of 1) is even simpler than in Theorem 4.2 since now $p > 2$ and additional regularization in ε is superfluous. One should take $\varphi = f_t|f_t|^{p-2}$ instead of φ_ε and v instead of v_ε. So we restrict ourselves by proving statement 3).

Since $|\Omega| < \infty$ then for any $\lambda \in \rho(-T_p(n))$ $(\lambda + T_p(n))^{-1}[L^\infty] \subset Q(A)$ and this holds in particular for $\lambda_p > \frac{c(\beta)}{p\sqrt{\beta}} \vee \|V_2\|_\infty$. Let $u_n = (\lambda_p + T_p(n))^{-1}f$, $f \in L^\infty$. Then $(\lambda_p + T_p(n))u_n = f$ and $T_p(n)u_n = T(n)u_n$. Therefore

$$\lambda_p \|u_n\|_2^2 + \langle d\bar{u}_n \cdot a \cdot du_n\rangle + \langle d\bar{u}_n \cdot b_{(n)} u_n\rangle = \langle f, u_n\rangle.$$

$$\langle d\bar{u}_n \cdot a \cdot du_n\rangle \leqslant -\lambda_p\|u_n\|_2^2 + \langle d\bar{u}_n \cdot a \cdot du_n\rangle^{\frac{1}{2}} \langle b \cdot a^{-1} \cdot b|u_n|^2\rangle^{\frac{1}{2}} + \frac{1}{4\varepsilon}\|f\|_2^2 + \varepsilon\|u_n\|_2^2.$$

$$\langle d\bar{u}_n \cdot a \cdot du_n\rangle \leqslant \|V_1\|_1 \cdot \|f\|_\infty^2 + \frac{1}{2\varepsilon}\|f\|_2^2 \ \ (\text{ for } \varepsilon \leqslant \lambda_p - \frac{c(\beta)}{p\sqrt{\beta}} \vee \|V_2\|_\infty).$$

Since $u_n \longrightarrow u$ strongly in L^p and hence in L^2 and the form t_A is closed it follows that $u = (\lambda_p + T_p)^{-1} f \in Q(A)$ and

$$\langle T_p u, g\rangle = \langle d\bar{u} \cdot a \cdot dg\rangle + \langle d\bar{u} \cdot bg\rangle, \ g \in Q(A). \qquad \square$$

Remark 4.1 For the detailed discussion of the results of this section as well as for the application to well-posedness of the Cauchy problem for the parabolic equation with non-divergent elliptic part and for the corresponding examples we refer to [L], [Se5].

5 Perturbations of symmetric Markov semigroups. II

Let A be a symmetric Markov generator and B be an operator both acting in L^{p_0} for some $p_0 \in [1, \infty)$ with $\mathcal{D}(B) \supset \mathcal{D}(A_{p_0})$. Define

$$H_{p_0} := A_{p_0} + B, \ \mathcal{D}(H_{p_0}) := \mathcal{D}(A_{p_0}).$$

The following theorems guarantee that H_{p_0} generates C_0-semigroup on L^{p_0} and

$$\|e^{-tH_{p_0}}\|_{p_0 \to p_0} \leqslant M e^{\omega t}, \ t > 0. \tag{5.1}$$

Theorem 5.1 *Let $p_0 \in (1,\infty)$. Suppose that for some $\lambda_B > 0$*

$$\|B(\lambda_B + A_{p_0})^{-1}\|_{p_0 \to p_0} < \frac{1}{2}.$$

Then H_{p_0} is the generator of a holomorphic semigroup and (5.1) holds with $\omega = \lambda_B$.

Theorem 5.2 ([KoSe1]) *Let $p_0 \in (1,\infty)$. Let A be a symmetric Markov generator and B be the operator of multiplication by a measurable function $V(\cdot) : M \to \mathbb{C}$. Suppose that*

$$\inf_{\lambda>0} \|B(\lambda + A_{p_0})^{-1}\|_{p_0 \to p_0} < 1.$$

Then H_{p_0} is the generator of a holomorphic semigroup and (5.1) holds. If in addition we assume that A_1 is the generator of a bounded holomorphic semigroup then the statement holds true for $p_0 = 1$ too.

Theorem 5.3 ([Vo]) *Let $p_0 = 1$. Let A be a symmetric Markov generator and B be the operator of multiplication by measurable function $V(\cdot) : M \to \mathbb{C}$. Suppose that*

$$\inf_{\lambda>0} \|B(\lambda + A_1)^{-1}\|_{1 \to 1} < 1.$$

Then H_1 is the generator of a C_0-semigroup.

Remark 5.1 Only Theorem 5.1 is of "general" nature. The others use that the perturbation is a multiplication operator.

Remark 5.2 Concrete Markov semigroups are far from being always holomorphic. So we emphasize that in Theorem 5.3 the semigroup e^{-tA_1} is not assumed to be holomorphic.

Denote by $\mathcal{H}(M,\theta)$ the set of all densely defined closed operators T acting in some Banach space X with the norm $\|\cdot\|$ such that

$$\rho(-T) \supset \bar{\Gamma}_{\frac{\pi}{2}+\theta} := \{z \in \mathbb{C};\ |\arg z| \leqslant \frac{\pi}{2} + \theta,\ \theta > 0\}$$

$$\|(z+T)^{-1}\| \leqslant M|z|^{-1},\ \forall z \in \Gamma_{\frac{\pi}{2}+\theta} \text{ and some } M \geqslant 1.$$

Every generator of a bounded holomorphic semigroup of angle θ belongs to $\mathcal{H}(M,\theta_1)$ for $\forall \theta_1 \in (0,\theta)$ and $M = M_{\theta-\theta_1}$.
Proof of Theorem 5.1 By Stein's interpolation theorem

$$\rho(-A_{p_0}) \supset \Gamma_{\frac{\pi}{2}+\theta_{p_0}},\ \theta_{p_0} = \frac{\pi}{2}(1-|\frac{2}{p_0}-1|),\ \|(z+A_{p_0})^{-1}\|_{p_0 \to p_0} \leqslant |z|^{-1} \text{ for } z \in \Gamma_{\frac{\pi}{2}+\theta_{p_0}},$$

so that $A_{p_0} \in \mathcal{H}(1,\theta)\ \forall \theta < \theta_{p_0}$. Using the second resolvent equation and the estimate $\|(z+\varepsilon+A_{p_0})^{-1}\|_{p_0 \to p_0} \leqslant |z|^{-1}$ for $z \in \Gamma_{\frac{\pi}{2}+\theta_{p_0}}$ and $\varepsilon > 0$ we have

$$\|B(z+\lambda_B+A_{p_0})^{-1}\|_{p_0 \to p_0} \leqslant 2\|B(\lambda_B+A_{p_0})^{-1}\|_{p_0 \to p_0}.$$

Therefore the Neumann series for the resolvent $(z+\lambda_B+H_{p_0})^{-1}$ converges and yields the estimate

$$\|(z+\lambda_B+H_{p_0})^{-1}\|_{p_0\to p_0} \leqslant (1-2\gamma_B)^{-1}|z|^{-1}, \text{ where } \gamma_B = \|B(\lambda_B+A_{p_0})^{-1}\|_{p_0\to p_0}, \text{ i.e.}$$

$$\lambda_B + H_{p_0} \in \mathcal{H}((1-2\gamma_B)^{-1},\theta)\ \forall\theta < \theta_{p_0}.\square$$

Proof of Theorem 5.2 According to the conditions of the theorem there exists $\lambda_B > 0$ such that $\|B(\lambda_B + A_{p_0})^{-1}\|_{p_0\to p_0} \equiv \gamma < 1$. Since for $f \in L^{p_0}$ and z with $\mathrm{Re}\ z > 0$ $|(z + A_{p_0})^{-1}f| \leqslant (\mathrm{Re}\ z + A_{p_0})^{-1}|f|$ pointwise μ-a.e. one obtains

$$\sup\{\|B(z+\lambda_B+A_{p_0})^{-1}\|_{p_0\to p_0};\ z \in \{s : \mathrm{Re}\, s > 0\}\} \leqslant \|B(\lambda_B+A_{p_0})^{-1}\|_{p_0\to p_0} = \gamma < 1.$$

Fix $\theta_1 \in (0, \theta_{p_0})$ by means of the condition $\gamma(1 + \tan\theta_1) < 1$. Let δ be an arbitrary number from $(0, \theta_1]$. Let $|\arg z| = \frac{\pi}{2} + \delta$ and let $\tilde{z} := i\ \mathrm{Im}\ z$. By virtue of the second resolvent equation $(z + \lambda_B + A_{p_0})^{-1} = (\tilde{z} + \lambda_B + A_{p_0})^{-1}(1 + (\tilde{z} - z)(z + \lambda_B + A_{p_0})^{-1})$ we get $(\lambda_B + A_{p_0}) \in \mathcal{H}(1, \theta_1)$

$$\|B(z + \lambda_B + A_{p_0})^{-1}\|_{p_0\to p_0} \leqslant \gamma(1 + \frac{|z-\tilde{z}|}{|z|}) \leqslant$$

$$\gamma(1 + \frac{|\mathrm{Re}\ z|}{|\mathrm{Im}\ z|}) \leqslant \gamma(1 + \tan\delta) \leqslant \gamma(1 + \tan\theta_1) < 1.$$

The last means that the Neumann series for $(z + \lambda_B + A_{p_0} + B)^{-1}$ converges and

$$\|(z + \lambda_B + A_{p_0} + B)^{-1}\|_{p_0\to p_0} \leqslant (1 - \gamma(1 + \tan\theta_1))^{-1}|z|^{-1},\ \forall z \in \Gamma_{\frac{\pi}{2}+\theta_1}.$$

The case $p_0 = 1$ needs slight modifications. According to the assumption $A_1 \in \mathcal{H}(M, \theta_1)$ for some $\theta_1 > 0$.

Fix an angle $\theta_2 \in (0, \theta_1]$ by asking $\gamma(1 + M\tan\theta_2) < 1$. Finally,

$$\|B(z + \lambda_B + A_1)^{-1}\|_{1\to 1} \leqslant \gamma(1 + M\tan\theta_2),\ z \in \Gamma_{\frac{\pi}{2}+\theta_2}.\square$$

6 Perturbations of symmetric Markov semigroups by potentials

In the previous section we have already seen that if the perturbation of a generator is a potential that is the operator of multiplication by a measurable function then the value of the operator norm bound can be enlarged at least twice. In this section the reader will find some other remarkable features of this case.

Theorem 6.1 ([Vo]) *Let A be a symmetric Markov generator and $V \geqslant 0$ be a potential such that $Q(A) \cap Q(V)$ is dense in L^2, and consequently $H := A\dot{+}V$ is a symmetric Markov generator. Then*

$$A_1 + V_1 \subset H_1 = \Lambda_1 + V_1,$$

where Λ *is the generator of the* C_0*-semigroup* $e^{-t\Lambda} := \lim_{\varepsilon \downarrow 0} e^{-t(A \dot{+} \varepsilon V)}$.

Proof Let V_n be the truncation of V, i.e. $V_n(x) = V(x) \wedge n$, $H(V) \equiv H$. Since $(1 + H(V_n))^{-1} \underset{L^2}{\overset{s}{\to}} (1 + H(V))^{-1}$ and $(1 + H(V_{n+1}))^{-1}|f| \leqslant (1 + H(V_n))^{-1}|f| \leqslant (1 + A)^{-1}|f|$ pointwise one obtains $(1+H_1(V_n))^{-1} \underset{L^1}{\overset{s}{\to}} (1+H_1(V))^{-1}$. Now using the same arguments as in the proof of Theorem 3.1 we arrive at $A_1 + V_1 \subset H_1$.
For $f \in \mathcal{D}(A_1) = \mathcal{D}(H_1(V_n))$

$$\|(1 + H_1(V_n))f\|_1 \geqslant \text{Re } \langle (1 + H_1(V_n))f, \text{sgn } f \rangle$$

$$= \|(1 + V_n)f\|_1 + \text{Re } \langle A_1 f, \text{sgn } f \rangle \geqslant \|V_n f\|_1.$$

From this and from the pointwise inequality $(1 + H_1(V))^{-1}|f| \leqslant (1 + H_1(V_n))^{-1}|f|$ it follows that $\|V_n(1 + H_1)^{-1}|f|\|_1 \leqslant \|f\|_1$, $\forall f \in L^1$. Using the B.Levi theorem yields $\|V(1 + H_1)^{-1}f\|_1 \leqslant \|V(1 + H_1)^{-1}|f|\|_1 \leqslant \|f\|_1$. Therefore $\mathcal{D}(H_1) \subset \mathcal{D}(V_1)$.

Now we prove that

$$H_1(\varepsilon V) + (1 - \varepsilon)V_1 \supset H_1(V), \quad \forall \varepsilon \in (0, 1). \tag{6.1}$$

Denote $F_n := (1+H_1(\varepsilon V)+(1-\varepsilon)V_n)^{-1}$, $F := (1+H_1)^{-1}$. Note that $F_n \underset{L^1}{\overset{s}{\to}} F$, $F_n f \in \mathcal{D}(H_1(\varepsilon V))$, $Ff \in \mathcal{D}(H_1(V))$ $\forall f \in L^1$, $n = 1, 2, \ldots$ It is clear that $H_1(\varepsilon V)F_n f = f - F_n f - (1 - \varepsilon)V_n F_n f$, $\forall f \in L^1$. Thus in order to prove (6.1) it suffices to check

$$\|V_n F_n f - V F f\|_1 \longrightarrow 0 \ (n \to \infty) \ \forall f \in L^1_+.$$

The last holds since $VF_n f \searrow VFf$, $0 \leqslant (V - V_n)F_n f \leqslant (V - V_n)(1 + H_1(\varepsilon V))^{-1} f \equiv g_n \leqslant V(1 + H_1(\varepsilon V))^{-1} f \equiv g$, $L^1 \ni g \geqslant g_n \geqslant 0$ and $g_n \searrow 0$, $g \geqslant VF_n f$.
For $f \in \mathcal{D}(H_1)$ (6.1) yields $H_1(\varepsilon V)f \underset{L^1}{\overset{s}{\to}} H_1 f - V_1 f$ $(\varepsilon \downarrow 0)$. Thus $\mathcal{D}(H_1) \subset \mathcal{D}(\Lambda_1)$ and $H_1 \subset \Lambda_1 + V_1$. Repeating the first argument with Λ instead of A one has $\Lambda_1 + V_1 \subset (\Lambda \dot{+} V)_1$. Hence $H_1 \subset (\Lambda \dot{+} V)_1$. Since these operators are generators of C_0-semigroups on L^1 the result follows. □

Remark 6.1 For the detailed analysis of conditions sufficient for $\Lambda = A$ and hence for $A_1 + V_1 = H_1$ see [Vo]. For further developments see also [Is].

Remark 6.2 In the case $(M, \mu) = (\mathbb{R}^d, dx)$, $A = -\Delta$ and $V \in L^1_{loc}(\mathbb{R}^d)$ Theorem 6.1 is proved also in [K3] and in [O]. If $V \in L^1_+(M, \mu)$ then using the inequality $\|V(1 + A_1 + V_n)^{-1}g\|_1 \leqslant \|V\|_1 \|g\|_\infty$, $g \in L^1 \cap L^\infty$ one can readily check that $\lim_n \|A_1(1 + A_1 + V_n)^{-1}f - f - (1 + V_1)(1 + H_1)^{-1}f\|_1 = 0 \ \forall f \in L^1$. Then $\mathcal{D}(H_1) \subset \mathcal{D}(A_1)$ follows. Therefore in this case $H_1 = A_1 + V_1$.

Remark 6.3 The ε-trick leading to the crucial (6.1) was used for a similar purpose in [PSe] when studying the operator $H(\vec{a}) = (i\nabla + \vec{a})^2$ with $\vec{a} \in L^1(\mathbb{R}^d \setminus S)$, $S = \bar{S}$, $|S| = 0$.

Definition 6.1 Let A be a symmetric Markov generator. Let W be the operator of multiplication by a measurable function $W : M \to \mathbb{C}$. We will write $W \in \Phi_p(A)$ for some $p \geqslant 1$ if and only if

$$\sup_n \|\mathbb{1}_{\{|W|<n\}} W(1+A_p)^{-1}\|_{p\to p} < \infty \text{ and } \inf_{\lambda>0} \|W(\lambda+A_p)^{-1}\|_{p\to p} < 1.$$

Proposition 6.1 *$W \in \Phi_p(A) \Longrightarrow W \in PK_\beta(A)$ for some $\beta < 1$.*

The proof follows from the next lemma.

Lemma 6.1 *Let $\lambda > 0$ and $p \in [1,\infty)$ be fixed. If $\|W(\lambda+A_p)^{-1}\|_{p\to p} \leqslant M_p < \infty$ then*

$$\||W|^s(\lambda+A_{p_s})^{-1}|W|^{1-s}\|_{p_s\to p_s} \leqslant M_p,\ \forall s \in [0,1], \text{ where } \frac{1}{p_s} = \frac{s}{p} + \frac{1-s}{p'}.$$

Proof Without loss of generality we can take $W \in L^\infty$. Let us introduce the operator-valued function
$F(z) = |W|^z(\lambda+A_{p_s})^{-1}|W|^{1-z}$, $z = s+it$, $0 \leqslant s \leqslant 1$, $t \in \mathbb{R}^1$ acting from the space $\mathcal{N}$ of simple functions from L^1 into the space of all measurable functions. Let $f, h \in \mathcal{N}$. Then by duality $\|(\lambda+A_{p'})^{-1}|W|\|_{p'\to p'} \leqslant M_p$ and

$$\|F(it)f\|_{p'} \leqslant \|(\lambda+A_{p'})^{-1}|W|^{1-it}f\|_{p'} \leqslant M_p\|f\|_{p'},$$

$$\|F(1+it)f\|_p \leqslant \|W(\lambda+A_p)^{-1}|W|^{it}f\|_p \leqslant M_p\|f\|_p.$$

Furthermore, setting $f = \sum_{k=1}^m c_k \mathbb{1}_{E_k}$, $h = \sum_{k=1}^m b_k \mathbb{1}_{B_k}$ we have

$$|\langle f, F(z)h\rangle| \leqslant c_{f,h,\lambda}\|W\|_{L^1(G)}.$$

The mapping $z \longmapsto \langle f, F(z)h\rangle$ is holomorphic in the strip $0 < s < 1$ so that $F(z)$ is an admissible family of linear operators in the sense of E.M.Stein [StW, p.250]. Consequently by Stein's interpolation theorem
$\|F(s)\|_{p_s\to p_s} \leqslant M_p$, $0 \leqslant s \leqslant 1$, $1/p_s = s/p + (1-s)/p'$.□

Let $W \in \Phi_p(A)$. Putting in Lemma 6.1 $s = 1/2$, $M_p < 1$ and $\lambda > \lambda_W$, where $\lambda_W > 0$ such that $\|W(\lambda_W + A_p)^{-1}\|_{p\to p} < 1$ we get $\||W|^{1/2}(\lambda+A)^{-1}|W|^{1/2}\|_{2\to 2} < 1$. This proves Proposition 6.1.

Theorem 6.2 *Let $V = V^+ - V^-$, $V^\pm \geqslant 0$. Suppose that*

1) $Q(V^+) \cap Q(A)$ is dense in L^2,

2) $V^- \in \Phi_1(H^+)$, $H^+ := A\dot{+}V^+$.

Let $H := A\dot{+}V = H^+\dot{-}V^-$ and $H_p = (A\dot{+}V)_p$ (according to Proposition 6.1 the operator H is well defined and by Theorem 5.3 H_p, $1 \leqslant p < \infty$, generates a C_0-semigroup). Then for all sufficiently large $\lambda > 0$, $\lambda \in \rho(-H_p)$,

$$1)\ (V^\pm)^{1/p}(\lambda+H_p)^{-1/p} \in \mathcal{L}(L^p),\ \forall p \in [1,2].$$

$$2)\ (V^\pm)^{1/p}(\lambda+H_p)^{-1} \in \mathcal{L}(L^p),\ \forall p \in [2,\infty],$$

where $(\lambda+H_\infty)^{-1} := ((\lambda+H_1)^{-1})^*$.

Proof Let $B = \lambda_0 + H_p$, where $\lambda_0 > 0$ is chosen by the condition

$$\|(t+B)^{-1}\|_{p\to p} \leqslant \frac{M}{1+t}, \quad \forall t \geqslant 0,\ \forall p \in [1,2]. \tag{6.2}$$

Define the following two operator-valued functions

$$f(z) := \int_0^\infty t^{1-z}(t+B)^{-2}dt, \quad 1/2 \leqslant \text{Re } z \leqslant 1$$

and

$$F(z) := (V^\pm)^{1-\frac{z}{2}} f(1-\frac{z}{2}), \quad 0 \leqslant \text{Re } z \leqslant 1.$$

The integral is convergent with respect to the operator norm for $0 < \text{Re } z < 2$. $f(z)$ is even analytic in the strip $0 < \text{Re } z < 2$, $f(1) = B^{-1}$ and $V^\pm f(1) \in \mathcal{L}(L^1)$, $(V^\pm)^{1/2} f(1/2) \in \mathcal{L}(L^2)$.

Indeed, $f(1) = B^{-1}$ follows from (6.2) and $\frac{d}{dt}(t+B)^{-1} = -(t+B)^{-2}$. $V^- \in \Phi_1(H^+)$ implies $V_1^- f(1) \in \mathcal{L}(L^1)$. By virtue of Theorem 6.1 $\mathcal{D}(H_1^+) \subset \mathcal{D}(V_1^+)$ so $V^+ f(1) \in \mathcal{L}(L^1)$. Finally, from $V^- \in \Phi_1(H^+)$ it follows that $Q(V^-) \supset Q(H^+) = Q(H)$ and $Q(H) = Q(A) \cap Q(V^+) \subset Q(V^+)$. Hence

$$(V^\pm)^{1/2} f(1/2) \equiv \frac{\pi}{2}(V^\pm)^{1/2} B^{-1/2} \in \mathcal{L}(L^2).$$

From $F(0+is) = (V^\pm)^{1-i\frac{s}{2}} \int_0^\infty t^{i\frac{s}{2}}(t+B)^{-2}dt,\ \ s \in \mathbb{R}^1$, and from positivity of $(t+B)^{-2}$ one has

$$\|F(0+is)\|_{1\to1} \leqslant \|V^\pm B^{-1}\|_{1\to1} < \infty.$$

Now $F(1+is) = (V^\pm)^{\frac{1}{2}-i\frac{s}{2}} \int_0^\infty t^{\frac{1}{2}+i\frac{s}{2}}(t+B)^{-2}dt,\ \ s \in \mathbb{R}^1$. By the spectral theorem for self-adjoint operators in a Hilbert space

$$\|F(1+is)\|_{2\to2} \leqslant \frac{\pi}{2}\|(V^\pm)^{1/2}(\lambda_0 + H_2)^{-1/2}\|_{2\to2} < \infty.$$

Therefore, $F(z)$, $0 \leqslant \text{Re } z \leqslant 1$ is an admissible family of linear operators in the sense of E.M.Stein so that

$$F(s) \in \mathcal{L}(L^{p_s}), \ 0 \leqslant s \leqslant 1, \ 1/p_s = 1 - s/2.$$

Statement 1) of the theorem now follows from the representation $B^{-z} = \frac{\sin \pi z}{\pi(1-z)} f(z)$ (see, e.g. [KrZPuSo, p.270]
Statement 2) also follows from Stein's interpolation theorem setting $F(z) = (V^\pm)^z B^{-1}$, $0 \leqslant \text{Re } z \leqslant 1$ and using the facts $V_1^\pm(\lambda_0 + H_1)^{-1} \in \mathcal{L}(L^1)$ and $(\lambda_0 + H_\infty)^{-1} \in \mathcal{L}(L^\infty)$.□

Theorem 6.3 *Let* $V = V^+ - V^-$, $V^\pm \geqslant 0$. *Suppose that*
1) $Q(V^+) \cap Q(A)$ *is dense in* L^2,

2) $V^- \in PK_\beta(H^+)$ *for some* $\beta \in (0,1)$. *Let* $H := A \dot{+} V = H^+ \dot{-} V^-$. *Then for all sufficiently large* $\lambda > 0$, $\lambda \in \rho(-H_p)$, *(e.g.,* $\lambda > c_{V^-}(\beta)$*)*

$$(V^+)^{1/p}(\lambda + H_p)^{-1/q} \in \mathcal{L}(L^p),\ \forall p \in \left[\frac{2}{1+\sqrt{1-\beta}}, \frac{2}{1-\sqrt{1-\beta}}\right],\ \forall q \in (0,p).$$

$$(V^-)^{1/p}(\lambda + H_p)^{-1/q} \in \mathcal{L}(L^p),\ \forall p \in \left(\frac{2}{1+\sqrt{1-\beta}}, \frac{2}{1-\sqrt{1-\beta}}\right),\ \forall q \in (0,p).$$

$$(V^\pm)^{1/p}(\lambda + H_p)^{-1/q}(V^\pm)^{1/p'} \in \mathcal{L}(L^p),\ \forall p \in \left(\frac{2}{1+\sqrt{1-\beta}}, \frac{2}{1-\sqrt{1-\beta}}\right), \forall q \in (0,p).$$

Proof First we prove the inequalities

$$\|(V^+)^{1/p}(\lambda + H_p)^{-1} g\|_p \leqslant a_p(1-\beta)^{-1/p}(\lambda - c(\beta))^{-1/p'}\|g\|_p, \tag{6.3}$$

$$\|(V^-)^{1/p}(\lambda + H_p)^{-1} g\|_p \leqslant \tilde{a}_p[\frac{1}{\beta}(4\frac{p-1}{p^2} - \beta)]^{-1/p}(\lambda - 4\frac{p-1}{p^2}\frac{c(\beta)}{\beta})^{-1/p'}\|g\|_p, \tag{6.4}$$

$\forall p \in \left[\frac{2}{1+\sqrt{1-\beta}}, \frac{2}{1-\sqrt{1-\beta}}\right]$, $\forall \lambda > c(\beta) \equiv c_{V^-}(\beta)$. The first two statements of the theorem follow from this by virtue of the formula

$$(\lambda + H_p)^{-\alpha} = \frac{\sin \pi\alpha}{\pi}\int_0^\infty t^{-\alpha}(t + \lambda + H_p)^{-1}dt,\ 0 < \alpha < 1.$$

It is enough to prove (6.3) and (6.4) for $0 \leqslant g \in L^1 \cap L^\infty$ and for the operator H_p with the truncations of V_m^+ and V_n^- instead of V^+ and V^- themselves since the rest follows from the positivity property and the Fatou lemma.

Let $f = (\lambda + H_p)^{-1}g$. Then $0 \leqslant f \in L^1 \cap L^\infty$. Moreover $f \in \mathcal{D}(H_p)$ $\forall p \in [1, \infty)$, in particular, $f \in Q(A) \cap Q(V^+)$. By Theorem 3.2 $f_p = f^{p/2} \in Q(A) \cap Q(V^+)$ and

$$\langle (H_p + \lambda)f, f^{p-1}\rangle \geqslant$$

$$(4\frac{p-1}{p^2} - \beta)\|A^{1/2}f_p\|_2^2 + (1-\beta)\|(V^+)^{1/2}f_p\|_2^2 + (\lambda - c(\beta))\|f_p\|_2^2. \tag{6.5}$$

Obviously, $\langle (H_p + \lambda)f, f^{p-1}\rangle \leqslant \|g\|_p\|f\|_p^{p-1}$.
(6.3) follows now from the inequality

$$(1-\beta)\|(V^+)^{1/p}f\|_p^p + (\lambda - c(\beta))\|f\|_p^p \leqslant \|g\|_p\|f\|_p^{p-1}$$

by using the Young inequality in its left-hand side. By means of the inequality

$$\frac{1}{\beta}\|(V^-)^{1/2}f_p\|_2^2 \leqslant \|A^{1/2}f_p\|_2^2 + \|(V^+)^{1/2}f_p\|_2^2 + \frac{c(\beta)}{\beta}\|f_p\|_2^2$$

one obtains from (6.5)

$$\frac{1}{\beta}(4\frac{p-1}{p^2}-\beta)\|(V^-)^{1/2}f_p\|_2^2+(\lambda-4\frac{p-1}{p^2}\frac{c(\beta)}{\beta})\|f\|_p^p \leqslant \|g\|_p\|f\|_p^{p-1}.$$

To get (6.4) one should use the Young inequality.
The last statement is proved in just the same way if instead of g one takes $(V_k^{\pm})^{1/p'}g$, where $V_k^{\pm}$ are the truncations of $V^{\pm}$. Observe only that in this case

$$\langle (H_p+\lambda)f, f^{p-1}\rangle = \langle (V_k^{\pm})^{1/p'}g, f^{p-1}\rangle \leqslant \|g\|_p\|(V_k^{\pm})^{1/2}f_p\|_2^{2/p'}.\square$$

Remark 6.4 Note the following interpolation result.
Let $V^- = V_1^- + V_2^-$, $V_{1,2}^- \geqslant 0$. Suppose that for some numbers $\beta_1 > 0$, $\beta_2 > 0$, $\beta_1+\beta_2 < 1$

$$\beta_1^{-1}V_1^- \in \Phi_1(H^+),\ \beta_2^{-1}V_2^- \in PK_1(H^+).$$

Then $\forall p \in (t(\beta_1+\beta_2), 2]$

$$(V_1^-)^{1/p}(\lambda+H_p)^{-1}(V_2^-)^{1/p'} \in \mathcal{L}(L^p).$$

Now we give the following result which in fact is due to T.Kato [K3] (see also [AB]).

Theorem 6.4 *Let A be a symmetric Markov generator and $V^+ \geqslant 0$ such that $Q(A)\cap Q(V^+)$ is dense in L^2. Suppose that A_1 is the generator of a bounded holomorphic semigroup of angle θ. Then there exists $\varepsilon \in (0,\theta)$ such that $-H_1^+$ is the generator of a bounded holomorphic semigroup of angle ε.*

Proof Let V_n^+ be the truncation of V^+. Then $\forall f \in \mathcal{D}(H_1(V_n^+)) = \mathcal{D}(A_1)$ and $\forall z$ with $\mathrm{Re}\, z \geqslant 0$

$$\|(z+H_1(V_n^+))f\|_1 \geqslant \mathrm{Re}\,\langle (z+H_1(V_n^+))f, \mathrm{sgn}\ f\rangle \geqslant \mathrm{Re}\, z\cdot\|f\|_1 + \|V_n^+f\|_1 \geqslant \|V_n^+f\|_1.$$

Since e^{-tA_1} is holomorphic then $A_1 \in \mathcal{H}(M,\theta_1)$ for some $\theta_1 \in (0,\theta)$. In particular $\|(z+A_1)^{-1}\|_{1\to 1} \leqslant M|z|^{-1}$, $\forall z$ with $\mathrm{Re}\, z>0$ and $|z| \geqslant \lambda_0 > 0$. Therefore for such z and $f \in \mathcal{D}(A_1)$

$$|z|\|f\|_1 \leqslant M\|(z+A_1)f\|_1 \leqslant M(\|(z+A_1+V_n^+)f\|_1+\|V_n^+f\|_1) \leqslant 2M\|(z+H_1(V_n^+))f\|_1$$

or

$$\|(z+H_1(V_n^+))^{-1}\|_{1\to 1} \leqslant 2M|z|^{-1}.$$

Using the monotone approximation theorem we conclude

$$\|(z+H_1^+)^{-1}\|_{1\to 1} \leqslant 2M|z|^{-1},\ \forall z \text{ with } \mathrm{Re}\, z>0 \text{ and } |z| \geqslant \lambda_0 > 0.$$

To finish the proof it is enough to use the Yosida-Solomyak theorem (see [KrZPuSo, p. 270]).$\square$

7 (L^p, L^q)-estimates for the resolvent and the semigroup, their integral properties and related problems

In many problems one meets "raising of smoothness" of solutions for equations with a given operator. This possibility is determined by Sobolev's imbedding property of the operator.

Let us consider the following problem:

Let $A \geqslant 0$ be a self-adjoint operator in L^2 which satisfies Sobolev's imbedding property and the semigroup e^{-tA}, $t > 0$ is L^p-bounded for some $p \neq 2$:

$$\|u\|_{2j}^2 \leqslant a\|A^{1/2}u\|_2^2 + b\|u\|_2^2, \; u \in Q(A)$$

for some numbers $j > 1$, $a > 0$ and $b \geqslant 0$;

$$\|e^{-tA}f\|_p \leqslant Me^{\omega t}\|f\|_p, \; f \in L^2 \cap L^p, \; t > 0$$

for some numbers $M \geqslant 1$, $\omega \in \mathbb{R}^1$ and $p \neq 2$, say $p > 2$. It is required to find the maximal exponent $q > \max\{2j, p\}$ such that

$$e^{-tA}[L^2 \cap L^p] \subset L^q$$

(L^q could be replaced by $L^{q,r}$ etc.). J.Nash [N] proved the following inequality for the concrete but important case of an elliptic operator in divergence form :

$\|u\|_2^{2+4/\nu} \leqslant a\|A^{1/2}u\|_2^2 \cdot \|u\|_1^{4/\nu}$, $\nu > 0$, $u \in Q(A) \cap L^1$.

He established the L^1-contractivity of e^{-tA} and from these two facts almost immediately proved the imbedding $e^{-tA_1}[L^1] \subset L^\infty$.

Considerably later B.Simon and one of the authors (Yu.S.) independently found other approaches leading to the proof of $e^{-tA_1}[L^1] \subset L^\infty$ in the case $A = -\Delta + V$ with $V \in \Phi_1(-\Delta)$(see [CyFrKiSi]).

It was N.Varopoulos who first showed that the Markov property and Sobolev's property of the operator A lead to maximal possible (and known for $A = -\Delta$) (L^p, L^q)-estimates for the (fractional) powers of the resolvent.

In the next theorem we do not assume the semigroup e^{-tA} to be Markov but only L^1-bounded (this is the only distinguishing feature of the following Theorems 7.1–7.3 from the corresponding results in [V],[CKuS], see also [D]). For the case of semigroups which are equicontinuous on L^1 and on L^∞ see [Co].

Theorem 7.1 *Let $A \geqslant 0$ be a self-adjoint operator in $L^2(M, \mu)$. Suppose that*

1) $\|f\|_{2j}^2 \leqslant c\; t_A[f]$, $f \in Q(A)$

for some numbers $j > 1$ and $c > 0$.

2) $\|e^{-tA}f\|_1 \leqslant M\|f\|_1$, $t > 0$, $f \in L^2 \cap L^1$

for some number $M \geqslant 1$.

Let $\lambda > 0$ and $\alpha \in (0, j'/p_0)$ for some $p_0 \geqslant 1$. Then

$$(\lambda + A_p)^{-\alpha} \in \mathcal{L}(L^p, L^q)$$

for all p, q such that $p_0 < p < q < \infty$ and $1/p - 1/q \leqslant \alpha/j'$.
For $\alpha \in (0, j')$ $(\lambda + A)^{-\alpha}$ can be extended by continuity to a bounded mapping from L^1 to $L^{q,\infty}$ for $q = (1 - \alpha/j')^{-1}$.

Proof Let $f \in Q(A) \cap L^1$ and let $\mu = 2j$. By Hölder's inequality

$$\|f\|_2^2 \leqslant \|f\|_\mu^{\mu'} \cdot \|f\|_1^{2-\mu'} \leqslant (c\, t_A[f])^{\mu'/2} \cdot \|f\|_1^{2-\mu'}$$

or

$$\|f\|_2^{2+4/\nu} \leqslant c\, t_A[f] \cdot \|f\|_1^{4/\nu}, \ \nu = 2j'.$$

Set $u(t) = \|f_t\|_2^2$, where $f_t = e^{-tA} f$. Then

$$-u'(t) = 2t_A[f_t] \geqslant 2c^{-1} u(t)^{1+2/\nu} \cdot \|f_t\|_1^{-4/\nu}$$

and since $\|f_t\|_1 \leqslant M \|f\|_1$ then

$$(u(t)^{-2/\nu})' \geqslant \frac{4}{\nu c M^{\frac{4}{\nu}}} \|f\|_1^{-4/\nu}.$$

Therefore

$$u(t)^{-2/\nu} \geqslant u(t)^{-2/\nu} - u(0)^{-2/\nu} \geqslant \frac{4t}{\nu c M^{\frac{4}{\nu}}} \|f\|_1^{-4/\nu}$$

or

$$\|e^{-tA} f\|_2 \leqslant c_1 t^{-\nu/4} \|f\|_1, \text{ where } c_1 = (\frac{\nu c}{4})^{\frac{\nu}{4}} M.$$

Choose $f = e^{-A/n} g$, $g \in L^1 \cap L^2$ and note that $f \in Q(A) \cap L^1$. Then from the last inequality we have $\|e^{-(t+1/n)A} g\|_2 \leqslant c_1 M t^{-\nu/4} \|g\|_1$. Since $e^{-(t+1/n)A} \xrightarrow[L^2]{s} e^{-tA}$ $(n \to \infty)$ then

$$\|e^{-tA} f\|_2 \leqslant c_1 t^{-\nu/4} \|f\|_1 \ \forall f \in L^1 \cap L^2, \text{ where } c_1 = (\frac{\nu c}{4})^{\frac{\nu}{4}} M^2.$$

By duality and semigroup property $\|e^{-tA} f\|_\infty \leqslant 2^{\frac{\nu}{2}} c_1^2 t^{-\nu/2} \|f\|_1$. By the Riesz–Thorin interpolation theorem

$$\|e^{-tA} f\|_\infty \leqslant M^{1-1/p} 2^{\nu/2p} c_1^{2/p} t^{-\nu/2p} \|f\|_p \ (t > 0, \ p \geqslant 1, \ f \in L^2 \cap L^p).$$

Let $p \in (1, j'/\alpha)$. Then $(\lambda + A_p)^{-\alpha} f \equiv R^\alpha f = g + h$, where

$$f \in L^p, \ g := \frac{1}{\Gamma(\alpha)} \int_0^\delta t^{\alpha-1} e^{-\lambda t} e^{-tA_p} f dt, \ h := \frac{1}{\Gamma(\alpha)} \int_\delta^\infty t^{\alpha-1} e^{-\lambda t} e^{-tA_p} f dt.$$

It is clear that

$$\|h\|_\infty \leqslant \frac{1}{\Gamma(\alpha)} \int_\delta^\infty t^{\alpha-1-\nu/2p} M^{1-1/p} 2^{\nu/2p} c_1^{2/p} \|f\|_p dt = c_2 \|f\|_p \cdot \delta^{\alpha-\nu/2p}.$$

For given $s > 0$ let us choose δ by the condition

$$s/2 = c_2 \|f\|_p \cdot \delta^{\alpha-\nu/2p}.$$

By Chebyshev's inequality and by L^p-boundedness of e^{-tA_p}

$$|\{x : |R^\alpha f(x)| \geqslant s\}| \leqslant |\{x : |g(x)| \geqslant s/2\}| \leqslant \left(\frac{2}{s}\right)^p \|g\|_p^p$$

$$\leqslant \left(\frac{2}{s}\right)^p \left(\frac{1}{\Gamma(\alpha)} \int_0^\delta t^{\alpha-1} M \|f\|_p dt\right)^p \leqslant \left(2 \frac{1}{\Gamma(\alpha)} \alpha^{-1} s^{-1} \delta^\alpha M \|f\|_p\right)^p.$$

Putting $1/q = 1/p - 2\alpha/\nu$ we obtain

$$|\{x : |R^\alpha f(x)| \geqslant s\}| \leqslant c_3 s^{-q} \|f\|_p^q, \text{ i.e. } R^\alpha : L^p \to L^{q,\infty}, \ \forall p \in (1, j'/\alpha).$$

Using the Marcinkiewicz interpolation theorem yields

$$R^\alpha \in \mathcal{L}(L^p, L^q), \ \forall p \in (1, j'/\alpha). \square$$

Remark 7.1 Let us consider the operator $\Delta^2 + V$ with $\inf_{\lambda>0} \|V(\lambda+\Delta^2)^{-1}\|_{1\to1} = 0$. Then by virtue of Theorem 2.4 [K1, p. 496] the conditions of Theorem 7.1 are satisfied. It should be pointed out that the corresponding semigroup is not positivity preserving and it is not quasi-contraction in L^p except for the case $p = 2$.

The above theorem can be given in a symmetric form.

Theorem 7.2 *Let $A \geqslant 0$ be a self-adjoint operator in L^2, and*

$$\|e^{-tA} f\|_1 \leqslant M \|f\|_1, \ f \in L^2 \cap L^1, \ t > 0$$

for some $M \geqslant 1$.
The following conditions are equivalent if $\nu > 2$:

$$\|e^{-tA} f\|_\infty \leqslant c_1 t^{-\nu/4} \|f\|_2, \ f \in L^2, \ t > 0. \tag{i}$$

$$\|f\|_{2j}^2 \leqslant c_2 \ t_A[f], \ j = \frac{\nu}{\nu-2}, \ f \in Q(A). \tag{ii}$$

$$\|f\|_2^{2+4/\nu} \leqslant c_3 \ t_A[f] \cdot \|f\|_1^{4/\nu}, \ f \in Q(A) \cap L^1. \tag{iii}$$

Moreover, if $\nu > 0$ then (i) and (iii) are equivalent.

Proof In the proof of Theorem 7.1 we have already proved that (ii) $\Longrightarrow$ (iii) $\Longrightarrow$ (i) $\Longrightarrow$ (ii) for $\nu > 2$ and (iii) $\Longrightarrow$ (i) for $\nu > 0$. So it remains only to prove (i) $\Longrightarrow$ (iii) for $\nu > 0$. We have (i) $\Longrightarrow \|e^{-tA}f\|_2 \leqslant c_1 t^{-\nu/4}\|f\|_1,\ f \in L^1 \cap L^2$, hence

$$c_1^2 t^{-\nu/2}\|f\|_1^2 \geqslant \langle e^{-2tA}f, f\rangle = \langle f, f\rangle - \int_0^{2t} t_A[f_{s/2}]ds$$
$$\geqslant \langle f, f\rangle - 2tM \cdot t_A[f]$$

for all $f \in Q(A) \cap L^1$. Consequently

$$\|f\|_2^2 \leqslant 2Mt \cdot t_A[f] + c_1^2 t^{-\nu/2}\|f\|_1^2$$

and (iii) follows by putting $t = [\|f\|_1^2/t_A[f]]^{2/2+\nu}$.□

A simple analysis of the proof of Theorem 7.1 shows that the following theorem is valid.

Theorem 7.3 *Let A and B be symmetric Markov generators. Suppose that*

1) $B \in PK_\beta(A)$ for some $\beta < 1$ (so that the form-difference $H = A \dot{-} B$ is well-defined).

2) $\|e^{-tH}f\|_1 \leqslant Me^{\omega t}\|f\|_1$ for some numbers $M \geqslant 1$ and $\omega \geqslant 0$ and all $t > 0$ and $f \in L^2 \cap L^1$.

If A satisfies the conditions of Theorem 7.2 then there exist constants $c_1,\ c_2,\ c_3 > 0$ and $b_1,\ b_2,\ b_3 \geqslant 0$ such that

$$\|H^\sigma e^{-tH}f\|_\infty \leqslant c_1 e^{b_1 t} t^{-\sigma-\nu/2p}\|f\|_p,\ f \in L^2 \cap L^p,\ t > 0,\ \sigma = 0, 1, 2, \ldots,\ 1 \leqslant p < \infty \qquad (i_\sigma)$$

for $\nu > 0$.

$$\|f\|_{2j}^2 \leqslant c_2\ t_{H+b_2}[f],\ f \in Q(A),\ j = \frac{\nu}{\nu-2},\ \text{for } \nu > 2. \qquad (ii)$$

$$\|f\|_2^{2+4/\nu} \leqslant c_3\ t_{H+b_3}[f] \cdot \|f\|_1^{4/\nu},\ f \in Q(A) \cap L^1 \text{ for } \nu > 0. \qquad (iii)$$

Moreover, (iii) $\Longleftrightarrow$ (ii) $\Longleftrightarrow$ (i$'_0$) *for $\nu > 2$, where*

$$\|e^{-tH}f\|_\infty \leqslant c_1\ t^{-\nu/4}\|f\|_2,\ f \in L^2,\ t \in (0, 1] \qquad (i'_0)$$

and (iii) $\Longleftrightarrow$ (i$'_0$) *for $\nu > 0$.*

Remark 7.2 Perhaps, only (i_σ) needs an explanation. For $2 < p < \infty$ by Stein's interpolation theorem $\|H_p e^{-tH_p}\|_{p\to p} \leqslant \frac{c'}{t}e^{b't}$. Therefore

$$\|He^{-2tH}f\|_\infty \leqslant c_1 e^{b_1 t} t^{-\nu/2p}\|He^{-tH}f\|_p \leqslant ce^{bt}t^{-1-\nu/2p}\|f\|_p.$$

For $1 \leqslant p \leqslant 2$

$$\|He^{-3tH}f\|_\infty \leqslant c_1 e^{b_1 t} t^{-\nu/4}\|He^{-tH}\|_{2\to 2} \cdot \|e^{-tH}f\|_2 \leqslant ce^{bt}t^{-1-\nu/2p}\|f\|_p.$$

Remark 7.3 The conditions of Theorem 7.3 can be checked, for example, for $B = V^- \in \Phi_1(A)$ by using Theorem 5.3.

Remark 7.4 The meaningful results of Theorem 7.3 are

$$H^\sigma e^{-H}[L^1 \cap L^2] \subset L^\infty;$$

$$(\lambda + H_p)^{-\alpha} \in \mathcal{L}(L^p, L^q),\ 1 < p < q < \infty,\ 1/p - 1/q \leqslant \alpha/j',$$

$0 < \alpha < j'/p$, $\lambda > 0$ is large enough.

Remark 7.5 In Theorem 7.3 we need not assume that A and B are Markov (obviously in this case remark 7.3 is not valid). We want to emphasize that Theorems 7.1 – 7.3 are not connected directly with the Beurling–Deny theory.

For arbitrary $B \in PK_\beta(A)$ with $\beta \in (0,1)$ one should not expect the validity of results similar to Theorems 7.1 or 7.3. Nevertheless the following result holds which essentially uses the Markov property in contrast to Theorem 7.1.

Theorem 7.4 *Let A and B be symmetric Markov generators. Suppose that*
1) *$B \in PK_\beta(A)$ for some $\beta \in (0,1)$.*
2) *$\|f\|_{2j}^2 \leqslant c\ t_A[f]$ for some $j > 1$ and all $f \in Q(A)$.*
Let $H = A \dot{-} B$ and $t(\beta), t'(\beta)$ be the corresponding roots of the equation $\beta\varkappa(q) = 4\frac{q-1}{q^2}$, $1 < q < \infty$. Fix $j_1 \in [1, j]$ and $\lambda_0 > \varkappa(p)c(\beta)$. Then
(i) $(\lambda_0 + H)^{-1}$ can be extended by continuity to a mapping from L^q to L^{pj_1}, for $\frac{1}{q} = \frac{1}{pj_1} + \frac{1}{j_1'}$ and for all $p \in (t(\beta), t'(\beta))$.
(ii) Moreover if $g \in L^2 \cap L^{t'(\beta)}$ then

$$|\{x : |(\lambda_0 + H)^{-1} g(x)| \geqslant s\}| \leqslant c \cdot s^{-jt'(\beta)} \cdot \ln^j s,\ \forall s > \exp\frac{1}{t'(\beta) - t(\beta)}.$$

Proof By Theorem 3.2(iv) we have

$$(4\frac{p-1}{p^2} - \varkappa(p)\beta)\|A^{1/2}|f|^{p/2}\|_2^2 + (\lambda_0 - \varkappa(p)c(\beta))\|f\|_p^p \leqslant \langle |g|, |f|^{p-1}\rangle,$$

where $f = (\lambda_0 + H_p)^{-1} g$, $g \in L^p$, $p \in (t(\beta), t'(\beta))$. By condition 2) $\|A^{1/2}|f|^{p/2}\|_2^2 \geqslant c^{-1}\|f\|_{pj}^p$. Since $\langle |g|, |f|^{p-1}\rangle \leqslant \|g\|_q \cdot \|f\|_{pj_1}^{p-1}$ and $\|f\|_{pj} + \|f\|_p \geqslant \|f\|_{pj_1}$ we obtain

$$\|f\|_{pj_1} \leqslant c_1\|g\|_q,\quad g \in L^p \cap L^q. \tag{7.1}$$

(7.1) implies (i) immediately.
Since

$$\langle |g|, |f|^{p-1}\rangle \leqslant \|g\|_p \cdot \|f\|_p^{p-1} \leqslant (\lambda_0 - \varkappa(p)c(\beta))\|f\|_p^p + p^{-p}(p-1)^{p-1}(\lambda_0 - \varkappa(p)c(\beta))^{1-p}\|g\|_p^p$$

we get

$$(4\frac{p-1}{p^2} - \varkappa(p)\beta)\|f\|_{pj}^p \leqslant K\|g\|_p^p. \tag{7.2}$$

Putting in (7.2) $p = t'(\beta) - \varepsilon$, $\varepsilon > 0$ and using the condition $g \in L^2$ we have from (7.2)

$$\varepsilon^j\|f\|_{pj}^{pj} \leqslant K_1, \ K_1 < \infty, \ K_1 \neq K_1(\varepsilon).$$

From this by Chebyshev's inequality

$$|\{x : |f(x)| > s\}| \leqslant K_1 s^{-jt'(\beta)} \cdot \varepsilon^{-j} \cdot s^{\varepsilon j}, \ s > \exp\frac{1}{t'(\beta) - t(\beta)}.$$

It remains to set $\varepsilon = \frac{1}{\ln s}$ and (ii) is proved.□

Remark 7.6 Statements (i) and (ii) of Theorem 7.4 are sharp in the following sense: simple examples show that for arbitrary $B \in PK_\beta(A)$ the imbedding $(\lambda + H)^{-1}[L^1 \cap L^\infty] \subset L_{loc}^{j \cdot t'(\beta)}$, $\lambda \in \rho(-H)$ fails. On the other hand, one can prove that if $M = \mathbb{R}^d$, $d\mu = dx$, $A = -\Delta$ and $B = V^- \in PK_\beta(-\Delta) \cap L^{d/2,\infty}(\mathbb{R}^d)$ then

$$(\lambda + H_p)^{-1} \in \mathcal{L}(L^p, L^{j \cdot t'(\beta), \infty}), \ \lambda \in \rho(-H_p), \ p \in (t(\beta), t'(\beta)), \ \frac{1}{p} - \frac{1}{jt'(\beta)} < \frac{1}{j'}.$$

The last leads to the conjecture that the estimate for the distribution function of $(\lambda_0 + H)^{-1}g$ obtained in Theorem 7.4 can be improved (on $\ln^j s$).

Simple arguments based on Remark 7.3 and on the Dunford–Pettis theorem lead to the following consequence.

Theorem 7.5 *Let A be symmetric Markov generator satisfying the conditions of Theorem 7.1 and let $V^- \in \Phi_1(A)$. Let $H = A \dot{-} V^-$. Then e^{-tH_p} is an integral operator for $\forall p \in [1, \infty)$ with non-negative integral kernel $K_H(t, x, y)$ and for all $q \in [1, \infty]$*

$$ess \sup_{y \in M} \|K_H(t, \cdot, y)\|_q = \|e^{-tH_1}\|_{1 \to q}.$$

The analogous consequences can be obtained also for the resolvent of the operator H_p.

8 Strong and Markov uniqueness

Let A be a closed densely defined operator acting in a Banach space. A linear manifold $\mathcal{E}_A \subset \mathcal{D}(A)$ is called a core or uniqueness domain of A if the closure $(A \restriction \mathcal{E}_A)^\sim = A$. If A is a symmetric operator bounded from below acting in a Hilbert space $\mathcal{K}$ and the linear manifold $\mathcal{E}_A \subset \mathcal{D}(A)$ is dense in $\mathcal{K}$ ($\mathcal{E}_A^a = \mathcal{K}$) and such that all bounded from below self-adjoint extensions of $A \restriction \mathcal{E}_A$ coincide then $A^\sim$ is a self-adjoint operator and $\mathcal{E}_A$ is a core or uniqueness domain of $A^\sim$(J. von Neumann).

Definition 8.1 ([Fu]) Let H be a symmetric Markov generator in $L^2(M,\mu)$ and let the linear manifold $\mathcal{E}_H \subset \mathcal{D}(H)$, $\mathcal{E}_H^a = L^2(M,\mu)$ be such that *all* Markov extensions of $H \restriction \mathcal{E}_H$ coincide. Such a domain $\mathcal{E}_H$ is called a domain of Markov L^2-uniqueness of the operator H.

Note that if $\mathcal{E}_H$ is a core of the operator H then $\mathcal{E}_H$ is a domain of Markov L^2-uniqueness of H. The converse is not true in general since the class of Markov extensions is a proper subclass of semibounded extensions. The corresponding examples will be given below.

Definition 8.2 Let H be a symmetric Markov generator. If $\mathcal{E}_p \subset \mathcal{D}(H_p)$ is such that whenever $\hat{H}$ is a symmetric Markov generator with $\hat{H}_p \restriction \mathcal{E}_p = H_p \restriction \mathcal{E}_p$ then $\hat{H} = H$, then $\mathcal{E}_p$ is called a domain of Markov L^p-uniqueness for H.

Observe that for example in the case $\mu(M) < \infty$ $H_p \subset H_1$ and $\mathcal{D}(H_p) = \mathcal{D}(H_1) \cap L^p$. So it is possible that $\mathcal{E}_1$ can be a domain of Markov L^1-uniqueness of the operator H and at the same time $\mathcal{E}_1 \not\subset \bigcup_{p>1} \mathcal{D}(H_p)$.

Example 8.1 Let A be a symmetric Markov generator and $V \in L^1_+(M,\mu)$. The algebraic sum $\hat{H} = A + V \equiv A_2 + V_2$ is not densely defined in L^2 in general (and in L^p, $p > 1$). Let $\mathcal{E}_1 = (1+A)^{-1}[L^1 \cap L^\infty]$. Then $\mathcal{E}_1 \subset L^\infty \cap \mathcal{D}(A_1) \cap \mathcal{D}(A_2) \subset \mathcal{D}(V_1)$ but $\mathcal{E}_1 \not\subset \mathcal{D}(V_2)$ in general. We claim that $\mathcal{E}_1$ is a domain of Markov L^1-uniqueness of the operator $\hat{H}$ and $H = A \dotplus V$ is L^1-unique Markov extension of $\hat{H}$. Indeed, according to Remark 6.2 after Theorem 6.1 $H_1 = A_1 + V_1$, $\mathcal{D}(H_1) = \mathcal{D}(A_1) \cap \mathcal{D}(V_1)$ and one can easily show that $\mathcal{E}_1$ is a core of H_1.

It is also possible to give a direct proof:

Let $V_{2,n}$ be the truncation of V and let $H_{2,n} = A_2 + V_{2,n}$, $\mathcal{D}(H_{2,n}) = \mathcal{D}(A_2)$. Since $(1+H_{2,n})^{-1} = (1+A)^{-1}[1 - V_{2,n}(1+H_{2,n})^{-1}]$ one has $(1+H_{2,n})^{-1}[L^1 \cap L^\infty] \subset \mathcal{E}_1$. If $\mathfrak{T}$ is an arbitrary symmetric Markov generator such that $\mathfrak{T}_1 f = A_1 f + V_1 f$, $\forall f \in \mathcal{E}_1$ then $\forall g \in L^1 \cap L^\infty$

$$(1+\mathfrak{T})^{-1}g - (1+H_{2,n})^{-1}g = (1+\mathfrak{T}_1)^{-1}(H_{1,n} - \mathfrak{T}_1)(1+H_{1,n})^{-1}g = \\ (1+\mathfrak{T}_1)^{-1}(V_{2,n} - V)(1+H_{1,n})^{-1}g \longrightarrow 0 \quad \text{in } L^1 \quad \text{for} \quad n \to \infty.$$

Therefore $(1+\mathfrak{T})^{-1} = (1+H)^{-1}$.□

The situation described in Example 8.1 is typical for other symmetric Markov generators too. Here we only dwell on the original statement of the problem [Fu].

Example 8.2 Let $(M,\mu) = (\mathbb{R}^d, \varphi^2 dx)$, where $\varphi : \mathbb{R}^d \longrightarrow \mathbb{R}^1$, $\varphi \neq 0$ a.e., $\int_{\mathbb{R}^d} \varphi^2 dx = 1$ and $\varphi \in \mathbb{H}^1(\mathbb{R}^d)$ (Sobolev space). In the space $L^2(M,\mu)$ let us consider the Dirichlet form $\langle \nabla u, \nabla u \rangle = \int_{\mathbb{R}^d} |\nabla u|^2 \varphi^2 dx$, $u \in C_0^\infty(\mathbb{R}^d)$. This form is closable in $L^2(M,\mu)$ since the adjoint to $\nabla \restriction C_0^\infty(\mathbb{R}^d)$ is densely defined. Let H be the operator associated with the closure of this form. Let $\mathcal{S}$ be Schwartz space. Then $\mathcal{S} \subset \mathcal{D}(H)$ and

$$\forall f \in \mathcal{S} \quad Hf = -\Delta f - \beta \cdot \nabla f, \quad \text{where} \quad \beta := 2\varphi^{-1}(\nabla \varphi).$$

We claim that $\mathcal{S}$ is a domain of Markov L^2-uniqueness of H. To prove this let us consider in $L^2(\mathbb{R}^d, dx)$ the operator $H_n = -\Delta - \beta^n \cdot \nabla$, $\beta^n \in [C_b^\infty(\mathbb{R}^d, dx)]^d$, $\mathcal{D}(H_n) = \mathbb{H}^2(\mathbb{R}^d)(= (1-\Delta)^{-1}L^2(\mathbb{R}^d, dx))$. $(e^{-tH_n}, t \geqslant 0)$ is a non-symmetric Markov semigroup on $L^2(\mathbb{R}^d)$ and moreover it is an analytic semigroup on $C_\infty(\mathbb{R}^d) := \{f \in C_b(\mathbb{R}^d); \lim_{|x|\to\infty} f(x) = 0\}$, where $C_b(\mathbb{R}^d)$ is the space of continuous bounded functions with the usual sup-norm and $e^{-tH_n}[\mathcal{S}] \subset \mathcal{S}$, $t > 0$. The mapping

$$(0,\infty) \ni t \longrightarrow e^{-tH_n} f \in L^2(M,\mu)$$

is strongly continuous and differentiable $\forall f \in \mathcal{S}$. Therefore for an arbitrary Markov extension $\mathfrak{T} \supset H \restriction \mathcal{S}$

$$e^{-t\mathfrak{T}}g - e^{-tH_n}g = \int_0^t e^{-(t-s)\mathfrak{T}}(H_n - \mathfrak{T})e^{-sH_n}g\,ds = \int_0^t e^{-(t-s)\mathfrak{T}}(\beta - \beta^n)\cdot\nabla u_n(s)ds,$$

$$\text{with } u_n(s) := e^{-tH_n}g,\ g \in C_0^\infty(\mathbb{R}^d),$$

$$\|e^{-t\mathfrak{T}}g - e^{-tH_n}g\|_1 \leqslant \|\beta - \beta^n\|_2 \int_0^t \|\nabla u_n(s)\|_2 ds.$$

So if we prove the estimate $\sup_n \int_0^t \|\nabla u_n(s)\|_2 ds < \infty$ for β^n such that $\|\beta - \beta^n\|_2 \longrightarrow 0$ $(n \longrightarrow \infty)$ then taking into account the density of $C_0^\infty(\mathbb{R}^d)$ in $L^2(M,\mu)$ we conclude that all such $\mathfrak{T}$ coincide and the desired statement is proved.

The following estimate holds

$$\int_0^t \|\nabla u_n(s)\|_2 ds \leqslant \sqrt{t}\|g\|_\infty \sqrt{1 + t\|\beta - \beta^n\|_2^2}.$$

Indeed, multiplying both sides of the equation $(\frac{d}{dt} + H_n)u_n(t) = 0$ by $\bar{u}_n = \bar{u}_n(t)$ we have after integrating

$$\frac{1}{2}\frac{d}{dt}\|u_n\|_2^2 + \|\nabla u_n\|_2^2 = \mathrm{Re}\,\langle(\beta^n - \beta)\cdot\nabla u_n, u_n\rangle \leqslant \frac{1}{2}\|u_n\|_\infty^2\|\beta - \beta^n\|_2^2 + \frac{1}{2}\|\nabla u_n\|_2^2.$$

Integrating over t the following inequality

$$\|\nabla u_n\|_2^2 \leqslant -\frac{d}{dt}\|u_n\|_2^2 + \|g\|_\infty^2\|\beta - \beta^n\|_2^2$$

we get

$$\int_0^t \|\nabla u_n(s)\|_2^2 ds \leqslant \|g\|_\infty^2(1 + t\|\beta - \beta^n\|_2^2).$$

Thus we have proved that $\mathcal{S}$ is a domain of Markov L^2-uniqueness of the operator H.

Now it is not difficult to show $C_0^\infty(\mathbb{R}^d)$ is also a domain of Markov L^2-uniqueness of the operator H. To this end it is enough to approximate an arbitrary element f from $\mathcal{S}$ by elements from $C_0^\infty(\mathbb{R}^d)$ in the graph norm of the operator H.

Let $f \in \mathcal{S}$ and $\chi(\cdot) \in C_0^\infty(\mathbb{R}^d)$, $0 \leqslant \chi(\cdot) \leqslant 1$. Set $\chi_n(\cdot) = \chi(\frac{\cdot}{n})$. Then obviously $\|(-\Delta - \beta\cdot\nabla)(1-\chi_n)f\|_2 \longrightarrow 0$ $(n \longrightarrow \infty)$.

Thus the following theorem is proved

Theorem 8.1 *Let $d \in \mathbb{N}$. Suppose that $\varphi : \mathbb{R}^d \longrightarrow \mathbb{R}^1$ satisfies the conditions*

$$\int_{\mathbb{R}^d} \varphi^2 dx = 1, \; \varphi \in \mathbb{H}^1(\mathbb{R}^d) \; \text{ and } \; \varphi \neq 0 \; \text{ a.e.} \tag{8.1}$$

Then the operator $\mathbb{D} = -\Delta - 2\varphi^{-1}(\nabla\varphi)\cdot\nabla$, $\mathcal{D}(\mathbb{D}) = C_0^\infty(\mathbb{R}^d)$, has exactly one self-adjoint extension in $L^2(\mathbb{R}^d, \varphi^2 dx)$ which generates a symmetric Markov semigroup on $L^2(\mathbb{R}^d, \varphi^2 dx)$.

The above proof is due to [Se2]. For other proofs see also [RöZ],[So].

Remark 8.1 Theorem 8.1 can be generalized in several directions. In particular, instead of (8.1) one can assume $\varphi \in \mathbb{H}^1_{loc}(\mathbb{R}^d)$, $\varphi \neq 0$ a.e. For this purpose one can use Povzner's or Weinholtz's methods.

Remark 8.2 For the proof of strong uniqueness the estimate $\sup_n \int_0^t \|u_n(s)\|_4^4 ds < \infty$ was proved in [LSe3] provided $|\beta| \in L^4(\mathbb{R}^d, \varphi^2 dx)$. This estimate is much deeper than the estimate $\sup_n \int_0^t \|\nabla u_n(s)\|_2 ds < \infty$ used in the proof of Markov uniqueness.

Remark 8.3 In the given proof of Theorem 8.1 the semigroup e^{-tH_n} is used. It should be pointed out that this semigroup cannot be extended to a bounded mapping on $L^2(M,\mu)$ except for some simplest cases like $|\beta| \in L^\infty_{loc}$; the estimate

$$\|e^{-tH_n} f\|_2 \leqslant e^{\omega_n t}\|f\|_2, \; f \in \mathcal{N}, \; \mathcal{N}^a = L^2(M,\mu)$$

is not valid since the operator H_n is not even quasi-accretive in $L^2(M,\mu)$.

Example 8.3 Let $(M,\mu) = (\mathbb{R}^d, \varphi^2 dx)$, where $\varphi(x) = |x|^\alpha e^{-|x|}/k$, $k := (\int |x|^{2\alpha} e^{-2|x|} dx)^{1/2}$. Let $\alpha > -\frac{d}{2}$ so that $k < \infty$. Let us consider the operator $\mathbb{D} = -\Delta - \beta\cdot\nabla$, $\mathcal{D}(\mathbb{D}) = C_0^\infty(\mathbb{R}^d)$, in $L^2(M,\mu)$. According to [LSe3] if $|\beta| \in L^4(M,\mu)$ then $\mathbb{D}^* = \mathbb{D}^\sim$ and in this case it gives the condition $\alpha > -\frac{d}{2} + 2$. Moreover, it is shown in [LSe3] that $\mathbb{D}^* \neq \mathbb{D}^\sim$ for $\alpha < -\frac{d}{2} + 2$ and $d \geqslant 4$. On the other hand Theorem 8.1 guarantees the Markov L^2-uniqueness of $\mathbb{D}$ for all $\alpha > -\frac{d}{2} + 1$. Hence for $\forall \alpha \in (-\frac{d}{2} + 1, -\frac{d}{2} + 2)$ the set $C_0^\infty(\mathbb{R}^d)$ is not a domain of strong uniqueness of the operator $\mathbb{D}$ but is a domain of Markov L^2-uniqueness (see also [Rö], [T]).

Now we turn to the problem of strong uniqueness for the algebraic difference $A - B$ of two symmetric Markov generators in the case when B is an operator of multiplication by a measurable function $V(\cdot) : M \longrightarrow \mathbb{R}_+$. L^p-versions of the KLMN-theorem and the concepts of L^1-theory open new ways in studying the uniqueness problem.

Suppose that $V \in PK_\beta(A)$ for some $\beta \in (0,1)$. Then according to Theorem 3.2 there exists the family $\{-(A\dot{-}V)_p\}_{p\in[t(\beta),t'(\beta)]}$, $t(\beta) = 2/(1+\sqrt{1-\beta})$, of generators of quasicontractive C_0-semigroups

$$\|e^{-t(A\dot{-}V)_p}\|_{p\to p} \leqslant e^{c(\beta)t}, \ c(\beta) = c_V(\beta),$$

and $(A\dot{-}V)_p$ is an extension of the algebraic difference $A_p - V_p$ defined on its natural domain which (this is not forbidden) can be equal to $\{0\}$ only.

Let $(A-V)_{p,min}$ be the closure of $A_p - V_p$. In order to prove the equality

$$(A-V)_{p,min} = (A\dot{-}V)_p \quad \text{for some} \quad p \in [t(\beta), t'(\beta)]$$

it suffices to show that the subspaces

$$\mathcal{D}(A_p) \cap \mathcal{D}(V_p) \quad \text{and} \quad (\lambda + A_p - V_p)[\mathcal{D}(A_p) \cap \mathcal{D}(V_p)]$$

are dense in $L^p(M,\mu)$. Here $\lambda > 0 \vee c(\beta)$ so that automatically $-\lambda \in \rho((A\dot{-}V)_p)$. The first requirement is satisfied if

$$V \in L^p(M,\mu)$$

since $\mathcal{E} := (1+A)^{-1}[L^1 \cap L^\infty] \subset \mathcal{D}(A) \cap L^\infty$ and dense in $L^p(M,\mu)$ and the second requirement reduces to establishing the implication

$$(\lambda > 0 \vee c(\beta), \ u = \operatorname{Re} u \in L^{p'}, \ \langle(\lambda + A_p - V_p)v, u\rangle = 0 \ \forall v \in \mathcal{E}) \Longrightarrow u = 0 \tag{8.2}$$

in order to conclude $(A\dot{-}V)_p = (A_p - V_p \restriction \mathcal{E})^{\sim}_{L^p \to L^p}$.

Setting in (8.2) $v = (\lambda + A)^{-1}g$, $g \in L^1 \cap L^\infty$ and taking into account that $Vu \in L^1$ we have

$$\langle g, u\rangle = \langle(\lambda + A)^{-1}g, Vu\rangle = \langle g, (\lambda + A_1)^{-1}Vu\rangle.$$

Since g is an arbitrary element from $L^1 \cap L^\infty$ one obtains

$$u = (\lambda + A_1)^{-1}Vu \ (\text{ equality in } \ L^1). \tag{8.3}$$

Therefore $u \in L^{p'} \cap \mathcal{D}(A_1)$ and

$$(\lambda + A_1)u = Vu \ (\text{ equality in } \ L^1). \tag{8.4}$$ [2]

Assume now that

$$Vu \in L^q \ \text{ for some } \ q > 1. \tag{8.5}$$

This can be reached, for example, in two ways:

[2]The reader can immediately notice that if $V \in \Phi_1(A)$ then (8.4) $\Longrightarrow u = 0$ for the appropriate λ. However we do not supp ose now any relative operator boundedness of $V_{\hat{p}}$ with respect to $A_{\hat{p}}$ for some $\hat{p}$.

I. to assume

$$(\lambda + A_1)^{-1}[L^1] \subset L^r \quad \text{for some} \quad r > p'$$

so that according to (8.3) by Hölder's inequality $Vu \in L^q$, $1/q = 1/p + 1/r$;

II. to assume

$$V \in L^k \quad \text{for some} \quad k > p$$

so that $Vu \in L^q$, $1/q = 1/k + 1/p'$.

The first case is reached when the operator A satisfies the Sobolev imbedding property

$$\|f\|_{2j}^2 \leqslant c_j t_A[f], \; f \in Q(A), \; j = \nu/\nu - 2 \;\text{ for some }\; \nu > 2 \;\text{ and in addition }\; \nu < 2p$$

(see Theorem 7.1). In the "simplest" case $(M, \mu) = (\mathbb{R}^d, dx)$, $d \geqslant 3$, $A = -\Delta$, assumption I) is equivalent to the inequality $d < 2p$ and holds, for example, for $d = 3$ and $p = 2$. The second case *does not depend on* the Sobolev imbedding property of the operator A.

Thus in both indicated cases we can multiply both sides of the equality (8.4) by $|u|^{z-1} \cdot \operatorname{sgn} u$, where $z = 1 + \frac{p'}{q'}$ ($u \in L^1 \cap L^{p'} \subset L^z$). After integration we have

$$\lambda \|u\|_z^z + \langle A_q u, |u|^{z-1} \cdot \operatorname{sgn} u\rangle = \langle V|u|^{z/2}, |u|^{z/2}\rangle.$$

At this point we are exactly in the conditions of Theorem 2.2 which immediately gives (just take $\varphi(t) = t^{z-1}$, $b = 0$):

$$|u|^{z/2} \in Q(A) \quad \text{and} \quad \langle A_q u, |u|^{z-1} \cdot \operatorname{sgn} u\rangle \geqslant 4\frac{z-1}{z^2}\|A^{1/2}|u|^{z/2}\|_2^2.$$

Using the condition $V \in PK_\beta(A)$ implies

$$(\lambda - c(\beta))\|u\|_z^z + (4\frac{z-1}{z^2} - \beta)\|A^{1/2}|u|^{z/2}\|_2^2 \leqslant 0$$

and if $\beta \leqslant 4\frac{z-1}{z^2}$ then $\|u\|_z^z = 0$, therefore $u = 0$.

In the first case the inequality $\beta \leqslant 4\frac{z-1}{z^2}$ is valid if $1/\nu > (p' - \frac{2\sqrt{1-\beta}}{1+\sqrt{1-\beta}})/2p'$, in particular, if $p = 2$, $\nu < 2(1+\sqrt{1-\beta})$ or if $\beta < 1-(\nu-2/2)^2$. Note that for $(M, \mu) = (\mathbb{R}^3, dx)$ $V \in L^2(\mathbb{R}^3) \implies V \in PK_0(-\Delta)$ and therefore the latter requirement is trivially satisfied.

In the second case $z = t(\beta)$ and $k = k_p(\beta) = \frac{1+\sqrt{1-\beta}}{1+\sqrt{1-\beta}-2/p'}$.

Thus we have proved the following

In fact in the second case one can slightly modify the above arguments like in the proof of Theorem 7.4(iii) to obtain a more precise result.

Theorem 8.2 *Let* $V \in PK_\beta(A) \cap \overset{\circ}{L}{}^{k_p(\beta)} \cap L^p$ *for some* $\beta \in (0,1)$ *and* $p \in (t(\beta), t'(\beta))$,

where $\overset{\circ}{L}{}^{s} := \{f \in \bigcap_{s-\varepsilon \leqslant q < s} L^q;\ \lim_{\varepsilon \downarrow 0} \varepsilon \cdot \|f\|_{s-\varepsilon} = 0\}$.

Then $(A_p - V_p \restriction \mathcal{E})^{\sim}_{L^p \to L^p} = (A \dot{-} V)_p$.

Proof Define q_ε and z_ε by $1/q_\varepsilon = 1/(k-\varepsilon) + 1/p'$, $z_\varepsilon = z - \varepsilon p'/(k-\varepsilon)k$, $\varepsilon > 0$ is small enough (further ε will tend to zero) ,where $k = k_p(\beta)$, $z = t(\beta)$. Just as above we have

$$\lambda\|u\|_{z_\varepsilon}^{z_\varepsilon} + \langle A_{q_\varepsilon}u, |u|^{z_\varepsilon - 1}\text{sgn } u\rangle = (1-\varepsilon_1)\langle V|u|^{z_\varepsilon}\rangle + \varepsilon_1\langle V|u|^{z_\varepsilon}\rangle,$$

where ε_1 is chosen from the equation $(1-\varepsilon_1)\beta = 4\frac{z_\varepsilon - 1}{z_\varepsilon^2}$. Then $\varepsilon_1 = \Gamma\varepsilon + o(\varepsilon)$ for some constant $\Gamma = \Gamma(p,\beta)$. Using Theorem 2.2 and the condition $V \in PK_\beta(A)$ we get

$$(\lambda - c(\beta) + \varepsilon\Gamma c(\beta))\|u\|_{z_\varepsilon}^{z_\varepsilon} \leqslant \varepsilon\Gamma\|V|u|^{z_\varepsilon}\|_1 \leqslant \varepsilon\|V\|_{k-\varepsilon}\cdot\Gamma\|u\|_{p'}^{z_\varepsilon}.$$

Since $u \in L^z \cap L^{z_\varepsilon}$ and $r \mapsto \|f\|_r$ is a continuous mapping (Proposition 1.1) one obtains

$$(\lambda - c(\beta))\|u\|_z^z \leqslant 0 \quad \text{thus} \quad u = 0.\square$$

Remark 8.4 $\overset{\circ}{L}{}^s + L^\infty \supset L^{s,\infty}$. If $(M,\mu) = (\mathbb{R}^d, dx)$, $d \geqslant 5$ and $V = d(d-4)/4|x|^2$ then for $\beta = d(d-4)(d-2)^{-2}$ $k_2(\beta) = \frac{d}{2}$. Thus the abstract result obtained is sharp!

Remark 8.5 Theorem 8.2 was proved in [LPSe] using Theorem 6.3. Here we followed the original idea [Se3].

Now we would like to answer the following question: is it possible to set $\beta = 0$ and $k_p(0) = p$ in the conditions of Theorem 8.2? In other words, is the condition $V \in PK_0(A) \cap L_+^p$ sufficient for the equality $(A_p - V_p \upharpoonright \mathcal{E})^\sim_{L^p \to L^p} = (A \dot{-} V)_p$ to hold? The answer is positive only in the case $(1+A)^{-1}L^1 \subset L^r$, $r > p'$, but in general the answer is negative. The corresponding example can be easily seen from the case $(M,\mu) = (\mathbb{R}^d, dx)$, $d = 4$, $p = 2$, $A = -\Delta$. Apropos, in the latter case $(1-\Delta)^{-1}L^1 \subset L^{2,\infty}$ and $(1-\Delta)^{-1}L^1 \not\subset L^2$.

So this question should be formulated somewhat differently: what are the additional assumption to $V \in PK_0(A) \cap L_+^p$, for some $p \in (1,\infty)$, which are sufficient for the equality $(A_p - V_p \upharpoonright \mathcal{E})^\sim_{L^p \to L^p} = (A \dot{-} V)_p$ to hold? Can one omit the requirement (8.5) which was crucial for the approach above?

Again, just as above, it is enough to find $\lambda \in \mathbb{R}_+^1 \cap \rho(-(A \dot{-} V)_p)$ such that from $u = \text{Re } u \in L^{p'}$ and $\langle(\lambda + A_p - V_p)v, u\rangle = 0$ $\forall v \in \mathcal{E}$ it follows that $u = 0$. Again we deduce that $u \in \mathcal{D}(A_1) \cap L^{p'}$ and $(\lambda + A_1)u = Vu$ (equality in L^1). According to Theorem 2.2 one can try to carry out the idea of the proof of Theorem 8.2 with the only knowledge $Vu\varphi(|u|) \in L^1$ and $\varphi(\cdot)$ is such that $\lim_{t\to\infty}\varphi(t)t^{-\gamma} = 0$ $\forall\gamma > 0$. For example, $\varphi(t) = \ln_+ t$. Then the condition $V\ln_+ V \in L^p$ is sufficient to conclude $Vu\varphi(|u|) \in L^1$.

Indeed, if $F \equiv \ln_+ V$ and $\alpha > p-1$ then

$$\begin{aligned} V|u|\ln_+|u| = (\mathbb{1}_{\alpha F \geqslant \varphi} + \mathbb{1}_{\alpha F \leqslant \varphi})V|u|\ln_+|u| &\leqslant \alpha VF|u| + |u|^{1+1/\alpha}\ln_+|u| \\ &\leqslant \alpha VF|u| + |u|^{p'}\sup_{t>1} t^{1+1/\alpha - p'}\cdot\ln t, \end{aligned}$$

hence $V|u|\ln_+|u| \in L^1$. Then

$$\lambda\langle |u|, a\rangle + \langle A_1 u, a\cdot \operatorname{sgn} u\rangle = \langle V|u|, a\rangle, \ a \equiv a(|u|), \tag{8.6}$$

$$\text{where } a(t) = \begin{cases} t & \text{if } t \leqslant 1, \\ 1+\ln t & \text{if } t \geqslant 1. \end{cases}$$

Subsequently we will estimate the right-hand side of (8.6) with the conclusion that $u = 0$.

Indeed, suppose in addition, that

$$e^{-tA_1}1 = 1 \ \text{ and } \ FV \in PK_\beta(A) \ \text{ for some } \ \beta < 4(j-1)$$

and obviously we are considering the case $p' \geqslant j$. Dividing both sides of the equation $(\lambda + A_1)u = Vu$ by $\|u\|_j$ one can assume that $\|u\|_j = 1$.

Let $a_n(t) := a(t\wedge n)$, $a_n^c(t) := a(t) - a_n(t)$, $n \geqslant 1$.

1°. $a(t) = a_1(t) + a_1^c(t)$ and by Theorem 2.2

$$g_{a_1}(t) = t - t\vee 1, \ c_{a_1} = 1; \ g_{a_1^c} = \sqrt{t} - \sqrt{t}\wedge 1, \ c_{a_1^c} = 4;$$

$$\langle A_1 u, a\cdot \operatorname{sgn} u\rangle \geqslant \|A^{1/2} g_{a_1}\|_2^2 + 4\|A^{1/2} g_{a_1^c}\|_2^2. \tag{8.7}$$

2°. $\langle V|u|, a\rangle = \langle V|u|, a_R\rangle + \langle V|u|, a_R^c\rangle$, $R \geqslant 16$. Define $g_1(t) := \sqrt{2}a_1(t)$, $g_2(t) := \sqrt{R} g_{a_1^c}(t)$. Using the condition $e^{-tA_1}1 = 1$ we have by Theorem 2.2 $g_1 \in Q(A)$ and $\|A^{1/2}a_1\|_2 = \|A^{1/2}g_{a_1}\|_2$. Moreover, since $t\cdot a_R(t) \leqslant g_1^2(t) + g_2^2(t)$ then (taking into account the condition $V \in PK_0(A)$)

$$\langle V|u|, a_R\rangle \leqslant \langle Vg_1^2\rangle + \langle Vg_2^2\rangle \leqslant \varepsilon(\|A^{1/2}g_1\|_2^2 + \|A^{1/2}g_2\|_2^2) + c_V(\varepsilon)(\|g_1\|_2^2 + \|g_2\|_2^2)$$

and therefore

$$\langle V|u|, a_R\rangle \leqslant \varepsilon(2\|A^{1/2}g_{a_1}\|_2^2 + \|A^{1/2}g_2\|_2^2) + c_V(\varepsilon)(\|g_1\|_2^2 + \|g_2\|_2^2) \ \forall \varepsilon > 0. \tag{8.8}$$

3°. Choose $\alpha > 1/(j-1)$. Let us estimate $\langle V|u|, a_R^c\rangle$.

$$\langle V|u|, a_R^c\rangle = \langle (\mathbb{1}_{\alpha F \geqslant 1+\varphi} + \mathbb{1}_{\alpha F < 1+\varphi})V|u|, a_R^c\rangle \leqslant \alpha\langle FV|u|\frac{a_R^c}{1+\varphi}\rangle + e^{1/\alpha}\langle |u|^{1+1/\alpha} a_R^c\rangle.$$

Using the inequalities

$$t\cdot a_R^c(t)(1+\varphi(t))^{-1} = t\frac{\ln_+ \frac{t}{R}}{\ln et} \leqslant (\sqrt{t} - \sqrt{t}\wedge 1)^2,$$

$$t^{1+1/\alpha}\cdot a_R^c(t) = t^{1+1/\alpha}\ln_+ \frac{t}{R} \leqslant t^j \cdot \mathbb{1}_{t \geqslant R} \sup_{t > R} \frac{\ln \frac{t}{R}}{t^{j-1-1/\alpha}},$$

$$t\mathbb{1}_{t\geqslant R} \leqslant 2(\sqrt{t} - \sqrt{t}\wedge 1)^2$$

and setting $k(R) = 2^j e^{1/\alpha} \sup_{t>R} \frac{\ln \frac{t}{R}}{t^{j-1-1/\alpha}}$, we have

$$\begin{aligned}
&\langle V|u|, a_R^c\rangle \leqslant \alpha\langle FV(\sqrt{|u|} - \sqrt{|u|}\wedge 1)^2\rangle + k(R)\langle(\sqrt{|u|} - \sqrt{|u|}\wedge 1)^{2j}\rangle \\
\equiv\ &\alpha\langle FV g_{a_1^c}^2\rangle + k(R)\langle g_{a_1^c}^{2j}\rangle.
\end{aligned}$$

The conditions $FV \in PK_\beta(A)$ and $\|u\|_j = 1$ imply

$$\langle V|u|, a_R^c\rangle \leqslant \alpha\beta\|A^{1/2}g_{a_1^c}\|_2^2 + k(R)\|g_{a_1^c}\|_{2j}^2 + \alpha c_{FV}(\beta)\|g_{a_1^c}\|_2^2. \tag{8.9}$$

4°. Combine (8.8) and (8.9). Using $g_{a_1^c}^2(t) \leqslant t\cdot a(t)$ we get

$$\begin{aligned}
\langle V|u|, a\rangle &\leqslant 2\varepsilon\|A^{1/2}g_{a_1}\|_2^2 + (R\varepsilon + \alpha\beta)\|A^{1/2}g_{a_1^c}\|_2^2 \\
&+k(R)\|g_{a_1^c}\|_{2j}^2 + [c_V(\varepsilon)(2+R) + \alpha\ c_{FV}(\beta)]\langle|u|, a\rangle.
\end{aligned} \tag{8.10}$$

From (8.6) and (8.7) we obtain

$$\lambda\langle|u|, a\rangle + \|A^{1/2}g_{a_1}\|_2^2 + 4\|A^{1/2}g_{a_1^c}\|_2^2 \leqslant \langle V|u|, a\rangle.$$

Substitution of (8.10) implies

$$\begin{aligned}
&[\lambda - c_V(\varepsilon)(2+R) - \alpha\ c_{FV}(\beta)]\langle|u|, a\rangle + (1-2\varepsilon)\|A^{1/2}g_{a_1}\|_2^2 \\
&+(4-\alpha\beta - R\varepsilon)\|A^{1/2}g_{a_1^c}\|_2^2 \leqslant k(R)\|g_{a_1^c}\|_{2j}^2 \leqslant k(R)c_j\|A^{1/2}g_{a_1^c}\|_2^2.
\end{aligned} \tag{8.11}$$

Let us choose the parameters: first choose $\alpha > j' - 1$ such that $4 > \alpha\beta$, then choose R such that $4 - \alpha\beta > k(R)c_j$ (c_j is the constant of Sobolev imbedding). Since $k(R) \downarrow 0$ $(R \to \infty)$ then such R can be chosen. Finally fix $\varepsilon \in (0,1)$ by the condition $4 - \alpha\beta - k(R)c_j > R\varepsilon$. Set $\lambda = 1 + c_V(\varepsilon)(2+R) + \alpha\ c_{FV}(\beta)$.

$$(8.11) \Longrightarrow u = 0.$$

Thus we have proved the following criterion of strong uniqueness.

Theorem 8.3 *Let A be a symmetric Markov generator satisfying the Sobolev imbedding property $\|f\|_{2j}^2 \leqslant c_j t_A[f]$, $j > 1$, $j' \geqslant p > 1$ and $e^{-tA_1}1 = 1$. Suppose that $V \in PK_0(A) \cap L_+^p$, $V\ln_+ V \in PK_\beta(A) \cap L^p$ and $\beta < 4(j-1)$. Then*

$$(A_p - V_p \upharpoonright \mathcal{E})^{\sim}_{L^p\to L^p} = (A\dot{-}V)_p,$$

where $\mathcal{E} = (1+A)^{-1}[L^1 \cap L^\infty]$.

The method described above in the case of Schrödinger operator $-\Delta + V$ leads to the following deeper result.

[Fu] Fukushima, M.: *Dirichlet Forms and Markov Processes*, North-Holland, Amsterdam-Oxford-New York 1980.

[G] Goldstein, J.: *Semigroups of Linear Operators and Applications,* Oxford University Press, New York 1985.

[HSl] Herbst, I.W. and Sloan, A.D. : Perturbation of translation invariant positivity preserving semigroups on $L^2(\mathbb{R}^N)$, *Trans. Amer. Math. Soc.* **233** (1978), 325–360.

[IT] Ionescu Tulcea, A. and Ionescu Tulcea, C.: *Topics in the Theory of Lifting*, Springer-Verlag, Berlin-Heidelberg-New York 1969.

[Is] Ishikawa, M.: Admissible and regular potentials for positive C_0-semigroups and application to heat semigroups, *Semigroup Forum* **48** (1994), 96–106.

[K1] Kato, T.: *Perturbation Theory for Linear Operators.* 2nd edition, Springer-Verlag , Berlin-Heidelberg- New York 1980.

[K2] Kato, T. : Trotter's product formula for an arbitrary pair of self-adjoint contraction semigroups, in *Advance in Math. Suppl. Studies*, vol. 3, Academic Press, New York 1978, 185–195.

[K3] Kato, T.: L^p-theory of Schrödinger operators, in *Aspects of Positivity in Functional Analysis*, North-Holand, Amsterdam 1986, 63–78.

[KoSe1] Kovalenko, V.F. and Semenov, Yu.A. : Semigroups generated by elliptic operators of second order, in *Methods of Functional Analysis in Problems of Math.Phys.* (Russ.), Inst. Math. Ukrain. Acad. Sci., Kiev 1987, 17–36.

[KoSe2] Kovalenko, V.F. and Semenov, Yu.A: On L^p-theory of Schrödinger semigroups.I, *Ukrain. Math. J.* **41** (1989), 273–278.

[KoSe3] Kovalenko, V.F. and Semenov, Yu.A.: On L^1-theory of parabolic semigroups, *Ukrain. Math. J.* **41** (1989), 1375–1389.

[KoSe4] Kovalenko, V.F. and Semenov, Yu.A: C_0-semigroups in $L^p(\mathbb{R}^d)$ and $\hat{C}(\mathbb{R}^d)$ spaces generated by the differential expression $\Delta + b \cdot \nabla$, *Theory Probab. Appl.* **35** (1990), 443–453.

[KrZPuSo] Krasnoselskii, M.A., Zabreiko, P.P., Pustylnik, E.I. and Sobolevskii, P.E.: *Integral Operators in Spaces of Summable Functions*, Noordhoff Inter. Publish., Leyden 1976.

[L] Liskevich, V.A.: On C_0-semigroups generated by elliptic second order differential expressions on L^p-spaces *Differential and Integral Equations*, to appear.

[LP] Liskevich, V.A. and Perelmuter, M.A.: Summability properties of solutions of second order elliptic equations (Russ.), *Mat. Zametki* **43** (1988), 337–345.

[LP1] Liskevich, V.A. and Perelmuter, M.A.: Analyticity of submarkovian semigroups, *Proceedings AMS* **123** (1995), 1097–1104.

[LPSe] Liskevich, V.A., Perelmuter, M.A. and Semenov, Yu.A.: Form-bounded perturbations of generators of submarkovian semigroups, Preprint.

[LSe1] Liskevich, V.A. and Semenov, Yu.A.: Some inequalities for submarkovian generators and their applications to the perturbation theory, *Proceedings AMS* **119** (1993), 1171–1177.

[LSe2] Liskevich, V.A. and Semenov, Yu.A.: One criterion of the essential self-adjointness of Schrödinger operators, Preprint.

[LSe3] Liskevich, V.A. and Semenov, Yu.A.: Dirichlet operators: a priori estimates and the uniqueness problem, *J. Funct. Anal.* **109** (1992), 199–213.

[LSe4] Liskevich, V.A. and Semenov, Yu.A.: Some problems on submarkovian semigroup, in *Operator Theory: Advances and Applications*, vol.70, Birkhäuser 1994, 143–148.

[MaR] Ma, Z.-M. and Röckner, M.: *Introduction to the Theory of (Non-symmetric) Dirichlet Forms*, Springer-Verlag, Berlin-Heidelberg-New York-London-Paris-Tokyo 1992.

[MSo] Maz'ja, V.G. and Sobolevskii, P.E.: On generators of operators semigroups (Russ.), *Uspechi Mat. Nauk* **17** (1962), 151–154.

[M] Meyer, P.A.: *Probability and potentials*, Blaisdel Publ. Comp. 1966.

[N] Nash, J.: Continuity of solutions of parabolic and elliptic equations, *Amer. J. Math.* **80** (1958), 931–954.

[O] Oinarov, R.: On the separability of the Schrödinger operator in the space of summability functions (Russ.), *Dokl. Akad. Nauk SSSR* **285** (1985), 1062–1064.

[Ou] Ouhabaz E.-M.: L^∞-contractivity of semigroups generated by sectorial forms, *J. London Math. Soc.* **2** (46) (1992), 529–542.

[PSe] Perelmuter, M.A. and Semenov, Yu.A.: On decoupling of finite singularities in the scattering theory for the Schrödinger operators with a magnetic field, *J. Math. Phys.***22**(3) (1981), 521–533.

[RSi1] Reed, M. and Simon, B.: *Methods of Modern Mathematical Physics II: Fourier analysis. Self-adjointness*, Academic Press, New York 1975.

[RSi2] Reed, M. and Simon, B.: *Methods of Modern Mathematical Physics IV: Analysis of operators*, Academic Press, New York 1978.

[Rö] Röckner, M.: Potential theory on non-locally compact spaces via Dirichlet forms, in *Potential theory*, de Gruyter & Co, Berlin–New York 1991, 107–126.

[RöZ] Röckner, M. and Zhang, T.-S.: Uniqueness of generalized Schrödinger operators and applications II, *J. Funct. Anal.* **119** (1994), 455–467.

[Se1] Semenov, Yu.A.: Schrödinger operators with L^p_{loc}-potentials, *Commun. Math. Phys.* **53** (1977), 277–284.

[Se2] Semenov, Yu.A., unpublished.

[Se3] Semenov, Yu.A.: One criterion of essential self-adjointness of the Schrödinger operator with negative potential, Preprint (Russ.), 1987.

[Se4] Semenov, Yu.A.: Two classes of perturbations and uniqueness of minimal and maximal Schrödinger operators, Preprint.

[Se5] Semenov, Yu.A.: On the theory of the operator $L_t = \frac{\partial}{\partial t} - \Delta + b \cdot \nabla + V$, to appear.

[Si] Simon, B.: *Quantum Mechanics for Hamiltonians Defined as Quadratic Forms*, Princeton University Press, Princeton, N.J. 1971.

[So] Song, S.: Maximality, regularity and perturbation of Dirichlet forms, Preprint.

[Sta] Stampacchia, G.: Formes bilinéaires coercitives sur les ensembles convexes, *C. R. Acad. Sci. Paris* **258** (1964), 4413–4416.

[St] Stein, E.M.: *Topics in Harmonic Analysis related to Littlewood–Paley Theory*, Princeton Univ. Press , Princeton, N.J. 1970.

[StW] Stein, E.M. and Weiss, G.: *Introduction to Fourier Analysis on Euclidian Spaces*, Princeton Univ. Press , Princeton, N.J. 1971.

[StoVo] Stollmann, P. and Voigt, J.: Perturbation of Dirichlet forms by measures, Preprint 1992.

[S] Stroock, D.W.: *An Introduction to the Theory of Large Deviations* , Springer-Verlag, New York 1984.

[T] Takeda, M.: The maximum Markovian self-adjoint extension of generalized Schrödinger operators, *J. Math. Soc. Japan* **44** (1992), 113–130.

[V] Varopoulos, N.Th.: Hardy-Littlewood theory for semigroups, *J. Funct. Anal.* **63** (1985), 240–260.

[VSC] Varopoulos, N.Th., Saloff-Coste, L. and Coulhon, Th.: *Analysis and Geometry on Groups* , Cambridge Univ. Press, Cambridge 1992.

[Vo] Voigt, J.: Absorption semigroups, their generators and Schrödinger semigroups, *J. Funct. Anal.* **67** (1986), 167–205.

Wavelet Analysis of Partial Differential Operators

Matthias Holschneider

Abstract

This paper is a self-contained introduction to continuous wavelet analysis. We introduce the general concepts that allow the characterization of various function spaces and microlocal regularity classes through wavelet transforms. In particular, we shall show how the wavelet transform may be used to define directional regularity in distributions. This concept is well suited for domains with cusps. We construct an algebra of wavelet pseudo-differential operators based on Calderón Zygmund operator-valued symbols for which we develop a symbolic calculus modulo smoothness. This allows us to prove some elliptic regularity for the directional microlocal classes in domains with cusps. We discuss, furthermore, how diffeomorphisms act in wavelet space. This allows us to discuss the transformation behavior of the directional microlocal classes under local coordinate changes. In the last part we use wavelet techniques to derive a formula for the essential spectral radius of some Fourier integral operators. We prove that the essential spectral radius is linked to the small scale behavior of the operator.

1 Introduction

In this paper we want to outline the basic principles of the analysis of partial differential operators through continuous wavelet transforms. This text is intended to serve two purposes. For the reader not familiar with the field of wavelet analysis it should help to understand the basic principles of this flexible analysis tool. In particular, this text should ease the access to the more specialized literature as, e.g., the books of Y. Meyer [Mey90] and R. Coifman and Y. Meyer [CM90]. The reader familiar with wavelet techniques, might find some interest in the construction of directional microlocal classes and the generalization of the Calderón Zygmund calculus of smoothly varying singular integrals [CZ57]. In this paper we will exclusively focus on continuous wavelet analysis and we refer the reader interested in the construction and application of wavelet bases to the book of I. Daubechies [Dau92] or the book of C. Chui [Chu92]. We will not consider any question of numerical applications of wavelet techniques to the solution of partial differential equations. These aspects are for instance treated in [DPS95], [Jaf92], [Tch95].

Wavelet analysis has been a field of active research for several years. The basics of wavelet analysis go back to Littlewood and Paley but it was the only recently that the fruitful collaboration between a physicist, A. Grossmann and an engineer J. Morlet, [GM84], brought into light this flexible technique of analysis. Since then, wavelet techniques have found many application fields ranging from acoustics [KMG87], [Sar94] to signal analysis in general [Fla93], [Tor95] or quantum mechanics [Pau84], to cite only a few. For an introductory overwiew see also [CGT89].

We briefly outline the central ideas of wavelet analysis and its application to partial differential operators. Let $g \in \mathcal{S}(\mathbb{R}^n)$, the class of *Schwartz* of rapidly decreasing functions. In addition suppose that g has all moments vanishing

$$\int dx\, x^m g(x) = 0,$$

for all multi-indices m. Then, the wavelet transform of $s \in L^p(\mathbb{R}^n)$ with respect to g is defined as the following convolutions

$$\mathcal{W}_g s(b,a) = \mathcal{W}[g,s](b,a) = (\tilde{g}_a * s)(b) = \int dx \frac{1}{a^n} \overline{g}\left(\frac{x-b}{a}\right) s(x), \qquad (1.1)$$

with $a > 0$ and $b \in \mathbb{R}^n$.

Here we have introduced the following notations, that we shall use in the sequel

$$\tilde{g}(t) = \overline{g}(-t), \; g_a = g(\cdot/a)/a^n, \; g_{b,a} = g_a(\cdot - b).$$

The wavelet transform thus maps functions over the n-dimensional real space to functions over the open half-space $\mathbb{H}^n = \{(b,a) : b \in \mathbb{R}^n, a > 0\}$.

From the definition it is clear that the wavelet transform is a sort of mathematical microscope whose position is fixed by b and whose enlargement is given by $1/a$ or to put it differently $\mathcal{W}[g,s](b,a)$ is obtained by "looking at s at position b and at scale a". As a general statement one can say that local regularity of s is mirrored in a certain speed of decay of $\mathcal{W}_g s$. So, for instance, a uniform (in b) decay of $O(a^\infty)$ as $a \to 0$, of the wavelet coefficients is equivalent to C^∞ regularity of s. More quantitative information is available. So a uniform decay of $O(a^\alpha)$ with $\alpha \in (0,1)$ is equivalent to $s \in \Lambda^\alpha$, the space of Hölder continuous functions of exponent α. Many more function spaces can be described in terms of localization of wavelet coefficients.

As the Fourier transform shows, the wavelet analysis is also a position-frequency analysis

$$\mathcal{W}[g,s](b,a) = \frac{1}{(2\pi)^n} \int d\xi \overline{\widehat{g}}(a\xi)\, e^{ib\xi}\, \widehat{s}(\xi).$$

However, in some sense we have

$$a \leftrightarrow 1/\,|\xi|\,.$$

The frequency variable is "degenerate", since it is direction invariant. However, in elliptic situations, this is no handicap, since elliptic operators themselves are in some sense direction independent at small scale.

The applications to partial differential equations can be easily understood if we write the Laplace operator in wavelets using integration by parts

$$\mathcal{W}[g,\Delta s](b,a) = a^{-2}\, \mathcal{W}[\Delta g, s](b,a).$$

Therefore we can factorize

$$\Delta = M\, CZO,$$

where the first operator is a multiplication operator (here multiplication by a^{-2}) and the action of the second operator *CZO* consists in changing from one wavelet g to the wavelet Δg. As it turns out this class of operators are precisely the so called Calderón Zygmund operators. The important point is that these operators are continuous in all Banach spaces of functions that can be described through the localization of wavelet coefficients. Moreover, *CZO* has a realization as singular integral and it corresponds to a symbol $\sigma(x,\xi)$ of order 0. However, it satisfies the so-called forbidden estimates

$$\left|\partial_x^\alpha \partial_\xi^\beta \sigma(x,\xi)\right| \le O(1)\,|\xi|^{|\alpha|-|\beta|}\,.$$

This kind of factorization into a singular integral operator and a diagonal part is typical for the analysis of elliptic operators through wavelets.

A second way to approach pseudo-differential operators through wavelet techniques is the following. In the pseudo-differential calculus, the relation between the symbol σ and the operator $\psi[\sigma]$ is neatly expressed through

$$\psi[\sigma]e^{ix\xi} = e^{ix\xi}\sigma(x,\xi).$$

We will construct algebras of pseudo-differential operators that behave in an analogous way, if we replace the exponential function by some wavelet as test function. The symbols $\sigma(b,a)$ we consider will be operator-valued taking their values in the space of *CZO*-operators. With the notation $g_{b,a} = g([\cdot - b]/a)/a^n$ the relation between the symbol σ and the operator $op[\sigma]$ can be expressed through

$$op[\sigma]g_{b,a} = (\sigma(b,a)g)_{b,a}.$$

However, for the wavelet calculus, this holds modulo smoothness only. As we will see, this algebra allows a symbolic calculus of arbitrary order if some differential inequalities are imposed on σ.

In the last section, we will use the wavelet calculus to derive a formula for the essential spectral radius of some operators of Fourier integral type.

2 Wavelet analysis, the basic formulas

For the convenience of the reader we shall list here the basic formulas of wavelet analysis. We limit ourselves to formal expressions. They actually have a precise meaning when we consider the wavelet analysis in $\mathcal{S}_0(\mathbb{R}^n)$ or $\mathcal{S}_0'(\mathbb{R}^n)$ (see below). Actually we shall prove some of them in the next section and this section serves as a reminder for the hurry reader.

2.1 The formulas for wavelet analysis and synthesis

Let f be a complex-valued function over $\mathbb{R}^n$. Let g be another such function. The wavelet transform of f with respect to the analyzing wavelet g is defined through

$$\mathcal{W}_g f(b,a) = \int_{\mathbb{R}^n} dx \, \frac{1}{a^n} \overline{g} \left(\frac{x-b}{a} \right) f(x), \quad b \in \mathbb{R}^n, a > 0.$$

We also write $\mathcal{W}[g,f](b,a)$ instead of $\mathcal{W}_g f(b,a)$. Here $b \in \mathbb{R}^n$ is a position parameter and $a \in \mathbb{R}_+$ is a scale parameter. The wavelet transform of a function over $\mathbb{R}^n$ is thus a function over the position scale half-space $\mathbb{H}^n = \mathbb{R}^n \times \mathbb{R}_+$.

If we introduce the dilation (D_a) and translation operators (T_b)

$$D_a s(x) = s(x/a)/a^n, \quad T_b s(x) = s(x-b),$$

then we may also write the wavelet transform as a family of scalar products

$$\mathcal{W}_g f(b,a) = \langle g_{b,a}, f \rangle, \quad g_{b,a} = T_b D_a g,$$

or as a family of convolutions indexed by a scale parameter

$$\mathcal{W}_g f(b,a) = (\tilde{g}_a * f)(b), \quad \tilde{g}_a = D_a \tilde{g}, \quad \tilde{g}(x) = \overline{g}(-x).$$

Here by $\langle \cdot, \cdot \rangle$ we always denote the sesqui-linear pairing, antilinear in the first argument. The convolution product is defined as usual

$$(s * r)(x) = \int_{\mathbb{R}^n} dy \, s(x-y) r(y) = (r * s)(x).$$

If we introduce the Fourier transform

$$\widehat{s}(k) = \int_{\mathbb{R}^n} dx e^{-ikx} s(x), \quad s(x) = \frac{1}{(2\pi)^n} \int_{\mathbb{R}^n} dk e^{ikx} \widehat{s}(k),$$

then the wavelet transform may also be written as

$$\mathcal{W}_g f(b,a) = \frac{1}{(2\pi)^n} \int_{\mathbb{R}^n} dk\, \overline{\widehat{g}(ak)}\, e^{ibk}\, \widehat{f}(k). \tag{2.1}$$

The wavelet synthesis $\mathcal{M}$ maps functions over the position - scale half-space to functions over $\mathbb{R}^n$. Let $r = r(b,a)$ be a complex-valued function over $\mathbb{H}^n$ and h a function over $\mathbb{R}^n$. Then, the wavelet synthesis of r with respect to the synthesizing wavelet h is defined as

$$\mathcal{M}_h r(x) = \int_{\mathbb{H}^n} \frac{dbda}{a}\, r(b,a)\, \frac{1}{a^n}\, h\left(\frac{x-b}{a}\right).$$

In Fourier space we rather have

$$\mathcal{F}\mathcal{M}_h r(k) = \int_{\mathbb{H}^n} \frac{dbda}{a}\, r(b,a)\, \widehat{h}(ak)\, e^{-ibk}.$$

2.2 Relation between $\mathcal{W}$ and $\mathcal{M}$

We now list some relations between $\mathcal{W}$ and $\mathcal{M}$. The wavelet synthesis is the adjoint of the wavelet transform, both with respect to the same wavelet, $\mathcal{W}_g^* = \mathcal{M}_g$

$$\int_{\mathbb{H}^n} \frac{dbda}{a} \overline{\mathcal{W}_g s(b,a)} r(b,a) = \int_{\mathbb{R}^n} dx\, \overline{s(x)}\, \mathcal{M}_g r(x).$$

Or, if we write for the scalar product of two functions over the half-space

$$\langle s, r\rangle_{\mathbb{H}^n} = \int_{\mathbb{H}^n} \frac{dbda}{a} \overline{s}(b,a)\, r(b,a),$$

then

$$\langle s, \mathcal{W}_g r\rangle_{\mathbb{H}^n} = \langle \mathcal{M}_g s, r\rangle_{\mathbb{R}^n}.$$

The combination $\mathcal{M}_h\mathcal{W}_g$ reads in Fourier space

$$\mathcal{M}_h\mathcal{W}_g : \widehat{s}(k) \mapsto m_{g,h}(k)\widehat{s}(k), \quad m_{g,h}(k) = \int_0^\infty \frac{da}{a} \overline{\widehat{g}(ak)}\, \widehat{h}(ak).$$

Note that the Fourier multiplier $m_{g,h}$ depends only on the direction of k, $m_{g,h} = m_{g,h}(k/|k|)$. This is because the measure da/a is scaling invariant. In case that g and h are such that

$$m_{g,h}(k) = \int_0^\infty \frac{da}{a} \overline{\widehat{g}(ak)}\, \widehat{h}(ak) = c_{g,h},$$

with $0 < |c_{g,h}| < \infty$, we say that g, h are an analysis reconstruction pair, or that h is a reconstruction wavelet for g. In this case we have

$$\mathcal{M}_h\mathcal{W}_g = c_{g,h}\,\mathbb{1}.$$

We say that g is strictly admissible if g is its own reconstruction wavelet, or (what is the same) if

$$\forall k \in \mathbb{R}^n \backslash \{0\} : \quad \int_0^\infty \frac{da}{a} \left|\widehat{g}(ak)\right|^2 = c_g.$$

We usually require in that case that $c_g = 1$. A sufficient condition for g to have a reconstruction wavelet r is that for some $c \geq 1$ we have

$$c^{-1} \leq \int_0^\infty \frac{da}{a} \left|\widehat{g}(ak)\right|^2 \leq c.$$

If this condition holds, we call g admissible. In this case the following function r will be a reconstruction wavelet for g

$$\widehat{r}(k) = \widehat{g}(k) \Big/ \int_0^\infty \frac{da}{a} \left|\widehat{g}(ak)\right|^2. \tag{2.1}$$

If g and h are an analysis reconstruction pair, then the following formula holds

$$\int_{\mathbb{H}^n} \frac{dbda}{a} \overline{\mathcal{W}_g s(b,a)} \mathcal{W}_h u(b,a) = c_{g,h} \int_{\mathbb{R}^n} dx\, \overline{s}(x)\, u(x).$$

In particular, if g is strictly admissible, then we have conservation of energy ($c_g = c_{g,g}$)

$$\int_{\mathbb{H}^n} \frac{dbda}{a} \left|\mathcal{W}_g s(b,a)\right|^2 = c_g \int_{\mathbb{R}^n} dx \left|s(x)\right|^2.$$

2.3 The reproducing kernel and the cross kernel

Let P and r be functions over the half-space $\mathbb{H}^n$. We define a (non-commutative) convolution for functions over the half-space via

$$(P * r)(b,a) = \int_{\mathbb{H}^n} \frac{db'da'}{a'} \frac{1}{a'^n} P\left(\frac{b-b'}{a'}, \frac{a}{a'}\right) r(b',a'). \tag{2.1}$$

Formally we can write for any g and h

$$(\mathcal{W}_g \mathcal{M}_h s)(b,a) = P_{g,h} * s(b,a), \quad P_{g,h} = \mathcal{W}_g h. \tag{2.2}$$

In the case where g, and h are an analysis reconstruction pair with $c_{g,h} = 1$, we clearly have $(\mathcal{W}_g \mathcal{M}_h)^2 = \mathcal{W}_g \mathcal{M}_h$ and hence

$$P_{g,h} * P_{g,h} = P_{g,h}.$$

The mapping $r \mapsto P_{g,h} * r$ is a projector into the range of the wavelet transform $\mathcal{W}_g$

$$P_{g,h} * \mathcal{W}_g = \mathcal{W}_g.$$

In case that g is strictly admissible with $c_{g,g} = 1$, we have that $P_{g,g}$ is an orthogonal projector, since it is self-adjoint.

The relation between $\mathcal{W}_g$ and $\mathcal{W}_h$ is given by the following. Consider a function s over $\mathbb{R}^n$. Suppose that g is admissible. It thus has a reconstruction wavelet r. An explicit formula has been given in (2.1). Then, since $\mathcal{W}_h = (\mathcal{W}_h \mathcal{M}_r)\mathcal{W}_g$, the wavelet transform of s with respect to g and the one with respect to h are related via the so-called cross kernel equation

$$\mathcal{W}_h s = P_{g\to h} * \mathcal{W}_g s, \quad P_{g\to h} = \mathcal{W}_h r. \tag{2.3}$$

It is clear, that behind wavelet analysis is the theory of group representations. Here it is the affine group. We will not give details and we refer the interested reader to the overview paper [AAGM95].

3 Wavelet transform of distributions

3.1 Wavelet analysis of $\mathcal{S}_0(\mathbb{R}^n)$

We start by listing some theorems that shall be proved in the sequel. Let $\mathcal{S}(\mathbb{R}^n)$ denote the class of Schwartz consisting of those functions that together with their derivatives decay faster than any polynomial, so that the following semi-norms are all finite

$$\|s\|_{n,m} = \sup_{t\in\mathbb{R}} |t^n \partial_t^m s(t)| < \infty \tag{3.1}$$

The topology they induce makes $\mathcal{S}(\mathbb{R}^n)$ a Fréchet space. Let $\mathcal{S}_0(\mathbb{R}^n)$ be the set of all those functions in $\mathcal{S}(\mathbb{R}^n)$ for which the Fourier transform

$$\widehat{s}(\xi) = \int dx\, s(x)\, e^{-i\xi x}, \quad s(x) = \frac{1}{(2\pi)^n} \int d\xi\, \widehat{s}(\xi) e^{i\xi x}$$

vanishes of infinite order in $\xi = 0$

$$\widehat{s}(\xi) = O(|\xi|^\infty) \quad (\xi \to 0),$$

or, what is the same, s is orthogonal to all polynomials π

$$\int dx\, \pi(x)\, s(x) = 0.$$

This space is a closed subspaces of $\mathcal{S}(\mathbb{R}^n)$. The analogous space of functions over the half-space $\mathbb{H}^n$ shall be denoted by $\mathcal{S}(\mathbb{H}^n)$. It consists of those functions r for which the following norms are all finite

$$\|r\|_{k,l} = \sup_{(b,a)\in\mathbb{H}^n} \left|(a + 1/a)^k (1 + |b|)^l r(b, a)\right| < \infty. \tag{3.2}$$

Note that this means that r decays rapidly for large b and for large or small a. It sometimes is useful to use the following equivalent system of semi-norms

$$\|r\|_l = \sup_{(b,a)\in\mathbb{H}^n} \left|(a+1/a)^l(1+a)^n(1+|b|/[1+a])^l r(b,a)\right| < \infty. \tag{3.3}$$

This means that for some function $f \in \mathcal{S}(\mathbb{R}^n)$ and some $\rho(a) = O((a+1/a)^l \ \forall l$ we may write

$$|r(b,a)| \leq \frac{\rho(a)}{(1+a)^n} f\left(\frac{b}{1+a}\right).$$

It is useful to introduce the following functions that we shall use systematically in the sequel. We write

$$\kappa_\alpha(x) = (1+|x|)^{-\alpha},$$

and

$$\phi_{\alpha,\beta}(a) = \frac{a^\alpha}{(1+a)^{\alpha+\beta}}, \quad \phi_\alpha(a) = \phi_{\alpha,\alpha}(a) \sim (a+1/a)^{-\alpha}.$$

Then we have the following useful estimates that we will use systematically for $\alpha \in \mathbb{R}$

$$\begin{array}{ll} \text{(i)} & \kappa_\alpha(x+x') \leq O(1)\kappa_\alpha(x)\,\kappa_{-|\alpha|}(x'), \\ \text{(ii)} & \kappa_\alpha(ax) \leq O(1)\phi_{-|\alpha|}(a)\,\kappa_\alpha(x), \\ \text{(iii)} & \phi_\alpha(aa') \leq O(1)\phi_{-|\alpha|}(a)\,\phi_\alpha(a'). \end{array}$$

We leave the verification to the reader. That the families of semi-norms (3.2) and (3.3) are equivalent follows now easily from

$$\kappa_\alpha(b) \leq \kappa_\alpha(b/[1+a]) \leq O(1)\,(1+a)^\alpha \kappa_\alpha(b).$$

Theorem 3.1 *We have*

$$\mathcal{W} : \mathcal{S}_0(\mathbb{R}^n) \times \mathcal{S}_0(\mathbb{R}^n) \to \mathcal{S}(\mathbb{H}^n), \quad (g,s) \mapsto \mathcal{W}_g s,$$

is continuous.

Note that we did not specify any smoothness in the topology of $\mathcal{S}(\mathbb{H}^n)$. In fact we have that the wavelet coefficients are smooth functions. More precisely we have for $r = \mathcal{W}_g s$ with g, $s \in \mathcal{S}_0(\mathbb{R}^n)$ that

$$\sup_{(b,a)\in\mathbb{H}^n} \left|(a+1/a)^k(1+|b|)^l \partial_b^p \partial_a^m r(b,a)\right| < \infty, \tag{3.4}$$

for any multi-index $p \in \mathbb{N}^n$ and any $m \in \mathbb{N}$. Moreover, we have

Lemma 3.2 *On the image of the wavelet transform of $\mathcal{S}_0(\mathbb{R}^n)$ with respect to $g \in \mathcal{S}_0(\mathbb{R}^n)$ the topology induced by the semi-norms (3.4) and the one induced by $\mathcal{S}(\mathbb{H}^n)$ coincide.*

The elementary proof can be found, e.g., in [Hol95] or in [Hol93b]. However, for the convenience of the reader, we repeat it here in a concise form. It will be sequence of elementary lemmas.

We can equip $\mathcal{S}_0(\mathbb{R}^n)$ with a (a priori) different topology that controls only localization properties of s and of $\widehat{s}$. That is we set

$$\|s\|_m = \|(1+|x|)^m s(x)\|_\infty + \|(|\xi| + 1/|\xi|)^m \widehat{s}(\xi)\|_\infty . \tag{3.5}$$

Clearly on $\mathcal{S}_0(\mathbb{R}^n)$ these norms are finite. But we have more.

Lemma 3.3 *On $\mathcal{S}_0(\mathbb{R}^n)$ the topology given by* (3.1) *and the one given by* (3.5) *coincide.*

Proof. That the topology generated by (3.5) is weaker follows from the fact that the localization of the Fourier transform may be estimated in terms of the localization of the derivatives of s. In addition, since all moments of s vanish, the behavior of $\widehat{s}$ around 0 is of the form $|\widehat{s}(\xi)| \le c_m |\xi|^m$ and by Taylor's theorem, the c_m may be estimated in terms of the localization of a finite number of derivatives of $\widehat{s}$.

The proof, that the topology is stronger is postponed, since it uses the inverse wavelet transform (see Section 3.2 below). □

In view of this theorem, proved below, we only need to estimate the localization of $\mathcal{W}[g,s]$ over the half-space in terms of the localization of g and s and of their Fourier transforms.

The localization in b follows immediately from the localization of s and g thanks to the estimate below.

Lemma 3.4 *Suppose that α, $\beta \in \mathbb{R}$ satisfy either of the following estimates*

$$\begin{array}{ll} \text{(i)} & \alpha > n, \beta > n, \\ \text{(ii)} & \alpha + \beta > n, \beta < 0, \\ \text{(iii)} & \alpha + \beta > n, \alpha < 0. \end{array}$$

Then, we may estimate

$$\mathcal{W}[\kappa_\alpha, \kappa_\beta](b,a) \le c_{\alpha,\beta}\phi_{\min\{\alpha,0\}\,\min\{\beta-n,0\}}(a)\frac{1}{(1+a)^n}\kappa_{\min\{\alpha,\beta\}}\left(\frac{b}{1+a}\right),$$

with some $c_{\alpha,\beta} > 0$.

In particular, we have for $\alpha > n$,

$$\mathcal{W}[\kappa_\alpha, \kappa_\alpha](b,a) \le c_\alpha \frac{1}{(1+a)^n}\kappa_\alpha\left(\frac{b}{1+a}\right).$$

Proof. First suppose we are in case (i). Globally we have by Young's inequality

$$\mathcal{W}[\kappa_\alpha, \kappa_\beta](b,a) \le \|\kappa_\alpha\|_{L^1(\mathbb{R}^n)} \|\kappa_\beta\|_{L^\infty(\mathbb{R}^n)} .$$

First suppose $a \le 1$. We may suppose $|b| > 2$. Then, we have to estimate

$$\int_{\mathbb{R}^n} dx\, a^{-n}\, \kappa_\alpha([x-b]/a)\kappa_\beta(x) = \left\{ \int_{K_1} + \int_{K_2} \right\} dx a^{-n}\, \kappa_\alpha([x-b]/a)\kappa_\beta(x),$$

where K_1 is the half-space $b(x - b/2) > 0$ and K_2 its complement. On K_1 we have $|x| > |b|/2$ and thus $\kappa_\beta(x) \le \kappa_\beta(b/2) \le O(1)\kappa_\beta(b)$. The remaining function is integrable, since $\alpha > n$ its integral being uniformly bounded in a.

In the second integral we have $|x - b| > |b|/2$ and thus $\kappa_\alpha([b-x]/a) < \kappa_\alpha(b/2a)$. For $|b| > 2$ and $\alpha > n$ we have $a^{-n}\kappa_\alpha(b/2a) \le O(1)\kappa_\alpha(b)$. Again the remaining integral is finite and we are done for $a < 1$.

The remaining part $a \ge 1$ follows from the following useful formula

$$\mathcal{W}[g,s](b,a) = \frac{1}{a^n}\mathcal{W}[\overline{s},\overline{g}]\left(-\frac{b}{a}, \frac{1}{a}\right). \tag{3.6}$$

Now suppose we are in case (ii). Then, writing $\kappa_\beta(x+b) \le O(1)(1+|x|)^{|\beta|}\kappa_\beta(b)$ we have to estimate

$$\int_{\mathbb{R}^n} dx\, a^{-n}(1 + (|x|/a))^{-\alpha}(1+|x|)^{|\beta|} = a^{-n}\int_0^\infty dr r^{n-1}(1+(r/a))^{-\alpha}(1+r)^{|\beta|}.$$

In the limit $a \to 0$ the integral converges and we may suppose $a > 1$. We split the integral into two parts, $0 \le r < a$ and $a \ge r$. In the first we may estimate

$$\le O(1)a^{-n}\int_0^a dr r^{n-1}(1+r)^{|\beta|} \le O(a^{|\beta|}).$$

For the second integral we have

$$\le O(1)a^{-n}\int_a^\infty dr r^{n-1}(r/a)^{-\alpha} r^{|\beta|},$$

which is again $O(a^{|\beta|})$.

Case (iii) follows from (3.6). □

We now want to discuss the localization in a. For any two functions over $\mathbb{R}_+$ we introduce their logarithmic convolution or scale-convolution via

$$s *_l r(x) = \int_0^\infty \frac{dy}{y}\, s(x/y)\, r(y) = r *_l s(x).$$

We write again

$$\phi_{\alpha,\beta}(a) = \frac{a^\alpha}{(1+a)^{\alpha+\beta}}.$$

Then, we have

Lemma 3.5 *We have for α, β, α', $\beta' > 0$, $\alpha \neq \alpha'$, $\beta \neq \beta'$*

$$\phi_{\alpha,\beta} *_l \phi_{\alpha',\beta'} \sim \phi_{\min\{\alpha,\alpha'\},\min\{\beta,\beta'\}}.$$

In the case $\alpha = \alpha'$ or $\beta = \beta'$ logarithmic corrections arise.

This means the smallest of the absolute values of the localization exponents determines the localization of the scale convolution.

Proof. We introduce the inversion operator

$$s \mapsto Is, \qquad (Is)(t) = s\left(\frac{1}{t}\right).$$

The scale convolution satisfies the co-variance

$$Is *_l Ir = I\,(s *_l r), \tag{3.7}$$

as can be verified by direct computation.

We have to estimate

$$\phi_{\alpha,\beta} *_l \phi_{\alpha',\beta'}(t) = t^\alpha \left\{ \int_0^1 + \int_1^\infty \right\} \frac{da}{a} \frac{a^{\alpha'-\alpha}}{(1+t/a)^{\beta+\alpha}(1+a)^{\beta'+\alpha'}} = X_1 + X_2.$$

Because $I\phi_{\alpha,\beta} = \phi_{\beta,\alpha}$ and because of (3.7) it is enough to show the lemma for $t \leq 1$; that is we have to show that

$$X_1 + X_2 \sim t^{\min\{\alpha,\alpha'\}}, \qquad 0 \leq t \leq 1.$$

In X_1 we have $0 < a \leq 1$ and thus $1 + a \sim 1$. Changing $a \to t/a$ and calling it again a we obtain

$$X_1 \sim t^\alpha \int_0^1 \frac{da}{a} \frac{a^{\alpha'-\alpha}}{(1+t/a)^{\beta+\alpha}} = t^{\alpha'} \int_0^{1/t} \frac{da}{a} \frac{a^{\alpha'+\beta}}{(1+a)^{\beta+\alpha}}.$$

We distinguish two cases and obtain

$$X_1 \sim \begin{cases} t^\alpha & \alpha' > \alpha, \\ t^{\alpha'} & \alpha' < \alpha. \end{cases}$$

In X_2 we have $a > 1$ and thus $1 + a \sim a$ and $1 + t/a \sim 1$. Therefore

$$X_2 \sim t^\alpha \int_1^\infty \frac{da}{a} a^{-\alpha-\beta'} \sim t^\alpha,$$

since by hypothesis on the exponents, the integral converges. □

We now can prove the following result estimating the localization in a of the wavelet coefficients in term of the localization of g and s.

Lemma 3.6 *Suppose that*

$$|\widehat{g}(\xi)| \leq \phi_{\alpha,\beta}(|\xi|), \quad |\widehat{s}(\xi)| \leq \phi_{\alpha',\beta'}(|\xi|).$$

Suppose that α, β, $\alpha'+n$ and $\beta'-n > 0$ and that $\alpha'+n \neq \beta$ and $\beta'-n \neq \alpha$. Then

$$|\mathcal{W}_g s(b,a)| \leq c\phi_{\min\{\beta'-n,\alpha\},\min\{\beta,\alpha'+n\}}(a).$$

We write $\phi_\alpha = \phi_{\alpha,\alpha}$. Then, we have, in particular, for $\alpha = \beta = \alpha' = \beta' > n$ that

$$|\widehat{g}(\xi)|\,, |\widehat{s}(\xi)| \leq \phi_\alpha(|\xi|) \Rightarrow |\mathcal{W}_g s(b,a)| \leq O(1)\frac{1}{(1+a)^n}\,\phi_\alpha(a).$$

Proof. Passing into Fourier space we have

$$|\mathcal{W}[g,s](b,a)| \leq \frac{1}{(2\pi)^n}\int d\xi\; |\widehat{g}(a\xi)|\; |\widehat{s}(\xi)|\,.$$

By hypothesis this may be estimated by

$$\leq O(1)\int_0^\infty dr\, r^{n-1}\, \phi_{\alpha,\beta}(ar)\, \phi_{\alpha',\beta'}(r) = O(1)\, \phi_{\beta,\alpha} *_l \phi_{\alpha'+n,\beta'-n}(1/a),$$

and the localization over the scales follows easily from the previous lemma. □

We now can conclude.

Proof. (of Theorem 3.1) We use the equivalent norm (3.5) as described in Lemma 3.3. The two localization Lemmata 3.4 and 3.6 imply that

$$\begin{aligned} \sup_{(b,a)\in\mathbb{H}^n} |(1+a)^n\kappa_{-\alpha}(b/[1+a])\mathcal{W}_g s(b,a)| &\leq c_\alpha \,\|g\|_\alpha \,\|s\|_\alpha\,, \\ \sup_{(b,a)\in\mathbb{H}^n} |(1+a)^n\phi_{-\alpha}(a)\mathcal{W}_g s(b,a)| &\leq c_\alpha \,\|g\|_\alpha \,\|s\|_\alpha\,. \end{aligned}$$

Therefore

$$\begin{aligned} &(1+a)^{2n}\kappa_{-\alpha}^2(b/[1+a])\phi_\alpha^2(a)\,|\mathcal{W}_g s(b,a)|^2 \\ &\quad = c\,(1+a)^n\kappa_{-2\alpha}(b/[1+a])\,|\mathcal{W}_g s(b,a)|\,(1+a)^n\phi_{2\alpha}(a)\,|\mathcal{W}_g s(b,a)|\,, \end{aligned}$$

which is

$$\leq c'_\alpha \,\|g\|_\alpha^2 \,\|s\|_\alpha^2\,.$$

The result follows upon taking a square root. □

3.2 The wavelet synthesis

The wavelet synthesis is defined point-wise via

$$\mathcal{M}[h,r](x) = \int_{\mathbb{H}^n} \frac{dbda}{a} r(b,a) \frac{1}{a^n} h\left(\frac{x-b}{a}\right).$$

We also write $\mathcal{M}_h r$ for $\mathcal{M}[h,r]$. The name is justified, since as we shall see, this transform allows us to write arbitrary functions as a superposition of wavelets.

The wavelet synthesis has the following basic mapping property.

Theorem 3.7 *The following holds*

$$\mathcal{M} : \mathcal{S}_0(\mathbb{R}^n) \times \mathcal{S}(\mathbb{H}^n) \to \mathcal{S}_0(\mathbb{R}^n), \quad (h,r) \mapsto \mathcal{M}_h r,$$

is continuous.

The following elementary proof is again taken from [Hol95] and from [Hol93b].

Proof. We have to estimate the localization of $\mathcal{M}_h r$ in terms of the localization of h and s. From the standard estimates on κ_α and ϕ_α we have

$$\sup_{x\in\mathbb{R}^n} \kappa_\alpha([x-b]/a)\kappa_{-\alpha}(x) = \sup_{x\in\mathbb{R}^n} \kappa_\alpha(x)\kappa_{-\alpha}(ax+b) \leq O(1)\,\phi_{-\alpha}(a)\kappa_{-\alpha}(b).$$

Therefore we have

$$\mathcal{M}[\kappa_\alpha(x'), \phi_\beta(a)\kappa_\beta(b)](x) \leq O(1)\kappa_\alpha(x) \int_{\mathbb{H}^n} \frac{dbda}{a} a^{-n} \phi_{\beta-\alpha}(a)\kappa_{\beta-\alpha}(b).$$

This last integral is finite for $\beta > \alpha + n$.

The localization of $\mathcal{F}\mathcal{M}_g s$ follows easily from

$$\mathcal{F}\mathcal{M}[h,r](\xi) = \int_{\mathbb{H}^n} \frac{dbda}{a} r(b,a)\, e^{-i\xi b}\, \widehat{h}(a\xi).$$

Note that we may exchange the integrations thanks to the localization of all functions involved. Thus if $\left|\widehat{h}(\xi)\right| \leq \phi_\alpha(|\xi|)$ we have

$$|\mathcal{F}\mathcal{M}[h, \phi_\beta(a)\kappa_\beta(b)]| \leq \phi_\alpha(|\xi|) \int_0^\infty \frac{d\lambda}{\lambda} \phi_{\beta-\alpha}(\lambda) \int_{\mathbb{R}^n} db\kappa_\beta(b).$$

All integrals are finite if $\beta > \alpha > n$ and we are done. □

Remark. It sometimes is useful to consider wavelets with compact support. Now they are clearly not in $\mathcal{S}_0(\mathbb{R}^n)$ since the Fourier transform of a compactly supported function is entire and hence cannot vanish of infinite order in 0 without vanishing everywhere. However, as the proof of the theorem shows, to show some localization we only need some vanishing moments and some smoothness, which is compatible with compact support.

3.3 The inverse wavelet transform

We now can justify all formal expression we have encountered in the previous section. Direct computation (upon exchanging the integrals) shows that we have that $\mathcal{M}_g$ is the adjoint of the wavelet transform $\mathcal{W}_g$. More precisely for r, s, $g \in \mathcal{S}_0(\mathrm{I\!R}^n)$ we have

$$\int_{\mathrm{I\!R}^n} dx\, \overline{(\mathcal{M}_g r)(x)}\, s(x) = \int_{\mathrm{I\!H}^n} \frac{dbda}{a}\, \overline{r}(b,a)\, \mathcal{W}_g s(b,a). \tag{3.1}$$

In certain cases the wavelet synthesis is also the inverse of the wavelet transform. Let $m \in L^\infty(\mathrm{I\!R}^n)$. We shall also denote by $M[m]$ or simply m, the operator "multiplication by m", $s \mapsto ms$, and by $T[m]$ we denote the Fourier multiplier given by

$$T[m]s = \mathcal{F}^{-1} M[m]\mathcal{F}s.$$

Theorem 3.8 *On $\mathcal{S}_0(\mathrm{I\!R}^n)$ we have that*

$$\mathcal{M}_h \mathcal{W}_g = T[m_{g,h}],$$

where

$$m_{g,h}(\xi) = m_{g,h}(\xi/\,|\xi|) = \int_0^\infty \frac{da}{a}\, \overline{\widehat{g}}(a\xi)\, \widehat{h}(a\xi).$$

Note that the multiplier is dilation invariant

$$m_{g,h}(\lambda\xi) = m_{g,h}(\xi), \quad \lambda > 0.$$

In case $m_{g,h} = 1$, this is precisely Calderóns reproducing formula.

Proof. Writing $s^\# = \mathcal{M}_h \mathcal{W}_g s$ and using the Fourier representation of the wavelet transform we may write

$$\begin{aligned} s^\#(x) &= \int_0^\infty \frac{da}{a} \int db\, \mathcal{W}_g s(b,a)\, \frac{1}{a^n}\, h\left(\frac{x-b}{a}\right) \\ &= \frac{1}{(2\pi)^n} \int_0^\infty \frac{da}{a} \int db \int d\xi\, \overline{\widehat{g}}(a\xi)\, e^{ib\xi}\, \widehat{s}(\xi)\, \frac{1}{a^n}\, h\left(\frac{x-b}{a}\right). \end{aligned}$$

The double integral over b and ξ converges absolutely by hypothesis on s, g and h and we may exchange the integration. Carrying out at first the integral over b we make appear the Fourier transform of $\frac{1}{a^n} h\left(\frac{x-b}{a}\right)$, and thus

$$s^\#(x) = \frac{1}{(2\pi)^n} \int d\xi \int_0^\infty \frac{da}{a}\, \overline{\widehat{g}}(a\xi)\, \widehat{s}(\xi)\, \widehat{h}(a\xi)\, e^{ix\xi}.$$

The double integral over the scales and ξ is absolutely convergent, since all functions are arbitrarily well polynomial strip-localized. Exchanging again the integration and using the scaling invariance of the measure da/a we obtain

$$s^\#(\xi) = m_{g,h}(\xi)\widehat{s}(\xi),$$

as claimed. □

In particular, it may happen that $m_{g,h} = c_{g,h}$ for some constant. In this case we have

$$\mathcal{M}_h \mathcal{W}_g = c_{g,h} \mathbb{1}. \tag{3.2}$$

Definition 3.9 *A function $h \in \mathcal{S}_0(\mathbb{R}^n)$ is called a reconstruction wavelet for $g \in \mathcal{S}_0(\mathbb{R}^n)$ if*

$$\int_0^\infty \frac{da}{a} \overline{\widehat{g}}(a\xi)\, \widehat{h}(a\xi) = 1, \quad \forall \xi \neq 0.$$

If g is its own reconstruction wavelet we say that g is strictly admissible.

Clearly the notion is symmetric in that if $g \in \mathcal{S}_0(\mathbb{R}^n)$ is a reconstruction wavelet for $h \in \mathcal{S}_0(\mathbb{R}^n)$, then h is a reconstruction wavelet for g and we say also that g, h is an analysis reconstruction pair. A necessary and sufficient condition for having a reconstruction wavelet is the following.

Proposition 3.10 *Suppose for $g \in \mathcal{S}_0(\mathbb{R}^n)$ we have*

$$c^{-1} \leq \int_0^\infty \frac{da}{a} \left|\widehat{g}(a\xi)\right|^2 \leq c,$$

with some $c \geq 1$. Then, r defined through

$$\widehat{r}(\xi) = \frac{\widehat{g}(\xi)}{\int_0^\infty \frac{da}{a} \left|\widehat{g}(a\xi)\right|^2},$$

is a reconstruction wavelet in $\mathcal{S}_0(\mathbb{R}^n)$. Vice versa, of g has a reconstruction wavelet, then the above inequality holds.

We then call such a g admissible. Recall that strictly admissible means that c may be chosen $= 1$.

Proof. The function

$$m(\xi) = \left(\int_0^\infty \frac{da}{a} \left|\widehat{g}(a\xi)\right|^2\right)^{-1}, \ \xi \neq 0,$$

is homogeneous of order 0, $m(\lambda\xi) = m(\xi)$, $\lambda > 0$ and C^∞. Its partial derivatives are therefore homogeneous too and they satisfy

$$\left|\partial^\alpha m(\xi)\right| \leq c_\alpha \left|\xi\right|^{|\alpha|}.$$

This shows that for $\widehat{r} = m\widehat{g}$ we have $r \in \mathcal{S}_0(\mathbb{R}^n)$. That it is a reconstruction wavelet is trivial.

On the other hand, suppose that the above integral is 0 for some ξ. Then, $g(\lambda\xi) = 0$ for all $\lambda > 0$ and g cannot have a reconstruction wavelet, since $m_{g,h}(\lambda\xi) = 0$ for all $h \in \mathcal{S}_0(\mathbb{R}^n)$. □

It might be interesting to note that we have quite a freedom in choosing the reconstruction wavelet for a given wavelet g in particular, if we leave the highly regular context. In fact g needs not even be regular as long as h compensates for this. This has been used in inverting the Radon transform which can be interpreted as a wavelet transform with respect to a singular wavelet [Hol93a]. A different application can be found in [Hol94], where a functional calculus based on wavelet transforms is constucted.

We now can prove Lemma 3.3.

Proof. It only remains to show that localization of s and $\widehat{s}$ simultaneously implies localization of all derivatives of s. For this take $g = h$ defined through

$$\widehat{g}(\xi) = c\, e^{-|\xi|-1/|\xi|}.$$

This function and all its derivatives have finite norm (3.5). We may choose c in such a way that g is strictly admissible and hence

$$\mathcal{M}_g \mathcal{W}_g = \mathbb{1}.$$

An integration by parts shows that the following formula holds for all multi-indices

$$\mathcal{W}[g, \partial^\alpha s](b, a) = (-a)^{-|\alpha|} \mathcal{W}[\partial^\alpha g, s](b, a),$$

and thus, since $\partial^\alpha g \in \mathcal{S}_0(\mathbb{R}^n)$ we have

$$\partial^\alpha s(x) = \mathcal{M}[g, (-a)^{-|\alpha|} \mathcal{W}[\partial^\alpha g, s]].$$

We thus may apply the previous theorem to conclude that all derivatives of s are rapidly decreasing. In addition their decay may be estimated in terms of the decay of s and of $\widehat{s}$. □

3.4 The reproducing kernel equation

For a given wavelet, the wavelet transform $\mathcal{W}_g$ maps functions over $\mathbb{R}^n$ to functions over the half-space. From this it is plausible that not every function over the half-plane is the wavelet transform with respect to g of some function over $\mathbb{R}^n$. In this section we will characterize the image of wavelet transforms. It turns out that there are internal correlations as expressed by the presence of a reproducing kernel.

For two functions over $\mathbb{H}^n$ we introduce the following convolution

$$s * r(b, a) = \int_{\mathbb{H}^n} \frac{db'\, da'}{a'} \frac{1}{a'^n} s\left(\frac{b - b'}{a'}, \frac{a}{a'}\right) r(b', a').$$

Note that this convolution is not commutative any more but it is still associative.

As we have seen, the wavelet transformation followed by an wavelet synthesis with respect to a reconstruction wavelet gives back the analyzed function. Let us consider this identity in wavelet space; that is we consider the operator on functions over the half-space $\mathcal{W}_g\,\mathcal{M}_h$

Proposition 3.11 *We have*

$$\mathcal{W}_g\,\mathcal{M}_h u = P * u,$$

with $P = \mathcal{W}_g h$.

Proof. We obtain exchanging the integrations

$$\begin{aligned}
\mathcal{W}_g\,\mathcal{M}_h u(b,a) &= \int_{\mathbb{R}^n} dx \frac{1}{a^n}\bar{g}\left(\frac{x-b}{a}\right) \int_{\mathbb{H}^n} \frac{db'da'}{a'}\frac{1}{a'^n} h\left(\frac{x-b'}{a'}\right) u(b',a) \\
&= \int_{\mathbb{H}^n} \frac{db'da'}{a'} \int_{\mathbb{R}^n} dx \frac{1}{a^n}\bar{g}\left(\frac{x-b}{a}\right) \frac{1}{a'^n} h\left(\frac{x-b'}{a'}\right) u(b',a) \\
&= \int_{\mathbb{H}^n} \frac{db'da'}{a'}\frac{1}{a'^n} \mathcal{W}_g h\left(\frac{[b-b']/a'}{[a/a']}, \frac{a}{a'}\right) u(b',a)
\end{aligned}$$

as claimed. □

The following theorem characterizes the image of the wavelet transform.

Theorem 3.12 *If* $h \in \mathcal{S}_0(\mathbb{R}^n)$ *is a reconstruction wavelet for* $g \in \mathcal{S}_0(\mathbb{R}^n)$, *then the operator* $\Pi_{g,h} = \mathcal{W}_g\mathcal{M}_g$ *is a continuous projector onto the image of* $\mathcal{W}_g$.

Proof. As composition of continuous maps it follows that $\Pi_{g,h}$ is a well-defined continuous map from $\mathcal{S}(\mathbb{H}^n)$ to $\mathcal{S}(\mathbb{H}^n)$. Clearly for $r \in \mathcal{S}(\mathbb{H}^n)$ we have that $\Pi_{g,h} r$ is in the image of $\mathcal{W}_g$, since it is the wavelet transform of $\mathcal{M}_h r$. On the other hand suppose that r is in the image of $\mathcal{W}_g$. Then there is a function $s \in \mathcal{S}_0(\mathbb{R}^n)$ such that $r = \mathcal{W}_g s$. It comes $\Pi_{g,h} r = \mathcal{W}_g(\mathcal{M}_h \mathcal{W}_g s) = \mathcal{W}_g s = r$. This shows that

$$\Pi_{g,h}\,\Pi_{g,h} = \Pi_{g,h},$$

and thus $\Pi_{g,h}$ is a projector on the image of $\mathcal{W}_g$. □

Therefore we have the following corollary

Corollary 3.13 *The image of* $\mathcal{S}_0(\mathbb{R}^n)$ *under the wavelet transform* $\mathcal{W}_g$ *with* $g \in \mathcal{S}_0(\mathbb{R}^n)$ *is the closed subspace of functions in* $r \in \mathcal{S}(\mathbb{H}^n)$ *that satisfy point-wise*

$$r(b,a) = \int_{\mathbb{H}^n} \frac{db'da'}{a'}\frac{1}{a'^n} P_{g,h}\left(\frac{b-b'}{a'}, \frac{a}{a'}\right) r(b',a') = P_{g,h} * r(b,a),$$

where h *is a reconstruction wavelet for* g *and* $P_{g,h} = \mathcal{W}_g h$.

Therefore the function $P_{g,h}$ is called the reproducing kernel.

From the localization theorems it follows that the reproducing kernel is highly localized over the half-space. Suppose now that $P_{g,h}$ is localized around the point $(0,1)$ in the upper half-space. Then, $\frac{1}{a'^n} P_{g,h}\left(\frac{b-b'}{a'}, \frac{a}{a'}\right)$ seen as a function of (b', a') for fixed (b, a) is localized around (b, a). Therefore the weighted mean with reproducing kernel localized around some point (b, a) gives back the value of $\mathcal{W}_g s$ at exactly this point. Therefore the localization of the reproducing kernel indicates how well the wavelet transform separates the scales and the positions of the analyzed function.

By direct computation we can verify that the following symmetry holds that we state for later reference

$$\frac{1}{a'^n} P_{g,h}\left(\frac{b-b'}{a'}, \frac{a}{a'}\right) = \frac{1}{a^n} \overline{P}_{h,g}\left(\frac{b'-b}{a}, \frac{a'}{a}\right).$$

3.5 The cross kernel

Given the wavelet transform of s with respect to some wavelet g how can we obtain the wavelet transform of s with respect to h? The first step is to recover s with the help of some reconstruction wavelet and then to analyze with h.

Proposition 3.14 *Suppose that $g \in \mathcal{S}_0(\mathbb{R}^n)$ satisfies at*

$$0 < c_1 \leq \int_0^\infty \frac{da}{a} |\hat{g}(a\xi)|^2 \leq c_2 < \infty.$$

Then, for all $h \in \mathcal{S}_0(\mathbb{R}^n)$ there is a function $P_{g\to h} \in \mathcal{S}(\mathbb{H}^n)$ such that

$$\mathcal{W}_h = P_{g\to h} * \mathcal{W}_g.$$

Proof. By Proposition 3.10 there is a reconstruction wavelet $r \in \mathcal{S}_0(\mathbb{R}^n)$ for g. We therefore may write

$$\mathcal{W}_h = \mathcal{W}_h(\mathcal{M}_r \mathcal{W}_g) = (\mathcal{W}_h \mathcal{M}_r)\mathcal{W}_g.$$

We now apply Proposition 3.11 to conclude that $P_{g\to h} = \mathcal{W}_h r$ will work. □

The convolution kernel $P_{g\to h}$ is called the cross-kernel from g to h. It clearly is again highly localized, $P_{g\to h} \in \mathcal{S}(\mathbb{H}^n)$.

3.6 The L^2-theory

We now can easily prove the following theorem.

Theorem 3.15 *For $g \in \mathcal{S}_0(\mathbb{R}^n)$ and $s \in L^2(\mathbb{R}^n)$ we have that*

$$\mathcal{W}_g s \in L^2(\mathbb{H}^n, db\,da/a).$$

For g admissible the converse holds, namely, if

$$\mathcal{W}_g s \in L^2(\mathbb{H}^n, dbda/a),$$

then $s \in L^2(\mathbb{R}^n)$. *Moreover, there is* $c \geq 1$ *such that*

$$c^{-1} \|s\|_{L^2(\mathbb{R}^n)} \leq \|\mathcal{W}_g s\|_{L^2(\mathbb{H}^n, dbda/a)} \leq c \|s\|_{L^2(\mathbb{R}^n)} .$$

If g *and* $h \in \mathcal{S}_0(\mathbb{R}^n)$ *are an analysis reconstruction pair, then we have in addition*

$$\int_{\mathbb{H}^n} \frac{dbda}{a} \overline{\mathcal{W}_g s}(b,a)\, \mathcal{W}_h r(b,a) = \int_{\mathbb{R}^n} dx\, \overline{s}(x)\, r(x).$$

Proof. First suppose that s, $r \in \mathcal{S}_0(\mathbb{R}^n)$. Then, we have

$$\langle \mathcal{W}_g s, \mathcal{W}_g s \rangle_{\mathbb{H}^n} = \langle s, \mathcal{M}_g \mathcal{W}_g s \rangle_{\mathbb{R}^n} .$$

Now $\mathcal{M}_g \mathcal{W}_g$ is a bounded Fourier multiplier, hence a continuous operator in $L^2(\mathbb{R}^n)$ and the first part of the proposition follows by a standard density argument.

In case that g is admissible, this multiplier is essentially bounded from below and hence has a bounded inverse. This shows the second statement of the theorem.

In case that h is a reconstruction wavelet for g we have $\mathcal{M}_h \mathcal{W}_g = \mathbb{1}$ and thus

$$\langle \mathcal{W}_h s, \mathcal{W}_g r \rangle_{\mathbb{H}^n} = \langle s, \mathcal{M}_h \mathcal{W}_g r \rangle_{\mathbb{R}^n} = \langle s, r \rangle_{\mathbb{R}^n} ,$$

as claimed. □

In view of the next section we rewrite the above equation in order to make it bi-linear in s and r

$$\int_{\mathbb{H}^n} \frac{dbda}{a} \mathcal{W}_{\overline{g}} s(b,a)\, \mathcal{W}_h r(b,a) = \int_{\mathbb{R}^n} dx\, s(x)\, r(x).$$

Recall that we have supposed that

$$\int_0^\infty \frac{da}{a} \overline{\widehat{g}(a\xi)}\, \widehat{h}(a\xi) = 1,$$

independent of $\xi \in \mathbb{R}^n \setminus \{0\}$.

It can easily be shown that the Theorems 3.12 and 3.14 carry over the the L^2 case. In addition we have that for g strictly admissible the convolution with the reproducing kernel $r \mapsto P_g * r$ is an orthogonal projection onto the image of the wavelet transform with respect to g.

3.7 Wavelet analysis of $\mathcal{S}'(\mathbb{R}^n)$

We write $\mathcal{S}'(\mathbb{R}^n)$, $\mathcal{S}'(\mathbb{H}^n)$ and of $\mathcal{S}_0'(\mathbb{R}^n)$ for the associated dual spaces. We shall equip them with the usual weak*-topologies. It is clear that $\mathcal{S}_0'(\mathbb{R}^n)$ can be canonically identified with

$$\mathcal{S}'(\mathbb{R}^n)/\mathcal{P}$$

where $\mathcal{P}$ is the space of polynomials in $x_1, x_2, \ldots, x_n$.

It is natural to define the wavelet transform with respect to any wavelet $g \in \mathcal{S}_0(\mathbb{R}^n)$ of any distribution $\eta \in \mathcal{S}'(\mathbb{R}^n)$ as the distribution in $\mathcal{S}'(\mathbb{H}^n)$, that has the following action on any function $r \in \mathcal{S}(\mathbb{H}^n)$

$$\mathcal{W}_g\eta(r) = \eta(\mathcal{M}_{\overline{g}}r).$$

In the same way we may define the wavelet synthesis by duality via $s \in \mathcal{S}_0(\mathbb{R}^n)$, $\chi \in \mathcal{S}'(\mathbb{H}^n)$.

$$(\mathcal{M}_h\chi)(s) = \chi(\mathcal{W}_{\overline{h}}s).$$

Theorem 3.16 *The wavelet transform and the wavelet synthesis with respect to some fixed wavelets $g, h \in \mathcal{S}_0(\mathbb{R}^n)$*

$$\mathcal{W}_g : \mathcal{S}_0'(\mathbb{R}^n) \to \mathcal{S}'(\mathbb{H}), \quad \eta \mapsto \mathcal{W}_g\eta,$$
$$\mathcal{M}_h : \mathcal{S}'(\mathbb{H}) \to \mathcal{S}_0'(\mathbb{R}^n), \quad \Omega \mapsto \mathcal{M}_h\Omega,$$

are continuous in the respective topologies.

This follows by standard duality arguments and we shall not repeat them here. Clearly for a fixed analyzed function the dependency on the wavelet is continuous, too. However, the bilinear map $(g, \eta) \mapsto \eta(g)$ is not continuous but rather hypocontinuous (e.g. [Tre67]).

The wavelet transform desingularizes the distributions.

Theorem 3.17 *Let $\eta \in \mathcal{S}_0'(\mathbb{R}^n)$, $g \in \mathcal{S}_0(\mathbb{R}^n)$. Then, $\mathcal{W}_g\eta$ may be identified with*

$$\Omega(b, a) = \eta(\overline{g}_{b,a}), \quad g_{b,a}(t) = a^{-n}g(a^{-1}[t-b]),$$

in that sense that for all $r \in \mathcal{S}(\mathbb{H}^n)$ we have

$$\mathcal{W}_g\eta(r) = \int_{\mathbb{H}^n} \frac{dbda}{a}\, \Omega(b, a)\, r(b, a). \tag{3.1}$$

The function Ω is together with all their derivatives of at most polynomial growth

$$\left|\partial_b^l (a\partial_a)^m \Omega(b, a)\right| \leq O(1)\,(1 + |b|^2)^{\alpha/2}\,(a + 1/a)^\beta,$$

with some $\alpha, \beta \geq 0$ only depending on l, m.

It can be shown that the analogous of Proposition 3.11 and 3.14 holds again. Explicitly Ω satisfies the reproducing kernel equation

$$\Omega(b,a) = \int_{\mathbb{H}^n} \frac{db'da'}{a'} \frac{1}{a'^n} P_{g,h}\left(\frac{b-b'}{a'}, \frac{a}{a'}\right) \Omega(b',a').$$

This equation holds point-wise. Vice versa any function of at most polynomial growth satisfying this reproducing kernel equation is in the image of the wavelet transform where we make again the identification (3.1).

The inversion formula may actually computed via an integral. Let $K_m \subset \mathbb{H}^n$ be a growing family of open sets, $K_m \subset K_{n+1}$ that tends to $\mathbb{H}^n$, $\cup_n K_m = \mathbb{H}^n$. Then, we have for any locally integrable function r of at most polynomial growth in b and $(a + 1/a)$ that

$$s_m = \int_{K_m} \frac{dbda}{a} r(b,a)\, h_{b,a} \to \mathcal{M}_h r,$$

where the convergence holds in $\mathcal{S}'_0(\mathbb{R}^n)$. This can be proved as follows. In wavelet space the approximations are given by the following

$$\mathcal{W}_g s_m = P * (K_m \cdot r),$$

where P is a reproducing kernel, and K_m stands for the multiplier with the characteristic function of the set K_m. The convolution with the highly localized kernel P defines a weak* continuous operator (it is clearly continuous is $\mathcal{S}(\mathbb{H}^n)$ and its transpose is again a convolution with some highly localized kernel). Now $\chi_{K_m} \cdot r$ tends to r weakly and thus the integral converges weakly, too.

3.8 Bounded sets in $\mathcal{S}_0(\mathbb{R}^n)$ and $\mathcal{S}'_0(\mathbb{R}^n)$

For further reference in later sections we note that the bounded sets of $\mathcal{S}_0(\mathbb{R}^n)$ and $\mathcal{S}'_0(\mathbb{R}^n)$ may be easily characterized in wavelet space by some uniform localization. Recall that in a topological vector space a set X is bounded iff it is eventually absorbed by all neighborhoods U of 0; in other words iff for all neighborhoods $U \ni 0$ we have $X \subset tU$ for all large enough t. Equivalently a set X is bounded iff for all sequences $x_n \in X$ we have

$$\alpha_m x_m \to 0,$$

for all numerical sequences $\alpha_m \to 0$ as $n \to \infty$. For a Fréchet space generated by the semi-norms p_m the bounded subsets $V \subset X$ are precisely those for which each p_m is bounded on V.

Theorem 3.18 *A set $X \subset \mathcal{S}_0(\mathbb{R}^n)$ is bounded iff there is some P over the half-space that is faster decreasing than any polynomial in b and $(a + 1/a)$ such that*

$$|\mathcal{W}_g x(b,a)| \leq P(b,a),$$

for all $x \in X$. A set $X \subset \mathcal{S}_0'(\mathbb{R}^n)$ is bounded iff there is some function P over the half-space of at most polynomial growth in b and $(a+1/a)$ such that

$$|\mathcal{W}_g x(b,a)| \leq P(b,a).$$

for all $x \in X$.

Proof. Since the wavelet transform is a topological isomorphism the bounded sets can be characterized in wavelet space. A set $X \subset \mathcal{S}_0(\mathbb{R}^n)$ is bounded iff for all $\alpha > 0$

$$\sup_{x \in X} \|x\|_{\mathcal{S}_0(\mathbb{R}^n);\alpha} < \infty.$$

This is equivalent to the existence of constants c_α such that

$$|\mathcal{W}_g x(b,a)| \leq c_\alpha \kappa_\alpha(b)\phi_\alpha(a),$$

for all $x \in X$ with $\phi_\alpha(a) = (a+1/a)^{-\alpha}$ and $\kappa_\alpha(b) = (1+|b|)^{-\alpha}$. This shows the first part of the theorem.

The proof of the second part shall be by contradiction. Let $X \subset \mathcal{S}_0'(\mathbb{R}^n)$ be bounded and suppose that for all α and all c we can find some point (b,a) and some $x \in X$ such that $|\mathcal{W}_g x(b,a)| > c\kappa_\alpha(b)\phi_\alpha(a)$. In particular, we may choose $\alpha = -m$ and $c = m$ obtaining that way a sequence of points (b_m, a_m) and distributions x_m such that

$$\mathcal{W}_g x_m(b_m, a_m) > m\kappa_{-m}(b_m)\phi_{-m}(a_m).$$

Consider now the sequence of translated and dilated wavelets $g_m = c_m g_{b_m,a_m}$ with $c_m = \kappa_m^(b_m)\phi_m(a_m)$. We claim that this set is bounded in $\mathcal{S}_0(\mathbb{R}^n)$. To see this let $F = |\mathcal{W}_g g|$. By the first part of the theorem we then have to show that the following expression is finite for all $l < 0$.

$$\sup_{m\in\mathbb{N}} \sup_{(b,a)\in\mathbb{H}^n} \kappa_m(b_m)\phi_m(a_m)\kappa_l(b)\phi_l(a)\frac{1}{a_m^n}F\left(\frac{b-b_m}{a_m}, \frac{a}{a_m}\right) < \infty.$$

Changing $a \to a_m a$ and $b \to a_m b + b_m$ and using the standard estimates for κ_α and ϕ_α, we see that this expression above is majorized by

$$O(1) \sup_{m\in\mathbb{N}} \kappa_{l+m}(b_m)\phi_{2l+m}(a_m)a_m^{-n}\kappa_m(a_m) \sup_{(b,a)\in\mathbb{H}^n} \phi_l(a)\kappa_l(b)F(b,a),$$

which actually is finite. But this is a contradiction. Indeed, suppose $Y \subset \mathcal{S}_0(\mathbb{R}^n)$ is a bounded set. Then, $\{x(y) : y \in Y, x \in X\}$ is bounded in $\mathbb{C}$. On the other hand we have found a bounded set in $\mathcal{S}_0(\mathbb{R}^n)$ on which we have $|x_m(\overline{g}_m)| = |\mathcal{W}_g x_m(b_m, a_m)| > m$. Therefore the theorem is proved. □

Corollary 3.19 *Let $\eta_m \to \eta$ in $\mathcal{S}'(\mathbb{R}^n)$. Then we have $\mathcal{W}_g\eta_m(b,a) \to \mathcal{W}_g\eta(b,a)$ uniformly on compact sets.*

where $\rho(a) \geq 0$ is bounded and satisfies

$$\forall K > 0 \; \rho(a) \leq O(a^K), (a \to 0), \quad \rho(a) \leq O(1), (a \to \infty). \tag{3.1}$$

Denote this space of functions by $\Sigma(\mathbb{H}^n)$ and equip it with the topology generated by the above semi-norms. It can be shown, that the mapping

$$\mathcal{W} : \mathcal{S}_0(\mathbb{R}^n) \times \mathcal{S}(\mathbb{R}^n) \to \Sigma(\mathbb{H}^n), \quad (g, s) \mapsto \mathcal{W}_g s,$$

is continuous. In particular, we have the following result that we state as a lemma for reference.

Lemma 3.21 *Let $\Omega \subset \mathcal{S}(\mathbb{R}^n)$ and $\Gamma \subset \mathcal{S}_0(\mathbb{R}^n)$ both be bounded sets. Then, there is $f \in \mathcal{S}(\mathbb{R}^n)$ and a function $\rho(a)$ satisfying (3.1) such that*

$$(g, s) \in \Gamma \times \Omega \Rightarrow |\mathcal{W}_g s(b, a)| \leq \rho(a) \frac{1}{(1+a)^n} f\left(\frac{b}{1+a}\right).$$

Proof. The localization in b is proved through Lemma 3.4. The decrease in $a \to 0$ follows from Lemma 3.5, since $|\widehat{s}(\xi)| \leq O(1)\phi_{0,m}(|\xi|)$ for all m. □

Note once more that there is no analogue for the wavelet synthesis. It is not true in general that a localization of the above type for r implies that $\mathcal{M}_h r$ is localized.

3.11 Wavelet analysis over $\mathbb{T}^n$

All results we have shown so far carry over in a natural way if we consider spaces of periodic functions. More precisely consider the n-dimensional torus $\mathbb{T}^n$ that we identify with $\mathbb{R}^n/\mathbb{Z}^n$. Accordingly we identify functions over $\mathbb{T}^n$ with periodic functions. From the translation covariance of the wavelet transform it follows that the wavelet transform of a periodic function is a function that is periodic in the first variable

$$\mathcal{W}_g s(b + k, a) = \mathcal{W}_g s(b, a), \quad k \in \mathbb{Z}^n.$$

Therefore, if we introduce the notation $\mathbb{M}^n = \mathbb{T}^n \times \mathbb{R}_+$, then we may identify wavelet transforms of functions over $\mathbb{T}^n$ with functions over the n-dimensional half-cylinder $\mathbb{M}^n$.

Accordingly we write $\mathcal{S}_0(\mathbb{T}^n)$ for the space of smooth functions together with the norms $(m \in \mathbb{N}^n)$

$$\|s\|_m = \|\partial^m s\|_{L^\infty(\mathbb{T}^n)},$$

and for which we have

$$\int_{\mathbb{T}^n} s = 0.$$

Its dual, the space of periodic distributions shall be denoted by $\mathcal{S}_0'(\mathbb{T}^n)$. On $\mathbb{M}^n$ let $\mathcal{S}(\mathbb{M}^n)$ denote the space of functions for which

$$\|s\|_k = \sup_{(b,a) \in \mathbb{M}^n} \left|(a + 1/a)^k s(b, a)\right| < \infty.$$

Then, it can be shown that as before

$$\mathcal{W} : \mathcal{S}_0(\mathbb{R}^n) \times \mathcal{S}_0(\mathbb{T}^n) \to \mathcal{S}(\mathbb{M}^n), \quad (g, s) \mapsto \mathcal{W}_g s,$$

is continuous.

The analogue holds for the wavelet synthesis defined as before upon identifying functions over $\mathbb{M}^n$ with periodic functions over $\mathbb{H}^n$. Again we have that

$$\mathcal{M} : \mathcal{S}_0(\mathbb{R}^n) \times \mathcal{S}(\mathbb{M}^n) \to \mathcal{S}_0(\mathbb{T}^n), \quad (h, r) \mapsto \mathcal{M}_h r,$$

is continuous.

The analogue statements hold for $\mathcal{S}_0'(\mathbb{T}^n)$ and $\mathcal{S}'(\mathbb{M}^n)$.

Note that the wavelet analysis and synthesis can also be formulated with the help of periodized wavelets. For this let us for $g \in \mathcal{S}(\mathbb{R}^n)$ denote by g° its periodization

$$g^\circ(x) = \sum_{k \in \mathbb{Z}^n} g(x + k).$$

Thus in particular $g^\circ \in \mathcal{S}_0(\mathbb{T}^n)$ for $g \in \mathcal{S}_0(\mathbb{R}^n)$. Then the periodized wavelet $g^\circ_{b,a} = (g_{b,a})^\circ$ may be used to write ($s \in \mathcal{S}_0(\mathbb{T}^n)$, $r \in \mathcal{S}(\mathbb{M}^n)$)

$$\mathcal{W}_g s(b, a) = \int_{\mathbb{T}^n} dx \, \overline{g^\circ_{b,a}}(x) \, s(x), \quad \mathcal{M}_h r(x) = \int_{\mathbb{M}^n} \frac{dbda}{a} \, r(b, a) \, h^\circ_{b,a}(x).$$

Since there is a natural maximal length scale (=the period) it is no surprise that the wavelet coefficients decay rapidly at large scale regardless of the distribution we analyzed

$$|\mathcal{W}_g s(b, a)| \leq O(a^{-K}), \quad (a \to \infty), \forall K > 0.$$

To see this note that

$$\mathcal{W}_g s(b, a) = \sum_{k \in \mathbb{Z}^n} \overline{\widehat{g}}(2\pi a k) \, e^{2i\pi k b} \, \widehat{s}_k,$$

where the Fourier coefficients of s read

$$\widehat{s}_k = \int_{\mathbb{T}^n} s(x) e^{-2i\pi k x}.$$

Since $\widehat{g}(0) = 0$, it follows that the sum is rapidly decaying as $a \to \infty$ thanks to the rapid decay of $\widehat{g}$. In particular we may choose g in such a way that for all $s \in \mathcal{S}'(\mathbb{T}^n)$ we have

$$a \geq 1 \Rightarrow \mathcal{W}_g s(b, a) = 0.$$

Indeed take $\widehat{g}$ compactly supported with support contained in $[-2\pi, 2\pi]^n$. This remark will play a role in some demonstrations.

4 Some Banach spaces

So far we have seen how $\mathcal{S}_0(\mathbb{R}^n)$, $L^2(\mathbb{R}^n)$ and $\mathcal{S}_0'(\mathbb{R}^n)$ can be characterized through wavelet coefficients. Many other function spaces can be characterized in terms of wavelet transforms. As a rule, the faster the wavelet coefficients decay, the more the analyzed function is regular. We will study this in more detail later. For the moment we give a simple proof of how to characterize the scale of Sobolev spaces through wavelets. As usual we define $H_\gamma(\mathbb{R}^n)$ as the Hilbert space of all distributions for which the following norms are finite

$$\|s\|^2_{H_\gamma(\mathbb{R}^n)} = \int d\xi\,(1+|\xi|^2)^{\gamma/2}\,|\widehat{s}(\xi)|^2 .$$

We also introduce the homogeneous Sobolev space $H^h_\gamma(\mathbb{R}^n)$ as the space of distributions for which

$$\|s\|^2_{H^h_\gamma(\mathbb{R}^n)} = \int d\xi\,|\xi|^{2\gamma}\,|\widehat{s}(\xi)|^2 .$$

is finite. Note that $H^h_\gamma(\mathbb{R}^n)$ is a Banach space only if we factorize by the polynomials of degree $\leq \gamma$.

Theorem 4.1 *Let $g \in \mathcal{S}_0(\mathbb{R}^n)$ be admissible. Then, there is a $c \geq 1$ such that*

$$c^{-1}\,\|s\|^2_{H_\gamma(\mathbb{R}^n)} \leq \int_{\mathbb{H}^n} \frac{dbda}{a}(1+a^{-2\gamma})\,|\mathcal{W}_g s(b,a)|^2 \leq c\,\|s\|^2_{H_\gamma(\mathbb{R}^n)},$$

and

$$c^{-1}\,\|s\|^2_{H^h_\gamma(\mathbb{R}^n)} \leq \int_{\mathbb{H}^n} \frac{dbda}{a} a^{-2\gamma}\,|\mathcal{W}_g s(b,a)|^2 \leq c\,\|s\|^2_{H_\gamma(\mathbb{R}^n)}.$$

This shows in a more qualitative way that localization of the wavelet coefficients at small scale is linked to local regularity. In the appendix we give a more detailed discussion of this point in terms of various moduli of continuity and their characterization through wavelet coefficients.

Proof. We have

$$\|s\|^2_{H_\gamma(\mathbb{R}^n)} = \left\|(1-\Delta)^{\gamma/2}s\right\|_{L^2(\mathbb{R}^n)}, \quad \|s\|^2_{H^h_\gamma(\mathbb{R}^n)} = \left\|(-\Delta)^{\gamma/2}s\right\|_{L^2(\mathbb{R}^n)}.$$

Now as integration by parts shows, we have

$$\mathcal{W}_h(-\Delta)^{\gamma/2}s(b,a) = a^{-\gamma}\mathcal{W}_g s(b,a), \quad \widehat{g}(\xi) = |\xi|^\gamma\,\widehat{h}(\xi).$$

Together with h, we have that g is admissible. Therefore we may use the characterization of $L^2(\mathbb{R}^n)$ in wavelet space to conclude

$$\|s\|^2_{H^h_\gamma(\mathbb{R}^n)} = \left\|(-\Delta)^{\gamma/2}s\right\|_{L^2(\mathbb{R}^n)} \sim \left\|\mathcal{W}_h(-\Delta)^{\gamma/2}s\right\|^2_{L^2(\mathbb{H}^n,dbda/a)} = \left\|a^{-\gamma}\mathcal{W}_g s\right\|_{L^2(\mathbb{R}^n)}.$$

This shows the second equivalence in the theorem. The first equivalence follows by cutting the half-space into the two regions $a < 1$ and $a \geq 1$ and we may leave the details to the reader. □

We now want to generalize this situation to more general Banach spaces. We have already seen that

$$\mathcal{S}_0(\mathbb{R}^n) \subset L^2(\mathbb{R}^n) \subset \mathcal{S}_0'(\mathbb{R}^n),$$

are characterized by the fact that the wavelet coefficients are in the following scale of spaces

$$\mathcal{S}_0(\mathbb{H}^n) \subset L^2(\mathbb{H}^n) \subset \mathcal{S}_0'(\mathbb{H}^n).$$

We generalize this situation and we recall the general principles of characterization of functions spaces through wavelet coefficients (see e.g. [FG89a],[FG89b] for a more abstract setting) For this consider a Banach space $B(\mathbb{H}^n)$ of locally integrable functions. We always require that it satisfies the following properties.

(A1) We have

$$\mathcal{S}(\mathbb{H}^n) \subset B(\mathbb{H}^n) \subset \mathcal{S}'(\mathbb{H}^n),$$

and the inclusions hold in the topological sense.

(A2) We have a Banach lattice

$$\|\,|s|\,\|_{B(\mathbb{H}^n)} = \|s\|_{B(\mathbb{H}^n)}\,.$$

(A3) It is tempered in the following sense. We suppose that there are $K > 0$ and $c > 0$ such that for all $s \in B(\mathbb{H}^n)$ we can estimate for $s_{\beta,\alpha}(b,a) = s(b + a\beta, \alpha a)$

$$\|s_{\beta,\alpha}\|_{B(\mathbb{H}^n)} \leq c\,\|s\|_{B(\mathbb{H}^n)}\,(\alpha + 1/\alpha)^K (1 + |\beta|)^K.$$

(A4) Its multiplier algebra contains the bounded functions

$$\mu \in L^\infty(\mathbb{H}^n), s \in B(\mathbb{H}^n) \Rightarrow \mu \cdot s \in B(\mathbb{H}^n),$$

and we have, if we denote by $M[\mu]$ the operator $s \mapsto \mu \cdot s$

$$\|M[\mu]\|_{B(\mathbb{H}^n)\to B(\mathbb{H}^n)} = \|\mu\|_{L^\infty(\mathbb{H}^n)}\,.$$

These are minimal requirements and we say that $B(\mathbb{H}^n)$ is an *admissible* Banach space.

From the first two conditions it follows easily that $B(\mathbb{H}^n)$ is invariant under convolutions with highly localized kernels.

$$r \in \mathcal{S}(\mathbb{H}^n),\ s \in B(\mathbb{H}^n) \quad \Rightarrow \quad r * s,\ s * r \in B(\mathbb{H}^n).$$

The convolution of two functions over $\mathbb{H}^n$ is defined through 2.1. More than that, we have the following lemma.

Lemma 4.2 *For fixed $P \in \mathcal{S}(\mathbb{H}^n)$ we have*

$$A[P] : B(\mathbb{H}^n) \to B(\mathbb{H}^n), \quad r \mapsto P * r,$$

is continuous.

Proof. By a simple change of variables, we may write ($s_{\beta,\alpha}(b,a) = s(b + a\beta, \alpha a)$)

$$P * s = \int_{\mathbb{H}^n} \frac{d\beta d\alpha}{\alpha} P(-\beta/\alpha, 1/\alpha) s_{\beta,\alpha}.$$

Upon passing to the absolute value and then to the norm on both sides we have using hypothesis (A3)

$$\|P * s\|_{B(\mathbb{H}^n)} \leq \|s\|_{B(\mathbb{H}^n)} \, c \int_{\mathbb{H}^n} \frac{d\beta d\alpha}{\alpha} \, |P(-\beta/\alpha, 1/\alpha)| \, (\alpha + 1/\alpha)^K (1 + |\beta|)^K.$$

The last integral is a finite constant. □

It therefore makes sense to pull back $B(\mathbb{H}^n)$ to some space of distributions over $\mathbb{R}^n$. More precisely, we define $B(\mathbb{R}^n)$ through the next proposition.

Proposition 4.3 *The following is equivalent for $s \in \mathcal{S}_0'(\mathbb{R}^n)$*

- *For some $g \in \mathcal{S}_0(\mathbb{R}^n)$ satsifying at*

$$0 < c^{-1} \leq \int_0^\infty \frac{da}{a} \, |\hat{g}(a\xi)|^2 \leq c < \infty, \quad \xi \neq 0,$$

  we have

$$\mathcal{W}_g s \in B(\mathbb{H}^n).$$

- *For all $h \in \mathcal{S}_0(\mathbb{R}^n)$ we have*

$$\mathcal{W}_h s \in B(\mathbb{H}^n).$$

We shall denote this space of distributions by $B(\mathbb{R}^n)$. More precisely we define for g admissible

$$B(\mathbb{R}^n) = \{\eta \in \mathcal{S}_0'(\mathbb{R}^n) : \mathcal{W}_g \eta \in B(\mathbb{H}^n)\}.$$

We can make $B(\mathbb{R}^n)$ a Banach space by setting

$$\|s\|_{B(\mathbb{R}^n)} = \|\mathcal{W}_g s\|_{B(\mathbb{H}^n)} \, .$$

Theorem 4.4 *The space $B(\mathbb{R}^n)$ is a well-defined Banach space. More precisely, as a set its definition does not depend on h. Its topology is well-defined, since for all g, $h \in \mathcal{S}_0(\mathbb{R}^n)$ there is a $c \geq 1$ such that*

$$\forall s \in B(\mathbb{R}^n) : c^{-1} \|\mathcal{W}_g s\|_{B(\mathbb{H}^n)} \leq \|\mathcal{W}_h s\|_{B(\mathbb{H}^n)} \leq c \|\mathcal{W}_g s\|_{B(\mathbb{H}^n)} \, .$$

Proof. As we have seen in Proposition 3.14 the transition from h to g is obtained via convolution with the help of a kernel in $\mathcal{S}(\mathbb{H}^n)$. But this a continuous map. □

We have the following easy convergence result that we state here for later reference.

Lemma 4.5 *For $B(\mathbb{H}^n)$ satisfying (A1) the following holds. Suppose that the family $s_m \in B(\mathbb{H}^n)$ is uniformly continuous on compact sets. Suppose further that $s_m(b,a) \to s(b,a)$ point-wise. Suppose further that $|s_m(b,a)| \le r(b,a)$ for some $r \in \mathcal{S}(\mathbb{H}^n)$. Then*

$$\lim_{m\to\infty} \|s_m - s\|_{B(\mathbb{H}^n)} = 0.$$

Proof. Consider the compact sets $P_w = \{(a + 1/a) + |b| \le w\}$ and denote the multiplier with its characteristic function with the same symbol. We have uniformly in m, that $(\mathbb{1} - P_w)(s_m - s) \to 0$ in the topology of $\mathcal{S}(\mathbb{H}^n)$ by localization. Therefore a forteriori by (A1)

$$\lim_{w\to\infty} \|P_w s_m\|_{B(\mathbb{H}^n)} = 0 \quad \text{uniformly in } m.$$

On the other hand, for fixed w we have $P_w s_m(b,a) \to P_w s(b,a)$ uniformly on the compact set P_w beause the s_m are uniformly continuous. Therefore, for fixed w we have $P_w s_m \to P_w s$ in $\mathcal{S}(\mathbb{H}^n)$ and a forteriori in $B(\mathbb{H}^n)$. □

We now list some additional conditions, that we shall need in order to prove the subsequent theorems. In Definition 4.8 below we shall summarize the different concepts we need.

We sometimes have to replace the third condition for the Banach space $B(\mathbb{H}^n)$ by the slightly stronger condition .

(ST) We say $B(\mathbb{H}^n)$ is *strongly tempered* in the following sense. We suppose that there are ϵ, K and $c > 0$ such that for all $s \in B(\mathbb{H}^n)$ we can estimate with $s_{\beta,\alpha}(b,a) = s(b + a\beta, \alpha a)$

$$\|s_{\beta,\alpha}\|_{B(\mathbb{H}^n)} \le c\,\|s\|_{B(\mathbb{H}^n)}\, \alpha^\epsilon (1+\alpha)^K (1+|\beta|)^K. \tag{4.1}$$

A homogeneity property is also needed.

(H) We say $B(\mathbb{H}^n)$ is homogeneous (of degree $\gamma \in \mathbb{R}$) if

$$\begin{aligned} \|s(\alpha\cdot+\beta,\cdot\alpha)\|_{B(\mathbb{H}^n)} &= \alpha^\gamma\,\|s\|_{B(\mathbb{H}^n)} \\ \Longleftrightarrow\ \|T_\beta D_\alpha r\|_{B(\mathbb{R}^n)} &= \alpha^{-n-\gamma}\,\|r\|_{B(\mathbb{R}^n)}, \end{aligned} \tag{4.2}$$

for some $\gamma \in \mathbb{R}$

We sometimes need the following density requirement.

second such family of equal cardinality. Let A *be a bounded, linear operator acting in* $B(\mathbb{H}^n)$*. Then we have*

$$\left\|\sum \Omega_k A\Gamma_k\right\| \leq \max_k \|\Omega_k A\Gamma_k\|.$$

In case that condition (CR2′) *holds, we have instead, with arbitrary sets* Ω_k, Γ_k,

$$\left\|\sum \Omega_k A\Gamma_k\right\| \leq \max_k \|\Omega_k A\Gamma_k\| (1+\eta).$$

Here η *is a function that satisfies* $\eta \to 0$ *as* $\delta_{cone}(\Omega_k, \Omega_{k'}) \to \infty$ *and* $\delta_{cone}(\Gamma_k, \Gamma_{k'}) \to \infty$*, for* $k \neq k'$.

Proof. We have

$$\left\|\sum \Omega_k A\Gamma_k\right\| = \sup_{\|s\|=1} \left\|\sum \Omega_k A\Gamma_k s\right\| \leq \sup_{\gamma_K((\|s_k\|))=1} \left\|\sum \Omega_k A\Gamma_k s_k\right\|,$$

where the s_k satisfy $\Gamma_k s_k = s_k$. The last inequality follows from the fact that $\|\sum_k \Gamma_k s\| \leq \|s\|$ because of (A4). Now

$$\sup_{\gamma_K((\|s_k\|))=1} \left\|\sum \Omega_k A\Gamma_k s_k\right\| = \sup_{\gamma_K((\|s_k\|))=1} \gamma_K((\|\Omega_k A\Gamma_k s_k\|)).$$

We may conclude by noting that

$$\gamma_K((\|\Omega_k A\Gamma_k s_k\|)) \leq \gamma_K((\|\Omega_k A\Gamma_k\| \, \|s_k\|)) \leq (\max_k \|\Omega_k A\Gamma_k\|)\gamma_K((\|s_k\|)),$$

thanks to (4.6).

The second part is similar and we may leave the details to the reader. □

To summarize we have the following definition for the different types of spaces we consider.

Definition 4.8 *We say a space* $B(\mathbb{H}^n)$ *is admissible if it satisfies conditions* (A1) - (A4)*. It is strongly tempered if it satisfies* (ST)*. It is deformable if it satisfies* (D)*. An admissible, deformable space for which* (R) *holds true is called regular. A regular space which is homogeneous* (H) *and for which in addition conditions* (CR1) *and* (CR2′) *hold is called completely regular.*

4.1 Admissible dual pairs

Of some interest are those spaces for which $B(\mathbb{H}^n)$ and its dual $B'(\mathbb{H}^n)$ are admissible. In this situation it is convenient to suppose that the duality between $B(\mathbb{H}^n)$ and $B'(\mathbb{H}^n)$ is implemented via the following: $s \in B(\mathbb{H}^n)$, $r \in B'(\mathbb{H}^n)$ implies $s \cdot r \in L^1(\mathbb{H}, dbda/a)$ and

$$r(s) = (r, s) = \int_{\mathbb{H}^n} \frac{dbda}{a} r(b,a)s(b,a).$$

We then have for $s \in B'(\mathbb{H}^n)$ and $r \in B(\mathbb{H}^n)$ that

$$|(s,r)_{\mathbb{H}^n}| \leq \|s\|_{B'(\mathbb{H}^n)} \, \|r\|_{B(\mathbb{H}^n)} \,. \tag{4.1}$$

Proposition 4.9 *If $\mathcal{S}(\mathbb{H}^n)$ is dense in $B(\mathbb{H}^n)$, then the dual $(B(\mathbb{R}^n))'$ of $B(\mathbb{R}^n)$ is canonically identified with $B'(\mathbb{R}^n)$.*

Proof. We may choose a $g \in \mathcal{S}_0(\mathbb{R}^n)$, radial and strictly admissible. Then for the reproducing kernel $P = \mathcal{W}_g g$ we have that $\Pi : s \to P * s$ is a self-adjoint projector on the image of $\mathcal{W}_g$. On the one hand, we have that $(B(\mathbb{R}^n))' \subset \mathcal{S}_0'(\mathbb{R}^n)$. On the other hand $B'(\mathbb{R}^n)$ is isomorph to $\mathcal{W}_g B'(\mathbb{R}^n)$, which is the closed supspace of $r \in B'(\mathbb{H}^n)$ for which $\Pi r = r$. Now for $s \in (B(\mathbb{R}^n))' \subset \mathcal{S}'(\mathbb{H}^n)$ and $r \in \mathcal{S}(\mathbb{H}^n)$ we have

$$(s, \Pi r) = (s, \Pi\Pi r) = (\Pi s, \Pi r).$$

Since by hypothesis $\mathcal{S}(\mathbb{H}^n)$ is dense in $B(\mathbb{H}^n)$ the proposition follows. □

Actually the duality can again be realized by a scalar product in the usual sense. Let g, $h \in \mathcal{S}_0(\mathbb{R}^n)$ be an analysis reconstruction pair. Then, we have for $s \in B_0'(\mathbb{R}^n)$ and $s \in B(\mathbb{R}^n)$ that

$\mathcal{W}_{\overline{h}} s \cdot \mathcal{W}_g r \in L^1(\mathbb{H}, dbda/a)$ and in addition

$$(s,r)_{\mathbb{R}^n} = \int_{\mathbb{H}^n} \frac{dbda}{a} \, \mathcal{W}_{\overline{h}} s(b,a) \, \mathcal{W}_g r(b,a) = (\mathcal{W}_{\overline{h}} s, \mathcal{W}_g r)_{\mathbb{H}^n} \,. \tag{4.2}$$

4.2 Spaces of periodic functions

For any admissible Banach space $B(\mathbb{H}^n)$ we can associate a second admissible Banach space $B_{\text{loc}}(\mathbb{H}^n)$ in the following way. Let χ denote the characteristic function of $\mathbb{M}^n = [0,1)^n \times \mathbb{R}_+$. Then, we may define a norm via

$$\sup_k \|\chi \cdot s(\cdot - k, \cdot)\|_{B(\mathbb{H}^n)} \,.$$

It is plain to see that this new Banach space is again admissible. Consider the subspace, denoted by $B(\mathbb{M}^n) \subset B_{\text{loc}}(\mathbb{H}^n)$ of periodic functions. On $B(\mathbb{M}^n)$, we have the following norm for $s \in B(\mathbb{M}^n) \subset B_{\text{loc}}(\mathbb{H}^n)$ which is equivalent to the induced norm

$$\|s\|_{B(\mathbb{M}^n)} := \|s\|_{B_{\text{loc}}(\mathbb{H}^n)} = \|\chi \cdot s\|_{B(\mathbb{H}^n)} \,.$$

This shows that $B(\mathbb{M}^n)$ is closed. Indeed, if $s_m \in B(\mathbb{M}^n)$ and $s_m \to s$ in $B_{\text{loc}}(\mathbb{H}^n)$ then $\chi \cdot s_m \to \chi s$ in $B(\mathbb{H}^n)$. In addition $s(b,a)$ is periodic in b since the lattice translation operator $s(\cdot,\cdot) \to s(\cdot - k, \cdot)$, $k \in \mathbb{Z}^n$ is continuous in $B_{\text{loc}}(\mathbb{H}^n)$. This space may be pulled back to a space of periodic functions over $\mathbb{T}^n$ via a wavelet synthesis. We denote this space by $B(\mathbb{T}^n)$.

For later reference we state the following density lemma.

Lemma 4.10 *If $\mathcal{S}(\mathbb{H}^n)$ is dense in $B(\mathbb{H}^n)$, then $\mathcal{S}(\mathbb{M}^n)$ is dense in $B(\mathbb{M}^n)$.*

Proof. Let $s \in B(\mathbb{M}^n)$. Then $\chi \cdot s \in B(\mathbb{H}^n)$. We therefore can find an approximating sequence $r_m \in \mathcal{S}(\mathbb{H}^n)$ that converges towards $\chi \cdot s$. But then $\chi \cdot r_m \to \chi \cdot s$ too since $M[\chi]$ is a continuous projector. If we set s_m the unique periodic function that satisfies at $\chi s_m = \chi r_m$ we have found an approximating sequence. □

4.3 Examples of admissible $B(\mathbb{H}^n)$

It is time to consider some concrete examples of admissible Banach spaces. Many function spaces are contained in the following two scales of functions. The following expressions define norms for functions over the half-space $1 \le p, q \le \infty$, $\gamma \in \mathbb{R}$

$$\|r\| = \left\{ \int_0^\infty \frac{da}{a} a^{-\gamma p} \left(\int_{\mathbb{R}^n} db\, |r(b,a)|^q \right)^{p/q} \right\}^{1/p},$$

or for $1 \le p \le \infty$, $1 \le q < \infty$

$$\|r\| = \left\{ \int_{\mathbb{R}^n} db \left(\int_{\mathbb{H}^n} \frac{db'da'}{a'} a'^{-\gamma q} \frac{1}{a'^n} \left(1 + \frac{|b-b'|^2}{a'^2} \right)^{-n} |r(b',a')|^q \right)^{p/q} \right\}^{1/p}.$$

The associated Banach spaces are denoted by $B^{p,q}_\gamma(\mathbb{H}^n)$ and $F^{p,q}_\gamma(\mathbb{H}^n)$, respectively. They are admissible. The first scale of spaces contains the Besov spaces, whereas the second scale are the Triebel Lizorkin spaces, which contains in e.g. the L^p-potential spaces and Sobolev spaces (see e.g. [Mey90], [CM90], [Tri84] for more details).

For $1 \le p, q < \infty$, both scales of spaces are regular (Definition 4.8). The Triebel Lizorkin scale $F^{p,q}_\gamma(\mathbb{H}^n)$ is even completely regular but the Besov scale $B^{p,q}_\gamma(\mathbb{H}^n)$ is not since the norm is not asymptotically free. For $p = q$, hovever, the scale $B^{p,p}_\gamma$ is assymptotically free, and these Besov spaces are completely regular for $p = q < \infty$. It even satisfies condition (CR2), whereas the Triebel Lizorkin norm above satisfies only (CR2′). For either p or $q = \infty$, $\mathcal{S}(\mathbb{H}^n)$ is not dense any more in any of these spaces, and the spaces are not regular any more. However, the closures $f^{p,q}_\gamma(\mathbb{H}^n)$ (respectively $b^{p,q}_\gamma(\mathbb{H}^n)$) of $\mathcal{S}(\mathbb{H}^n)$ in $F^{p,q}_\gamma(\mathbb{H}^n)$ (respectively $B^{p,q}_\gamma(\mathbb{H}^n)$) are regular, and in the case of $f^{p,q}_\gamma(\mathbb{H}^n)$ and $b^{p,p}_\gamma$, completely regular for all considered p, q.

We have already seen that $F^{2,2}_1(\mathbb{R}^n) = B^{2,2}_1(\mathbb{R}^n) = L^2(\mathbb{R}^n)$ and that the Sobolev spaces are

$$F^{2,2}_\gamma(\mathbb{R}^n) = H^h_\gamma(\mathbb{R}^n).$$

More general we have

$$F^{p,2}_1(\mathbb{R}^n) = L^p(\mathbb{R}^n).$$

and accordingly $F^{p,2}_\gamma(\mathbb{R}^n)$ are the homogeneous potential spaces. In the appendix we prove that $B^{\infty,\infty}_\gamma(\mathbb{R}^n)$ is the Hölder-Zygmund scale of spaces.

4.4 A priori estimates

We now want to give some a-priori growth estimates of wavelet coefficients and of their derivatives for functions in $B(\mathbb{R}^n)$. It illustrates once more that qualitatively regularity is linked to growth behavior of wavelet coefficients.

Lemma 4.11 *For $B(\mathbb{H}^n)$ admissible, we have for $P \in \mathcal{S}(\mathbb{H}^n)$ and $r \in B(\mathbb{H}^n)$ that*

$$|P * r(b,a)| \leq c\, \|r\|_{B(\mathbb{H}^n)} (a + 1/a)^M (1 + |b|)^M,$$

where $c > 0$ depends only on P and $B(\mathbb{H}^n)$ and $M > 0$ depends only on $B(\mathbb{H}^n)$.

Proof. By a simple change of variables we may write as before

$$P * r(b,a) = \int_{\mathbb{H}^n} \frac{dbda}{a} P(-b'/a', 1/a')\, r(b + ab', aa').$$

Now we have that $\tilde{P}(b,a) = \overline{P}(-b/a, 1/a)$ is in $\mathcal{S}(\mathbb{H}^n)$ and hence, by the first hypothesis on $B(\mathbb{H}^n)$, it is in the dual, $B'(\mathbb{H}^n)$ of $B(\mathbb{H}^n)$. We therefore may estimate using the temperdness of the norm in $B(\mathbb{H}^n)$, that

$$\begin{aligned} |P * r(b,a)| &\leq c\, \|r\|_{B(\mathbb{H}^n)} \left\|\tilde{P}\right\|_{B'(\mathbb{H}^n)} (a + 1/a)^M (1 + |b|)^M \\ &\leq c'\, \|r\|_{B(\mathbb{H}^n)} (a + 1/a)^M (1 + |b|)^M. \end{aligned}$$

This is the desired estimate. □

Corollary 4.12 *Suppose $\partial_b^\alpha \partial_a^m P \in \mathcal{S}(\mathbb{R}^n)$ for every mulit-index α and every $m \in \mathbb{N}$. Then we have that the growth behavior of the derivatives of $P * r$ may be estimated in terms of the norms of r as follows*

$$|\partial_b^\alpha P * r(b,a)| \leq c\, \|r\|_{B(\mathbb{H}^n)} (a + 1/a)^{M+|\alpha|} (1 + |b|)^M,$$

and

$$|\partial_a P * r(b,a)| \leq c\, \|r\|_{B(\mathbb{H}^n)} (a + 1/a)^{M+1} (1 + |b|)^M.$$

with some c depending on α and m only.

Proof. (of Lemma 4.11.) We have

$$\partial_b^\alpha \frac{1}{a'^n} P\left(\frac{b-b'}{a'}, \frac{a}{a'}\right) = a'^{-|\alpha|} \frac{1}{a'^n} (\partial_b^\alpha P)\left(\frac{b-b'}{a'}, \frac{a}{a'}\right),$$

and

$$\partial_a \frac{1}{a'^n} P\left(\frac{b-b'}{a'}, \frac{a}{a'}\right) = a'^{-1} \frac{1}{a'^n} (a\partial_a P)\left(\frac{b-b'}{a'}, \frac{a}{a'}\right).$$

□

In particular this shows thanks to the reproducing kernel equation $\mathcal{W}_g s = P_{g,h} * \mathcal{W}_g s$, (compare (2.2)) that for all bounded sets $\Omega \subset B(\mathbb{R}^n)$, the tempered growth behavior of $\mathcal{W}_g s(b,a)$ and all its derivatives can be estimated uniformly for in $s \in \Omega$.

4.5 Compact embedding

Let $\gamma \in \mathbb{R}$. Then, denote by $a^\gamma B(\mathbb{H}^n)$ the space of functions $a^\gamma r(b,a)$ with $r \in B(\mathbb{H}^n)$. This new space is a Banach space by choosing the final topology for the map $r \mapsto a^\gamma r$. It is plain to see that if $B(\mathbb{H}^n)$ is admissible, then the same holds true for $a^\gamma B(\mathbb{H}^n)$. In case $\gamma \geq 0$, the analogue holds if we require the stronger temperdness of condition (4.1). Note once more that partial integration shows that we have the following formula

$$\mathcal{W}_g(-\Delta)^{\gamma/2}s(b,a) = a^{-\gamma}\mathcal{W}_h s(b,a), \quad h = (-\Delta)^{\gamma/2}g.$$

Therefore multiplication with powers of a essentially acts like a smoothing operator.

More generally, let $\rho : \mathbb{H}^n \to \mathbb{R}_+$ satisfy

$$\rho(b+\beta, \alpha a) \leq O((\alpha + 1/\alpha)^m(1+|\beta|)^m \rho(b,a),$$

for some $m > 0$. We call such a ρ a tempered weight function. Examples for such ρ are obviously $r = a^\gamma$ and $\rho = (\alpha + 1/\alpha)^\gamma(1+|\beta|)^\gamma$. Note that the space of tempered weight functions is closed under addition, multiplication and formation of arbitrary real powers. Note that for all tempered ρ we have, in particular,

$$\rho(b,a) \leq c(a+1/a)^m(1+|b|)^m,$$

for some $c > 0$ and some $m > 0$.

Then, with the obvious notations we define $\rho \cdot B(\mathbb{H}^n)$ as the image on $B(\mathbb{H}^n)$ of the multiplication operator with ρ. We make it a Banach space through the norm

$$\|s\|_{\rho \cdot B(\mathbb{H}^n)} = \|s/\rho\|_{B(\mathbb{H}^n)}.$$

Proposition 4.13 *For tempered ρ and admissible $B(\mathbb{H}^n)$, we have that $\rho \cdot B(\mathbb{H}^n)$ is admissible, too.*

Proof. Obvious. □

We therefore can define $\rho B(\mathbb{R}^n)$ without ambiguity. Note already now, that this space is the image of $B(\mathbb{R}^n)$ under the Toeplitz operator

$$\mathcal{M}_h \rho \cdot \mathcal{W}_g,$$

where g and $h \in \mathcal{S}_0(\mathbb{R}^n)$ are admissible wavelets.

We now prove a fairly general theorem concerning compact embeddings. It shows that for weight functions that vanish at infinity the embedding $\rho B(\mathbb{H}^n) \to B(\mathbb{H}^n)$ is compact. Note however, that the infinity on $\mathbb{H}^n$ means small or large scales or large positions.

Theorem 4.14 *Let $\rho(b,a)$, be a tempered weight function that is $o(1)$ as $(a+1/a)(1+|b|) \to \infty$. Then, $\rho B(\mathbb{R}^n) \subset B(\mathbb{R}^n)$ and the embedding is compact for any admissible Banach space $B(\mathbb{R}^n)$.*

An analogous statement holds for spaces of periodic distributions as discussed in the appendix. In that case only the behavior of ρ at small scale $a \to 0$ counts.

Proof. Since ρ is bounded, it follows that $\rho B(\mathbb{R}^n) \subset B(\mathbb{R}^n)$. We now have to show that given a sequence $s_l(b,a) = \mathcal{W}_g r_l(b,a)$, such that $\|s_l\|_{\rho B(\mathbb{H}^n)} \leq 1$ we can extract a subsequence $\tilde{s}_m = s_{m(l)}$, that converges in $B(\mathbb{H}^n)$:

$$\lim_{k,l\to\infty} \|\tilde{s}_k - \tilde{s}_l\|_{B(\mathbb{H}^n)} = 0.$$

We first observe that thanks to Lemma 4.11 we have that for any multi-index $|\alpha| = 1$

$$|s_m(b,a)| + |\partial_b^\alpha s_m(b,a)| + |\partial_a s_m(b,a)| \leq c(a + 1/a)^M (1 + |b|)^M,$$

for some $c > 0$ and $M > 0$. Consider now the level sets

$$I_k = \overline{\{(b,a) \in \mathbb{H}^n : \rho(b,a) \leq k\}}.$$

These sets are compact, monotone $I_k \subset I_{k+1}$ and exhaustive $\mathbb{H}^n = \bigcup I_k$. Moreover, we have, with χ_{I_k} the characteristic function of I_k, that

$$\lim_{k\to\infty} \|(1 - \chi_{I_k})\rho\|_{L^\infty(\mathbb{H}^n)} = 0.$$

Now, on each I_k, the functions s_m are uniformly continuous. Therefore for every $k \in \mathbb{N}$ we may select a sub-sequence of the s_m that converges uniformly in I_k. Therefore using a diagonal sequence, we have that $\tilde{s}_m$ converges uniformly in I_k for all $k \in \mathbb{N}$ to some continuous function $\tilde{s}(b,a)$. Since $B(\mathbb{H}^n)$ is tempered, we have that

$$\forall K \in \mathbb{N}: \quad \|\chi_{I_k}(\tilde{s}_l - \tilde{s}_m)\|_{B(\mathbb{H}^n)} \to 0 \quad (l, m \to \infty).$$

On the other hand

$$\begin{aligned} \|(1 - \chi_{I_k})\tilde{s}_m\|_{B(\mathbb{H}^n)} &= \|\rho\,(1 - \chi_{I_k})\tilde{s}_m/\rho\|_{B(\mathbb{H}^n)} \\ &\leq \|\rho(1 - \chi_{I_k})\|_{L^\infty(\mathbb{H}^n)} \|\tilde{s}_m/\rho\|_{\rho B(\mathbb{H}^n)} \\ &\leq c_1\rho(1/k). \end{aligned}$$

In the last inequality we have used the boundedness of s_m in $\rho B(\mathbb{H}^n)$. Therefore we end up with

$$\begin{aligned} &\limsup_{l,m\to\infty} \|\tilde{s}_l - \tilde{s}_m\|_{B(\mathbb{H}^n)} \\ &\leq \limsup_{l,m\to\infty} \|\chi_{I_k}(\tilde{s}_l - \tilde{s}_m)\|_{B(\mathbb{H}^n)} + \limsup_{l,m\to\infty} \|(1 - \chi_{I_k})(\tilde{s}_l - \tilde{s}_m)\|_{B(\mathbb{H}^n)} \\ &\leq O(1)\,(\rho(1/k)). \end{aligned}$$

Now as k gets large this is arbitrary small by hypothesis on ρ and the theorem follows. □

5 More general microlocal classes

In this section we shall be concerned with the problem of constructing new local regularity spaces out of old ones. The idea is easily explained. Consider an arbitrary set $\Omega \subset \mathbb{H}^n$. Eventually we want to consider lines $\Omega = \{(b = \lambda\xi, a = \lambda^\gamma)\}$ in order to define regularity in the direction $\xi \in \mathbb{R}^n$. But for the moment we stay general since it renders the discussion more easy to understand. Now fix in addition to Ω, an admissible, local regularity space $B(\mathbb{R}^n)$. It is characterized by the fact that the wavelet transforms of its members are in some vector space $B(\mathbb{H}^n)$. The regularity classes we want to construct are roughly speaking distributions whose wavelet coefficients have on Ω a growth behavior governed by the class $B(\mathbb{H}^n)$.

A slightly more general concept is obtained if we take two local regularity spaces $B_1(\mathbb{R}^n)$ and $B_2(\mathbb{R}^n)$, both of the type considered in Section 3.11. We now want to cut the half-space $\mathbb{H}^n$ into two parts, say Ω and its complement Ω^c. In some sense, to be made precise below, we consider classes of distributions whose wavelet coefficients behave inside Ω like the wavelet coefficients of functions in $B_1(\mathbb{R}^n)$, whereas in Ω^c, these distributions have regularity governed by $B_2(\mathbb{R}^n)$.

A naive approach might be to require that the restriction of $\mathcal{W}_g\eta$ to Ω satisfies $\mathcal{W}_g\eta \in B_1(\mathbb{H}^n)$, whereas its restriction to the complement of Ω should correspond (via $\mathcal{M}_h$) to a function in $B_2(\mathbb{R}^n)$. However this definition might depend on the wavelets we use and thus it is not useful.

To get around this difficulty we first construct a suitable family of neighborhoods for Ω. With these neighborhoods it turns out that we can define vector spaces that are independent of the wavelets g and h.

5.1 A non-Euclidian distance

The first step of the construction is to introduce a non-euclidian distance function adapted to the geometry of the half-space. A suitable choice is given by dist as defined via

$$\operatorname{dist}((b,a),(b',a')) = \max\{a/a', a'/a\} + |(b-b')/a'| + |(b'-b)/a| \,.$$

This clearly is not a distance in the usual sense, since $\operatorname{dist}((b,a),(b,a)) = 1$. However, a kind of multiplicative triangular inequality holds (see Lemma 5.1 below), so that $\log \operatorname{dist}((b,a),(b',a'))$ actually is a distance. Note However, that the distance function is symmetric.

$$\operatorname{dist}((b,a),(b',a')) = \operatorname{dist}((b',a'),(b,a)).$$

Clearly the upper half-space carries a natural group structure. It is given by the following composition law

$$(b,a) \circ (b',a') = \delta_a \tau_b (b',a') = (ab' + b, aa'),$$

where τ_b and δ_a stand for the translation and dilation as left actions of $\mathbb{R}_+$ and $\mathbb{R}^n$ on $\mathbb{H}^n$ via

$$\delta_\alpha : (b,a) \mapsto (\alpha b, \alpha a), \quad \tau_\beta : (b,a) \mapsto (b+\beta, a'). \tag{5.1}$$

The inverse element reads

$$(b,a)^{-1} = (-b/a, 1/a),$$

and the neutral element clearly is $(0,1)$.

If we denote by $\Delta((b,a))$ the distance of a point (b,a) from the point $(0,1)$

$$\begin{aligned}\Delta((b,a)) &= \operatorname{dist}((b,a),(0,1)) = \exp(|\log a|) + |b|\,(1+1/a) \\ &= \max\{a, 1/a\} + |b|\,(1+1/a),\end{aligned}$$

then we have the following relation

$$\operatorname{dist}((b,a),(b',a')) = \Delta((b,a)^{-1}(b',a')).$$

Note also the following identity

$$\Delta((b,a)^{-1}) = \Delta((b,a)).$$

Moreover, the function Δ is a tempered weight function

$$\Delta((b+\beta, \alpha a)) \leq (\alpha + 1/\alpha)(1+|\beta|)\Delta((b,a)).$$

The next lemma shows that a kind of triangular inequality holds

Lemma 5.1 *We have the following triangular inequalities*

$$\begin{aligned}\max\{ \;& \Delta((b,a))/\Delta((b',a')), \Delta((b',a'))/\Delta((b,a))\} \\ & \leq \Delta((b,a)(b',a')) \leq \Delta((b,a))\Delta((b',a')).\end{aligned}$$

Proof. To prove this inequality note that an elementary direct computation shows that

$$\begin{aligned}aa'\Delta((b,a)^{-1}(b',a')) &= aa'\Delta(((b'-b)/a, a'/a)) \\ &= \max\{a^2, a'^2\} + a\,|b-b'| + a'\,|b-b'| \\ &\leq \max\{a^2, a'^2\} + a\,|b| + a\,|b'| + a'\,|b| + a'\,|b'|\,.\end{aligned}$$

On the other hand

$$aa'\Delta((b,a))\Delta((b',a')) = (\max\{1,a^2\} + |b| + a\,|b|)(\max\{1,a'^2\} + |b'| + a'\,|b'|).$$

Upon neglecting positive terms, we have

$$\begin{aligned}& aa'\Delta((b,a))\Delta((b',a')) \\ & \quad \geq \max\{1,a^2\}\max\{1,a'^2\} + a\,|b| + \max\{1,a'^2\}\,|b| + a'\,|b'| + \max\{1,a^2\}\,|b'|\,.\end{aligned}$$

Therefore we have by the triangular inequality

$$\begin{aligned}\Delta((b',a'))^{\epsilon_1} &< \operatorname{dist}((b',a'),(b'',a'')) \\ &\leq \operatorname{dist}((b',a'),(b,a)) \operatorname{dist}((b,a),(b'',a'')) \\ &\leq \Delta((b',a'))^{\epsilon_3}\Delta((b'',a''))^{\epsilon_2}.\end{aligned}$$

Now, relation 5.2 implies

$$\Delta((b,a)) \leq \Delta((b',a'))^{1+\epsilon_3}.$$

Moreover, (5.3) implies

$$\Delta((b'',a''))^{1-\epsilon_2} \leq \Delta((b,a)).$$

It follows that

$$\Delta((b'',a'')) < \Delta((b',a'))^{\frac{1+\epsilon_3}{1-\epsilon_2}}.$$

Therefore

$$1 < \Delta((b',a'))^{\epsilon_3-\epsilon_1+\epsilon_2\frac{1+\epsilon_3}{1-\epsilon_2}},$$

which is impossible by the choice of ϵ_3 since $\Delta((b',a')) \geq 1$. □

This immediately implies the following corollary that we shall use in the next section.

Lemma 5.5 *Let $\Sigma \supset \Omega$ be such that Σ^c is well separated from Ω. Then there is a set Ξ, $\Sigma \supset \Xi \supset \Omega$ such that Ξ is well separated from Σ^c and Ω is well separated from Ξ^c.*

Proof. Some $\Gamma_\epsilon(\Omega)$ with ϵ small enough will do. □

5.4 Cutting the half-space

Let us come back to our original goal of dividing the half-space into two sets of different regularity. As we said already, it is not possible to speak of regularity $B(\mathbb{H}^n)$ inside a given set $\Omega \subset \mathbb{H}^n$, since this notion is not independent under highly regular Calderón Zygmund operators, or to put it simpler, it might depend on the given wavelet we use for the definition.

However, if we require regularity $B(\mathbb{H}^n)$ in a region that that is slightly larger than Ω, it then follows that the same regularity holds true in Ω for any wavelet.

We denote by abuse of notation Σ, (respectively Ω) the operator that restricts functions over $\mathbb{H}^n$ to the set Σ (respectively Ω). That is we have

$$\Sigma : r \mapsto \chi_\Sigma r,$$

where χ_Σ is the characteristic function of Σ.

Theorem 5.6 *Consider two sets* Σ *and* Ω *and suppose that* $\Sigma \supset \Omega$ *in such a way that* Σ^c *and* Ω *are well separated. Let* g, g', h, $h' \in \mathcal{S}_0(\mathbb{R}^n)$ *satisfy for some* $c > 1$ $(s = g,\, g',\, h,\, h')$,

$$c^{-1} < \int_0^\infty \frac{da}{a} |\widehat{s}(ak)|^2 < c.$$

Suppose that $\eta \in \mathcal{S}'(\mathbb{R}^n)$ *satisfies*

$$\mathcal{M}_h \Sigma \mathcal{W}_g \eta \in B(\mathbb{R}^n).$$

Then

$$\mathcal{M}_{h'} \Omega \mathcal{W}_{g'} \eta \in B(\mathbb{R}^n).$$

Therefore it makes sense to separate $\mathbb{H}^n$ into regions of different regularity provided, the regions are well separated.

The proof is based on the following lemma. It estimates the influence under convolution operators over the half-space of a nasty function inside some region Ξ on Ξ' when both are well separated.

Lemma 5.7 *Suppose* Ξ' *and* Ξ *are well separated. Let* r *be a locally integrable function over* $\mathbb{H}^n$ *that is equal to* 0 *except on* Ξ', *where it satisfies for some* $M > 0$ *and some* $c > 0$

$$(b,a) \in \Xi' \quad \Rightarrow \quad |r(b,a)| \leq c\,\Delta((b,a))^M.$$

Then, for $P \in \mathcal{S}(\mathbb{H}^n)$ *we have that* $\tilde{r} = P * r$ *satisfies*

$$(b,a) \in \Xi \quad \Rightarrow \quad |\tilde{r}(b,a)| \leq c_k \Delta((b,a))^{-k},$$

for all $k \in \mathbb{N}$.

Proof. We have to estimate the localization of $P * r(b,a)$ for $(b,a) \in \Xi$. By definition $P * r(b,a)$ equals

$$\int_{\Xi'} \frac{db'da'}{a'} \frac{1}{a'^n} P\left(\frac{b-b'}{a'}, \frac{a}{a'}\right) r(b',a').$$

By dilation and translation we also may write using the action (5.1) of dilation and translation on $\mathbb{H}^n$

$$\int_{\delta_{1/a}\tau_{-b}\Xi'} \frac{db'da'}{a'} P(-b'/a', 1/a')\, r(ab'+b, aa').$$

Now by hypothesis on r we may write with some K and some $c > 0$

$$|r(ab'+b, aa')| \leq c\,\Delta((b,a))^K \Delta((b',a'))^K.$$

Plugging this estimate into the previous expression we have to estimate for $(b,a) \in \Xi$

$$\Delta((b,a))^K \int_{\delta_{1/a}\tau_{-b}\Xi'} \frac{db'da'}{a'} |P'(b',a')|, \tag{5.1}$$

with $P'(b',a') = \Delta((b',a'))^K P(-b'/a', 1/a')$. Together with P we have that P' is highly localized.

For $\lambda \geq 0$ let us look at the following integral running over the complement of a non-euclidian ball centered at $(0,1)$

$$F(\lambda) = \int_{\Delta((b',a'))>\lambda} \frac{db'da'}{a'} |P'(b',a')| .$$

Thanks to the high localization of P', this function is faster decreasing than any power of $1/\lambda$ as $\lambda \to \infty$. Now, since Ξ and Ξ' are well separated we may use characterization (5.1) to conclude that the integral in (5.1) is estimated by $F(\Delta((b,a))^\epsilon)$ for some $\epsilon > 0$. But this function is again rapidly decreasing as $\Delta((b,a)) \to \infty$ and the proof is finished. □

Note that the lemma we went to prove may be rephrased as follows: for all $P \in \mathcal{S}(\mathbb{H}^n)$ we have that

$$s \mapsto \Xi' \left(P * (\Xi s)\right))$$

is infinitely smoothing in the sense that it maps functions of polynomial growth into rapidly decreasing functions over the half-space.

Proof. (of theorem 5.6) These previous considerations imply the following: if we have that $\Sigma \mathcal{W}_g \eta \in B(\mathbb{H}^n)$, with g admissible, then for all $g' \in \mathcal{S}_0(\mathbb{R}^n)$ we have that $\Omega \mathcal{W}_{g'} \eta \in B(\mathbb{H}^n)$. To show this note the transition from $\mathcal{W}_g \eta$ to $\mathcal{W}_{g'} \eta$ is done by convoluting with a highly localized kernel P. Now we may write,

$$\Omega \mathcal{W}_{g'} \eta = \Omega(P * (\Sigma \mathcal{W}_g \eta)) + \Omega(P * (\Sigma^c \mathcal{W}_g \eta)).$$

Since by hypothesis $B(\mathbb{H}^n)$ is invariant under multiplications with bounded functions and convolutions with P the first term is again in $B(\mathbb{H}^n)$, whereas the second term is arbitrary smooth.

A slightly more complicated situation occurs in our theorem, since we can not conclude from $\mathcal{M}_h r \in B(\mathbb{R}^n)$ that $r \in B(\mathbb{H}^n)$, since the wavelet synthesis is not injective.

Now we can find a set Ξ between Σ and Ω

$$\Omega \subset \Xi \subset \Sigma,$$

such that Ξ is well separated from the complement of Σ and Ω is well separated from the complement of Ξ. This follows from Lemma 5.5. We may conclude that

$$\Xi \mathcal{W}_g \mathcal{M}_h \Sigma \mathcal{W}_g \eta = \Xi(P_1 * (\Sigma \mathcal{W}_g \eta)) \in B(\mathbb{H}^n).$$

Where $P_1 = \mathcal{W}_f h$ for any admissible $f \in \mathcal{S}_0(\mathbb{R}^n)$. In particular, we may choose f to be a reconstruction wavelet for g and thus it follows that $P_1 * \mathcal{W}_g \eta = \mathcal{W}_g \eta$. Now writing (as characteristic functions!) $\Sigma = 1 - \Sigma^c$ the last expression equals

$$\Xi(P_1 * \mathcal{W}_g \eta) - \Xi(P_1 * (\Sigma^c \mathcal{W}_g \eta)).$$

The set Ξ is well separated from Σ^c and thus the second term has rapid decay as $\Delta((b,a))$ gets large. Let us call this function u. Then, since u is well localized we have

$$r * u \in \mathcal{S}(\mathbb{H}^n),$$

for all $r \in \mathcal{S}(\mathbb{H}^n)$. We therefore obtain, up to a function of rapid decay

$$\Xi \mathcal{W}_g \eta \in B(\mathbb{H}^n).$$

Now $\mathcal{W}_{g'}\eta = P * \mathcal{W}_g \eta$ for some $P \in \mathcal{S}(\mathbb{H}^n)$. Therefore, since $\Omega \subset \Xi$ is well separated from the complement of Ξ we have as at the beginning of the proof that $\Omega \mathcal{W}_{g'}\eta \in B(\mathbb{H}^n)$ up to the well localized function Ωu. But then clearly $\mathcal{M}_{h'}\Omega\mathcal{W}_{g'}\eta \in B(\mathbb{R}^n)$. □

Let $\Sigma \supset \Omega$ be open and let again Ω be well separated from the complement of Σ as before. Consider the two Toeplitz operators

$$T_\Sigma = \mathcal{M}_h \Sigma \mathcal{W}_g, \quad T_\Omega = \mathcal{M}_h \Omega \mathcal{W}_g.$$

We then have proved the following

Corollary 5.8 *We have that*

$$[T_\Sigma, T_\Omega] = T_\Sigma T_\Omega - T_\Omega T_\Sigma,$$

is infinitely smoothing in the sense that it maps the tempered distributions in $\mathcal{S}_0'(\mathbb{R}^n)$ into smooth function in $\mathcal{S}_0(\mathbb{R}^n)$.

5.5 Microlocal classes

The theorems of the previous section may be used to define some very general microlocal classes. Suppose we are given two regularity spaces $A(\mathbb{R}^n)$ and $B(\mathbb{R}^n)$ and Suppose in addition that $B(\mathbb{R}^n) \subset A(\mathbb{R}^n)$. Consider a set $\Omega \subset \mathbb{H}^n$. Since we are only interested in local properties we may suppose that Ω is bounded in the euclidian norm. In order to avoid technicalities we suppose Ω is closed.

The first type of local regularity classes corresponds to the idea the globally, a distribution has a regularity described by $A(\mathbb{H}^n)$ whereas locally, in Ω we have some higher regularity of type $B(\mathbb{H}^n)$.

A dual idea would be to have the wavelet coefficients concentrated on the subset Ω. That is that outside of Ω, the wavelet coefficients are small, hence correspond to the higher regularity $B(\mathbb{H}^n)$, whereas inside Ω, the coefficients are in $A(\mathbb{H}^n)$.

We now want to make these statements more precise. Consider first the case of higher local regularity. Suppose that there is a sequence of closed sets $\{\Omega_k\}$, $k = 1, 2, \ldots$ with

$$\Omega_1 \subset \ldots \Omega_k \subset \Omega_{k+1} \cdots \subset \Omega.$$

We suppose that Ω_k converges to Ω in the sense that

$$\Omega = \bigcup_k \Omega_k.$$

Suppose that Ω_k and Ω_{k+1}^c are well separated for each k. Then clearly Ω_k and Ω_l^c are well separated for $k < l$. We then say that $\eta \in \mathcal{S}_0'(\mathbb{R}^n)$ belongs to the microlocal class $\Lambda(\Omega; A, B)$ iff for some admissible g and all k we have

$$\eta \in A(\mathbb{R}^n), \quad \text{and} \quad \mathcal{M}_g \Omega_k \mathcal{W}_g \eta \in B(\mathbb{R}^n).$$

By the results of the previous theorem it is clear that the definition does not depend on the specific wavelets nor on the family of approximating sets Ω_k. Indeed, by Lemma 5.4 we may take the family $\Gamma_{1/k}(\Omega)$ as universal family of approximating sets.

Note however, that for arbitrary Ω, the previous class might coincide with $A(\mathbb{R}^n)$. Indeed, in order to have an approximating sequence from the interior, of mutually well separated sets the "smoother" region can not be arbitrary thin. It must contain at least some non-euclidian neighborhood of some set.

Frequently one takes $A = \mathcal{S}_0'(\mathbb{R}^n)$ in which case one is only interested in the behavior of the wavelet coefficients around Ω. In the next section we shall use this kind of classes to define directional regularity in distributions.

Consider now the dual approach, where we want to formalize the idea of wavelet coefficients concentrated on Ω. Suppose now that a sequence of open sets $\{\Omega^k\}$, $k = 1, 2, \ldots$ with

$$\Omega^1 \supset \ldots \Omega^k \supset \Omega^{k+1} \cdots \supset \Omega.$$

converges to Ω in the sense that

$$\Omega = \bigcap_k \Omega^k.$$

Again we require that $(\Omega^k)^c$ and Ω^{k+1} are well separated for each k. We then say that $\eta \in \mathcal{S}_0'(\mathbb{R}^n)$ belongs to the microlocal class $\Sigma(\Omega; A, B)$ iff for some admissible g and all k we have

$$\eta \in A(\mathbb{R}^n), \quad \text{and} \quad \mathcal{M}_g (\Omega^k)^c \mathcal{W}_g \eta \in B(\mathbb{R}^n).$$

Note that in the case where $B(\mathbb{R}^n) = \mathcal{S}_0(\mathbb{R}^n)$, this corresponds to the idea having the wavelet coefficients concentrated on the set Ω, where they satisfy the less restrictive regularity estimate given by $A(\mathbb{H}^n)$.

Note again that the region that contains the wavelet coefficients corresponding to the smoother behavior can not be arbitrary thin. However, the set Ω on which the wavelet coefficients are concentrated is arbitrary.

We can equip $\Lambda(\Omega; A, B)$ and $\Sigma(\Omega; A, B)$ with a topology generated by the following family of semi-norms for g, $h \in \mathcal{S}_0(\mathbb{R}^n)$ both admissible

$$\begin{aligned} \|s\|_{\Lambda(\Omega;A,B)} &= \|s\|_{A(\mathbb{R}^n)} + \|\mathcal{M}_h \Omega_k \mathcal{W}_g\|_{B(\mathbb{R}^n)}, \\ \|s\|_{\Sigma(\Omega;A,B)} &= \|s\|_{A(\mathbb{R}^n)} + \left\|\mathcal{M}_h (\Omega^k)^c \mathcal{W}_g\right\|_{B(\mathbb{R}^n)}. \end{aligned}$$

The topologies do not depend on the specific choice of wavelets.

We clearly have

$$\Lambda(\Omega; A, B) = \Sigma(\Omega^c; A, B)$$

A second class of microlocal classes can be defined as follows. Denote by $\lambda(\Omega; A, B)$ the class of functions η for which there is some set $\Gamma \supset \Omega$, such that Γ^c is well separated from Ω for which

$$\eta \in A(\mathbb{R}^n), \quad \text{and} \quad \mathcal{M}_h \Gamma \mathcal{W}_g \eta \in B(\mathbb{R}^n).$$

Suppose now that the dual spaces A' and B' are admissible, too. We then have

$$\Lambda(\Omega; A, B)' = \lambda(\Omega; B', A').$$

We can thus equip the set $\lambda(\Omega; B', A')$ with the weak* topology induced by $\Omega_{A,B}$.

As before we may also consider the complementary situation. Here we consider the set of distributions for which there is some $\Gamma \subset \Omega$, such that Ω^c is well separated from Γ and for which

$$\eta \in A(\mathbb{R}^n), \quad \text{and} \quad \mathcal{M}_h \Gamma \mathcal{W}_g \eta \in B(\mathbb{R}^n).$$

The set of all such distributions will be denoted by $\sigma(\Omega; A, B)$. Again we have duality

$$\Sigma(\Omega; B, A)' = \sigma(\Omega; B', A').$$

5.6 Directional microlocal classes

We now propose to look at more specific examples of regularity classes. In particular, to those we mentioned in the beginning of the paper, that is to classes related to the notion of singular or regular directions in distributions.

The classical definition of local singular directions of a distribution η is given by the wavefront set (e.g. [Hör82]). At a given point x it is roughly speaking the cone of all directions in which the Fourier transform of the localized distribution $\phi \cdot \eta$ does not decay rapidly where ϕ is any smooth function that is supported by some neighborhood of x. More precisely for fixed $\phi \in C_0^\infty(\mathbb{R}^n)$, $x \in \operatorname{supp} \phi$, a direction $\xi \in \mathbb{R}^n \setminus \{0\}$ is regular if in some conic neighborhood $\gamma \ni \xi$ we have

$$k \in \gamma \Rightarrow |(\phi \cdot \eta)^\frown(k)| \leq c_N (1 + |k|)^{-N}, N = 0, 1, 2, \ldots,$$

The complement of the regular directions is denoted by $\Sigma_{x,\phi}$. The singular directions at x are then defined as

$$\Sigma_x = \cap_\phi \Sigma_{x,\phi},$$

where the the intersection is over all $\phi \in C_0^\infty(\mathbb{R}^n)$ with $x \in \operatorname{supp} \phi$. However, this concept of singular and regular directions does not always fit with what one would

with some $f \in \mathcal{S}(\mathbb{R}^n)$ *and some function* $\rho : \mathbb{R}_+ \to \mathbb{R}_+$ *satisfying for all* $m > 0$ *at*

$$\rho(a) \leq O(a^m(1+a)^{-m}).$$

Then $Op[E]$ *acts continuously in every admissible Banach space* $B(\mathbb{H}^n)$ *subject in addition condition* 4.1.

Proof. We have for $s = Op[E]r$

$$s(b,a) = \int_{\mathbb{H}^n} \frac{db'da'}{a'} E(b,a;b',a')r(b',a').$$

A simple change of variables leads to

$$s(b,a) = \int_{\mathbb{H}^n} \frac{db'da'}{a'} E(b,a;b'+b,aa')r(b'+b,a'a).$$

Using the various estimates on E and r we have

$$\|s\| \leq O(1)\,\|r\| \int_{\mathbb{H}^n} \frac{db'da'}{a'} \rho(1/a') \frac{1}{(1+a')^n} f\left(\frac{-b'}{1+a'}\right) a'^{\epsilon}(1+a')^M.$$

Changing b' to $(1+a')b'$ we may estimate the integral by

$$\int_{\mathbb{R}^n} db' f(b') \int_0^\infty \frac{da'}{a'} \rho(1/a') a'^{\epsilon}(1+a')^M < \infty.$$

This shows the lemma. □

Let $\gamma \in \mathbb{R}$. Then denote by $a^\gamma B(\mathbb{H}^n)$ the space of functions $a^\gamma r(b,a)$ with $r \in B(\mathbb{H}^n)$. This new space is a Banach space by choosing the final topology for the map $r \mapsto a^\gamma r$. It is plain to see that if $B(\mathbb{H}^n)$ is admissible, (regular, completely regular), then the same holds true for $a^\gamma B(\mathbb{H}^n)$. In case $\gamma \geq 0$, if $B(\mathbb{H}^n)$ is strongly tempered then the same holds true for $a^\gamma B(\mathbb{H}^n)$.

Remark. For all admissible $B(\mathbb{H}^n)$, there is $\gamma \in \mathbb{R}$ such that $a^\gamma B(\mathbb{H}^n)$ is strongly tempered. This will play a role in subsequent demonstrations.

Partial integration shows that we have the following formula

$$\mathcal{W}_g(-\Delta)^{\gamma/2} s(b,a) = a^{-\gamma}\mathcal{W}_h s(b,a), \quad h = (-\Delta)^{\gamma/2} g.$$

Therefore multiplication with powers of a essentially acts like a smoothing operator. More generally, let $\rho : \mathbb{R}_+ \to \mathbb{R}_+$ satisfy

$$\rho(\alpha a) \leq O((\alpha + 1/\alpha)^m)\rho(a),$$

for some $m > 0$. Then with the obvious notations we define $\rho \cdot B(\mathbb{H}^n)$ and norm it through

$$\|s\|_{\rho \cdot B(\mathbb{H}^n)} = \|s/\rho\|_{B(\mathbb{H}^n)}.$$

Then $\rho \cdot B(\mathbb{H}^n)$ is admissible if $B(\mathbb{H}^n)$ is. With these notations we have the following corollaries.

Corollary 6.3 *Keep the notations of Lemma* 6.1 *respectively of Lemma* 6.2. *Let* E *satisfy instead*

$$|E(b,a;b',a')| \leq \frac{1}{a'^n} F\left(\frac{b-b'}{a'}, \frac{a}{a'}\right) \rho(a'), \tag{6.3}$$

or

$$|E(b,a;b',a')| \leq \rho(a) \frac{1}{a'^n} F\left(\frac{b-b'}{a'}, \frac{a}{a'}\right). \tag{6.4}$$

If F *satisfies the localization conditions of Lemma* 6.1, *then* $Op[E]$ *maps* $B(\mathbb{H}^n)$ *continuously into* $\rho \cdot B(\mathbb{H}^n)$ *for every admissible Banach space. Suppose now that* F *only satisfies the conditions of Lemma* 6.2. *Then, if* E *satisfies* 6.3 *and if* $\rho \cdot B(\mathbb{H}^n)$ *is strongly tempered then again* $Op[E]$ *maps* $B(\mathbb{H}^n)$ *into* $\rho \cdot B(\mathbb{H}^n)$ *continuously. The same holds true if* E *satisfies* 6.4 *and if* $B(\mathbb{H}^n)$ *itself is strongly tempered.*

Proof. $Op[E]$ is the composition of two operators. The first (respectively second) maps $r(b,a) \to \rho(a)r(b,a)$. The second (first) has a Schwartz kernel that satisfies the estimates of the previous lemma. The first (second) operator is trivially continuous, and for the second Lemma 6.1 (respectively Lemma 6.2) applies. □

7 Operator algebras

As we have seen in the previous section, operators acting in functions spaces over $\mathbb{R}^n$ may be successfully analyzed through localization estimates of the their kernel acting in wavelet space. In this section now we want to construct an algebra of operators over $\mathbb{H}^n$ for which their continuity properties can be easily read from their symbols.

7.1 Convolution estimates

For two functions r, s over $\mathbb{H}^n$ we have defined their convolution (formally)

$$(s * r)(b,a) = \int_{\mathbb{H}^n} \frac{db'da'}{a'} \frac{1}{a'^n} s\left(\frac{b-b'}{a'}, \frac{a}{a'}\right) r(b',a').$$

We now give some estimates for convolutions over the half-space. The name convolution is justified, since we may write

$$(s * r)(b,a) = \int_{\mathbb{H}^n} \frac{db'da'}{a'^{n+1}} s((b',a')^{-1} \circ (b,a)) r(b',a').$$

Again we use constantly the following notations ($b \in \mathbb{R}^n$, $a > 0$),

$$\kappa_\alpha(b) = (1+|b|)^{-\alpha}, \quad \phi_{\alpha,\beta}(a) = \frac{a^\alpha}{(1+a)^{\alpha+\beta}}.$$

Lemma 7.1 *Let α, β, γ, α', β', $\gamma' \in \mathbb{R}$. Suppose that*

$$\alpha' + \beta > 0, \quad \alpha + \beta' + n > 0, \quad \gamma + \gamma' > n.$$

Suppose further that either $\gamma \leq 0$ or $\gamma' \leq 0$ or if both are positive that γ, $\gamma' > n$. For

$$P(b,a) = \phi_{\alpha,\beta}(a)\,\frac{1}{(1+a)^n}\kappa_\gamma\left(\frac{b}{1+a}\right) = \phi_{\alpha,\beta}(a) D_{1+a}\kappa_\gamma(b),$$

$$P'(b,a) = \phi_{\alpha',\beta'}(a)\,\frac{1}{(1+a)^n}\kappa_{\gamma'}\left(\frac{b}{1+a}\right) = \phi_{\alpha',\beta'}(a) D_{1+a}\kappa_{\gamma'}(b),$$

we have that

$$P * P'(b,a) = \int_{\mathbb{H}^n} \frac{db'da'}{a'}\frac{1}{a'^n} P\left(\frac{b-b'}{a'}, \frac{a}{a'}\right) P(b',a'), \tag{7.1}$$

is well-defined point-wise. Then, if $\alpha \neq \alpha'$ and $\beta \neq \beta'$ and $\alpha + \beta' + n \neq \min\{\gamma, \gamma'\}$, then

$$P * P'(b,a) \leq O(1)\,\phi_{\min\{\alpha,\alpha'\},\{\beta,\beta'\}}(a)\,\frac{1}{(1+a)^n}\kappa_{\min\{\alpha+\beta'+n,\gamma,\gamma'\}}\left(\frac{b}{1+a}\right).$$

The estimate of the exponents is optimal. In case that the above inequalities do not hold, logarithmic corrections appear (see the proof below for details).

Proof. Let us write

$$\Phi(b,a) = P * P'(b,a).$$

We have

$$\frac{1}{a'^n}P\left(\frac{b-b'}{a'}, \frac{a}{a'}\right) = \phi_{\alpha,\beta}(a/a')\,\frac{1}{(a+a')^n}\kappa_\gamma\left(\frac{b-b'}{a+a'}\right).$$

Now we may write for ξ, $\xi' > 0$

$$(D_\xi\kappa_\gamma) * (D_{\xi'}\kappa_{\gamma'}) = D_{\xi'}((D_{\xi/\xi'}\kappa_\gamma) * \kappa_{\gamma'}),$$

and thus using Lemma 3.4 we have

$$(D_\xi\kappa_\gamma) * (D_{\xi'}\kappa_{\gamma'}) \leq D_{\xi+\xi'}\kappa_{\min\{\gamma,\gamma'\}}.$$

This estimates the convolution in (7.1) and we obtain ($\rho = \min\{\gamma, \gamma'\}$)

$$\Phi(b,a) \leq \int_0^\infty \frac{da'}{a'}\,\phi_{\alpha,\beta}(a/a')\,\phi_{\alpha',\beta'}(a')\frac{1}{(1+a+2a')^n}\kappa_\rho\left(\frac{b}{1+a+2a'}\right).$$

This expression is equivalent ($\sim$) to

$$F^{\alpha',\beta'}_{\alpha,\beta;\rho}(b,a) = \int_0^\infty \frac{da'}{a'}\,\phi_{\alpha,\beta}(a/a')\,\phi_{\alpha',\beta'}(a')\frac{1}{(1+a+a')^n}\kappa_\rho\left(\frac{b}{1+a+a'}\right).$$

We have to show that

$$F^{\alpha',\beta'}_{\alpha,\beta;\rho}(b,a) \leq G^{\rho}_{\min\{\alpha,\alpha'\},\min\{\beta,\beta'\}}(b,a),$$

with

$$G^{\nu}_{\lambda,\mu}(b,a) = \phi_{\lambda,\mu}(a)\frac{1}{(1+a)^n}\kappa_\nu\left(\frac{b}{1+a}\right).$$

By direct computation using the scaling invariance of the measure da'/a' and the obvious fact that $\phi_{\alpha,\beta}(1/a) = \phi_{\beta,\alpha}(a)$, we see that

$$\frac{1}{a^n}F^{\alpha',\beta'}_{\alpha,\beta;\rho}\left(\frac{b}{a},\frac{1}{a}\right) = F^{\beta,\alpha}_{\beta',\alpha';\rho}(b,a).$$

In addition we have that

$$\frac{1}{a^n}G^{\nu}_{\lambda,\mu}\left(\frac{b}{a},\frac{1}{a}\right) = G^{\nu}_{\mu,\lambda}(b,a).$$

It therefore is enough to show the estimates for $a \leq 1$. Here we have to show that for $|b| < 2$ we have

$$F^{\alpha',\beta'}_{\alpha,\beta;\rho}(b,a) \leq a^{\min\{\alpha,\alpha'\}},$$

and that for $|b| > 2$ we have

$$F^{\alpha',\beta'}_{\alpha,\beta;\rho}(b,a) \leq a^{\min\{\alpha,\alpha'\}}\,|b|^{-\min\{\alpha+\beta'+n,\rho\}}.$$

We now split the integral over a' into three parts, $X_1 + X_2 + X_3$, if $|b| \leq 2$, or four parts, $X_1 + X_2 + X_4 + X_5$, if $|b| > 2$.

X_1. For $0 < a' < a \leq 1$ we may estimate $\phi_{\alpha,\beta}(a/a') \sim (a/a')^{-\beta}$, $\phi_{\alpha',\beta'}(a') \sim a'^{\alpha'}$ and $D_{1+a+a'}\kappa_\rho \sim \kappa_\rho$. Thus we have to compute

$$\int_0^a \frac{da'}{a'}\,(a/a')^{-\beta}\,a'^{\alpha'}\,\kappa_\rho(b) = \begin{cases} a^{\alpha'}\kappa_\rho(b) & \text{for } \alpha'+\beta > 0,\\ \infty & \text{else.}\end{cases}$$

X_2. For $0 < a \leq a' \leq 1$ we may estimate $\phi_{\alpha,\beta}(a/a') \sim (a/a')^{\alpha}$, $\phi_{\alpha',\beta'}(a') \sim a'^{\alpha'}$ and again $D_{1+a+a'}\kappa_\rho \sim \kappa_\rho$. Thus we have to compute

$$\int_a^1 \frac{da'}{a'}\,(a/a')^{\alpha}\,a'^{\alpha'}\,\kappa_\rho(b) = \begin{cases} a^{\alpha'}\kappa_\rho(b) & \text{for } \alpha'-\alpha < 0,\\ \log(1/a)a^{\alpha}\kappa_\rho(b) & \text{for } \alpha'-\alpha = 0,\\ a^{\alpha}\kappa_\rho(b) & \text{for } \alpha'-\alpha > 0.\end{cases}$$

X_3. For $0 < a \leq 1 < a'$ and $|b| \leq 2$ we may estimate $\phi_{\alpha,\beta}(a/a') \sim (a/a')^{\alpha}$, $\phi_{\alpha',\beta'}(a') \sim a'^{-\beta'}$ and $D_{1+a+a'}\kappa_\rho \sim a'^{-n}$. Thus we have to compute

$$\int_1^\infty \frac{da'}{a'}\,(a/a')^{\alpha}\,a'^{-\beta'}\,a'^{-n} = \begin{cases} a^{\alpha} & \text{for } \alpha+\beta'+n > 0,\\ \infty & \text{else.}\end{cases}$$

X_4. For $0 < a \leq 1 < a' \leq |b|$ and $|b| > 2$ we may estimate $\phi_{\alpha,\beta}(a/a') \sim (a/a')^\alpha$, $\phi_{\alpha',\beta'}(a') \sim a'^{-\beta'}$ and $D_{1+a+a'}\kappa_\rho \sim a'^{-n}(|b|/a')^{-\rho}$. Thus we have to compute

$$\int_1^{|b|} \frac{da'}{a'} (a/a')^\alpha a'^{-\beta'} a'^{-n}(|b|/a')^{-\rho} = \begin{cases} a^\alpha |b|^{-\rho} & \text{for } \alpha+\beta'+n-\rho > 0, \\ a^\alpha |b|^{-\rho} \log(1/|b|) & \text{for } \alpha+\beta'+n-\rho = 0, \\ a^\alpha |b|^{-\alpha-\beta'-n} & \text{for } \alpha+\beta'+n-\rho < 0. \end{cases}$$

X_5. For $0 < a \leq 1 < 2 \leq |b| < a'$ we may estimate $\phi_{\alpha,\beta}(a/a') \sim (a/a')^\alpha$, $\phi_{\alpha',\beta'}(a') \sim a'^{-\beta'}$ and $D_{1+a+a'}\kappa_\rho \sim a'^{-n}$. Thus we have to compute

$$\int_{|b|}^{\infty} \frac{da'}{a'} (a/a')^\alpha a'^{-\beta'} a'^{-n} = \begin{cases} a^\alpha |b|^{-\alpha-\beta'-n} & \text{for } \alpha+\beta'+n > 0, \\ \infty & \text{else.} \end{cases}$$

This shows the lemma. □

7.2 The highly regular Calderón Zygmund operators (CZO)

Let us denote by $G_{\alpha,\beta,\gamma}(\mathbb{H}^n)$ the set of all measurable complex-valued functions s over the half-space for which we have

$$|s(b,a)| \leq O(1)\phi_{\alpha,\beta}(a)\frac{1}{(1+a)^n}\kappa_\gamma\left(\frac{b}{1+a}\right).$$

Then clearly, the indices are hierarchical in that we have

$$\alpha' \geq \alpha, \beta' \geq \beta, \gamma' \geq \gamma \Rightarrow G_{\alpha',\beta',\gamma'}(\mathbb{H}^n) \subset G_{\alpha,\beta,\gamma}(\mathbb{H}^n).$$

Let $\Gamma_\alpha(\mathbb{H}^n)$ denote the set

$$\Gamma_\alpha(\mathbb{H}^n) = \bigcap_{\beta<\alpha} G_{\beta,\beta,\beta}(\mathbb{H}^n).$$

And in addition

$$\Gamma_\infty(\mathbb{H}^n) = \bigcap_\alpha \Gamma_\alpha(\mathbb{H}^n) = \mathcal{S}(\mathbb{H}^n).$$

We denote by $CZO_\alpha(\mathbb{H}^n \times \mathbb{H}^n)$ the set of all measurable complex valued functions σ over $\mathbb{H}^n \times \mathbb{H}^n$ for which we have

$$|\sigma(b,a;b',a')| \leq \frac{1}{a'^n} r\left(\frac{b-b'}{a'}, \frac{a}{a'}\right),$$

for some $r \in \Gamma_\alpha(\mathbb{H}^n)$. Instead of $CZO_\infty(\mathbb{H}^n \times \mathbb{H}^n)$ we write $CZO(\mathbb{H}^n \times \mathbb{H}^n)$.

Corollary 7.2 *The set $\Gamma_\alpha(\mathbb{H}^n)$ is an algebra with respect to the $*$-product. The set $Op[CZO_\alpha(\mathbb{H}^n \times \mathbb{H}^n)]$ is an algebra of operators. Here $\alpha > n/2$, or $\alpha = \infty$.*

Proof. By Lemma 7.1, we have that the following graduation holds

$$\Gamma_\alpha(\mathbb{H}^n) * \Gamma_\beta(\mathbb{H}^n) \subset \Gamma_{\min\{\alpha,\beta\}}(\mathbb{H}^n).$$

The analogous graduation holds for $Op[CZO_\alpha(\mathbb{H}^n \times \mathbb{H}^n)]$, since we can majorize by convolutions. □

The following theorem shows that the $CZO(\mathbb{H}^n \times \mathbb{H}^n)$ may be pulled pack to an algebra of operators acting on functions spaces in $\mathbb{R}^n$. Clearly an analogous theorem holds for $CZO_\alpha(\mathbb{H}^n \times \mathbb{H}^n)$ but we leave the details to the reader.

Theorem 7.3 *Let $K : \mathcal{S}_0(\mathbb{R}^n) \to \mathcal{S}_0'(\mathbb{R}^n)$ be linear. Then the following is equivalent*

(i) *The matrix elements*

$$\sigma[K](b,a;b',a') = \sigma_{g,h}[K](b,a;b',a') = \langle g_{b,a}, Kh_{b',a'}\rangle,$$

with respect to some functions g, $h \in \mathcal{S}_0(\mathbb{R}^n)$ satisfying

$$\int_0^\infty \frac{da}{a} |\widehat{g}(ak)|^2 \sim \int_0^\infty \frac{da}{a} \left|\widehat{h}(ak)\right|^2 \sim 1,$$

satisfy $\sigma \in CZO(\mathbb{H}^n \times \mathbb{H}^n)$.

(ii) *The matrix elements with respect to all functions g' and $h' \in \mathcal{S}_0(\mathbb{R}^n)$ satisfy $\sigma_{g',h'} \in CZO(\mathbb{H}^n \times \mathbb{H}^n)$.*

(iii) *The family of dilated and translated operators*

$$K_{b,a} = D_a^{-1} T_b^{-1} K T_b D_a, \quad (b,a) \in \mathbb{H}^n,$$

is uniformly continuous from $\mathcal{S}_0(\mathbb{R}^n)$ to $\mathcal{S}_0(\mathbb{R}^n)$.

(iv) *There are functions v, $w \in \mathcal{S}_0(\mathbb{R}^n)$ and some $F \in CZO(\mathbb{H}^n \times \mathbb{H}^n)$ such that*

$$K = \mathcal{M}_v Op[F] \mathcal{W}_w.$$

Before we come to the proof we list some obvious consequences. These operators are called highly regular Calderón Zygmund operators and we shall denote this set by *CZO*. As condition (iii) shows, *CZO* is an algebra with respect to the operator product. Indeed, this follows from

$$T_b\, D_a\, K_1 K_2\, D_a^{-1} T_b^{-1} = (T_b\, D_a\, K_1\, D_a^{-1} T_b^{-1})(T_b\, D_a\, K_2\, D_a^{-1} T_b^{-1}),$$

and the fact that the product of two uniformly continuous families of operators is again uniformly continuous.

Condition (i) and (ii) show that, roughly speaking, a highly regular Calderòn Zygmund operator maps a wavelet that is localized at some position b and at some scale a into something that is again wavelet like and localized at some position b' and scale a' such that $(b-b')/a'$ and a/a' stay close to 1.

Condition (iii) shows that these operators are approximately self-similar (in terms of continuity) at all scales and all positions: they have the same continuity properties at all positions and all scales.

Eventually condition (iv) states that these operators can be written in wavelet space in terms of the algebra $CZO(\mathbb{H}^n \times \mathbb{H}^n)$.

Proof. The implication (ii) $\Rightarrow$ (i) is obvious. Let us prove (i) $\iff$ (iv), then (i) $\Rightarrow$ (ii), and then (i) $\iff$ (iii).

(iv) $\Rightarrow$ (i). Suppose that we have

$$K : s \mapsto \mathcal{M}_v Op[F] \mathcal{W}_w s,$$

where F is in $CZO(\mathbb{H}^n \times \mathbb{H}^n)$. Then K maps $\mathcal{S}_0(\mathbb{R}^n)$ into itself. Indeed $Op[F]\mathcal{W}_w s \in \mathcal{S}(\mathbb{H}^n)$, since $\mathcal{W}_w s \in \mathcal{S}(\mathbb{H}^n)$ and we may estimate the action of $Op[F]$ by the convolution with some function in $\mathcal{S}(\mathbb{H}^n)$. In addition its matrix elements read

$$\langle g_{b,a}, K h_{b',a'} \rangle = \mathcal{W}_g(\mathcal{M}_v(Op[F]\mathcal{W}_w g_{b',a'}))(b,a).$$

Now set $P_1 = |\mathcal{W}_w g|$ and $P_2 = |\mathcal{W}_g v|$. Then P_1, $P_2 \in \mathcal{S}(\mathbb{H}^n)$. By hypothesis there is $P_3 \in \mathcal{S}(\mathbb{H}^n)$ such that

$$|F(b,a;b',a')| \leq \frac{1}{a'^n} P_3\left(\frac{b-b'}{a'}, \frac{a}{a'}\right).$$

We therefore may estimate upon passing to the modulus under the integral signs

$$\langle g_{b,a}, K h_{b',a'} \rangle \leq \frac{1}{a'^n} (P_2 * P_3 * P_1)\left(\frac{b-b'}{a'}, \frac{a}{a'}\right).$$

Since $\mathcal{S}(\mathbb{H}^n)$ is a convolution algebra (see Corollary 7.2), this shows that (iv) implies (i).

(i) $\Rightarrow$ (iv). Vice versa, suppose that (i) holds. Let f and r be defined through

$$\widehat{f}(k) = \frac{\widehat{g}(k)}{\int_0^\infty \frac{da}{a} \, |\widehat{g}(ak)|^2}, \quad \widehat{r}(k) = \frac{\widehat{h}(k)}{\int_0^\infty \frac{da}{a} \left|\widehat{h}(ak)\right|^2}.$$

Then f, $r \in \mathcal{S}_0(\mathbb{R}^n)$ and we have

$$\mathcal{M}_f \mathcal{W}_g = \mathcal{M}_h \mathcal{W}_r = \mathbb{1}.$$

Therefore, we may write

$$K = \mathcal{M}_f Op[\sigma_{g,h}[K]] \mathcal{W}_r.$$

Indeed, this follows upon exchanging the integrals in

$$K = \mathcal{M}_f(\mathcal{W}_g K \mathcal{M}_h)\mathcal{W}_r.$$

This shows that (i) implies (iv).

(i) $\Rightarrow$ (ii). Indeed, changing wavelets amounts to convolve the symbol σ with some highly localized kernel, either in the first component or in the second (see (6.2). This leaves invariant the functions in *CZO*$(\mathbb{H}^n \times \mathbb{H}^n)$.

(i) $\Rightarrow$ (iii). Let K satisfy (i). It then clearly is continuous from $\mathcal{S}_0(\mathbb{R}^n)$ to $\mathcal{S}_0(\mathbb{R}^n)$. The matrix elements of the family $K_{\alpha,\beta}$ of scaled and translated versions of K read $\beta \in \mathbb{R}^n$, $\alpha > 0$

$$\left\langle g_{b,a}, T_\beta\, D_\alpha K\,(D_\alpha)^{-1}\,(T_\beta)^{-1} h_{b',a'} \right\rangle = \alpha^{-n} \left\langle g_{(b-\beta)/\alpha, a/\alpha}, K\, h_{(b'-\beta)/\alpha, a'/\alpha} \right\rangle.$$

By hypothesis this is bounded in absolute value by

$$\frac{1}{a'^n}\, P\left(\frac{b-b'}{a'}, \frac{a}{a'}\right),$$

with some $P \in \mathcal{S}(\mathbb{H}^n)$ independently of β, α. This shows that the family above is equi-continuous thanks to Theorem 3.18.

(iii) $\Rightarrow$ (i). On the other hand suppose that the family above is equi-continuous. Then, for a fixed, admissible $g \in \mathcal{S}_0(\mathbb{R}^n)$ the set $T_\beta D_\alpha\, K\, D_\alpha^{-1}\, T_\beta^{-1} g$ is a bounded set in $\mathcal{S}_0(\mathbb{R}^n)$. Therefore, by Theorem 3.18 there is a highly localized function $P \in \Gamma_\infty(\mathbb{H}^n)$ such that for all $(\beta, \alpha) \in \mathbb{H}^n$ we have

$$\left|(\mathcal{W}_g\, T_\beta D_\alpha\, K\, D_\alpha^{-1}\, T_\beta^{-1} g)(b,a)\right| \le P(b,a).$$

By the co-variance of the wavelet transform we obtain once more the desired result. □

7.3 Fourier multipliers as highly regular CZO

In the next two sections, we focus on the case where K commutes with all the translations. In this simple situation all *CZO* can be easily characterized. We continue this analysis in Section 8.3.

Theorem 7.4 *Let K be a highly regular Calderón-Zygmund operator. Then K commutes with the translations*

$$T_b\, K = K\, T_b,$$

if and only if there is a smooth function $m : \mathbb{R}^n \to \mathbb{C}$ satisfying

$$|\partial^\alpha m(\xi)| \le \frac{c_\alpha}{|\xi|^{|\alpha|}}, \quad \xi \ne 0,$$

for all multi-indices α such that K is given by the Fourier multiplier

$$K : \widehat{s} \mapsto m \cdot \widehat{s}.$$

More precisely we shall show that for every $m \geq 0$ and some suitable admissible g, h in Λ_m we have

$$X(b,a) \leq C_\beta\,(a+1/a)^{-\beta}, \quad \text{and} \quad X(b,a) \leq C_\beta(1+|b|^2)^{-\beta/2},$$

with some localization exponent $\beta > 0$ and some constant C_β, that depends only on g and the constants c and c_α above. In addition we shall show that this exponent β may become arbitrary large with m thereby proving the theorem.

Now pick $g \in \Lambda_m$. The first condition on B ensures that X is localized over the scales

$$X(b,a) \leq C_\beta(a+1/a)^{-\beta},$$

with some exponent β that may become arbitrary large as $m \to \infty$ and a constant C_β depending only on g and c above. Indeed, the localization over the scales depends as we know only on the behavior of the Fourier transform of the analyzing wavelet and the analyzing function around 0 and at ∞. But by hypothesis on B this behavior is not altered by K.

The second condition on B ensures the localization in b. Let $I = \{|x| \leq 1\} \subset \mathbb{R}^n$ and pick a rotation invariant, non-negative $\phi \in C_c^\infty(I)$, $\phi = \phi(|x|) \geq 0$. Now set $g = (-\Delta)^k\phi$ with k the smallest integer $\geq m/2$. Clearly $g \in \Lambda_m$ and g is admissible. We now want to estimate the localization of

$$Kg(x) = \int_{\mathbb{R}^n} dy\, g(x-y)\, B(y).$$

Clearly by the first condition the Fourier transform of B we have globally

$$|Kg(x)| \leq c\,\|\widehat{g}\|_1\,,$$

and we may restrict ourselves to $|x| \geq 2$. Using integration by parts we have

$$|Kg(x)| = \left|\int_{\mathbb{R}^n} dy\,\phi(x-y)\,(-\Delta)^k B(y)\right| \leq \frac{C_\gamma}{|x|^{n+m}}\,\|\phi\|_1\,,$$

where in the last inequality we have used the fact that the integral actually only runs over $|x-y| \leq 1$. The constant C_γ depends only on the c_γ and g. In addition together with m the localization in b becomes arbitrary large, which shows the theorem. □

7.5 The smoothing operators

We have already encountered conditions on K under which $Op[K]$ acts like a smoothing operator. We now want to generalize this. Let $\Sigma_m(\mathbb{H}^n \times \mathbb{H}^n)$ denote the set of all measurable functions $s : \mathbb{H}^n \times \mathbb{H}^n \to \mathbb{C}$ for which we have

$$|s(b,a;b',a')| \leq c_s a^m\,\frac{1}{a'^n} r\left(\frac{b-b'}{a'}, \frac{a}{a'}\right),$$

with some $r \in \mathcal{S}(\mathbb{H}^n)$. An equivalent condition reads

$$|s(b,a;b',a')| \leq c'_s a'^m \frac{1}{a'^n} r'\left(\frac{b-b'}{a'}, \frac{a}{a'}\right) \tag{7.1}$$

with $r'(b,a) = r(b,a)a^m$, which is again in $\mathcal{S}(\mathbb{H}^n)$.

Proposition 7.6 *We have the following graduation of algebras*

$$Op[\Sigma_l(\mathbb{H}^n \times \mathbb{H}^n)]Op[\Sigma_m(\mathbb{H}^n \times \mathbb{H}^n)] \subset Op[\Sigma_{l+m}(\mathbb{H}^n \times \mathbb{H}^n)].$$

Proof. Indeed, suppose $s_1 \in \Sigma_l(\mathbb{H}^n \times \mathbb{H}^n)$ and $s_2 \in \Sigma_m(\mathbb{H}^n \times \mathbb{H}^n)$. Then, with some r_1, $r_2 \in \mathcal{S}(\mathbb{H}^n)$, we have

$$\begin{aligned} |s_1(b,a;b',a')| &\leq O(1)a^l \frac{1}{a'^n} r_1\left(\frac{b-b'}{a'}, \frac{a}{a'}\right), \\ |s_1(b'',a'';b',a')| &\leq O(1)a'^m \frac{1}{a'^n} r_2\left(\frac{b''-b'}{a'}, \frac{a''}{a'}\right). \end{aligned}$$

We thus have that $Op[s_1]Op[s_2] = Op[s_3]$, where s_3 may be estimated through

$$|s_3(b,a;b',a')| \leq a'^{l+m} \frac{1}{a'^n} (r_1 * r_2)\left(\frac{b-b'}{a'}, \frac{a}{a'}\right)\left(\frac{a}{a'}\right)^l.$$

Now $a^l r_1 * r_2 \in \mathcal{S}(\mathbb{H}^n)$ and we are done. □

Again, this algebra may be pulled back to an operator algebra acting on function spaces over $\mathbb{R}^n$. It is the analogue of Theorem 7.3

Theorem 7.7 *Let* $K : \mathcal{S}_0(\mathbb{R}^n) \to \mathcal{S}'_0(\mathbb{R}^n)$ *be linear and let* $m \in \mathbb{R}$. *Then the following is equivalent*

(i) *The matrix elements*

$$\sigma(b,a;b',a') = \sigma_{g,h}[K](b,a;b',a') = \langle g_{b,a}, K h_{b',a'} \rangle,$$

with respect to some functions g, $h \in \mathcal{S}_0(\mathbb{R}^n)$ *satisfying*

$$\int_0^\infty \frac{da}{a} |\widehat{g}(a\xi)|^2 \sim \int_0^\infty \frac{da}{a} \left|\widehat{h}(a\xi)\right|^2 \sim 1,$$

satisfy $\sigma \in \Sigma_m(\mathbb{H}^n \times \mathbb{H}^n)$.

(ii) *The matrix elements with respect to all functions* g' *and* $h' \in \mathcal{S}_0(\mathbb{R}^n)$ *satisfy* $\sigma_{g',h'} \in \Sigma_m(\mathbb{H}^n \times \mathbb{H}^n)$.

(iii) *The family of dilated, translated and rescaled operators*

$$K_{b,a} = a^m D_a^{-1} T_b^{-1} K T_b D_a, \quad (b,a) \in \mathbb{H}^n,$$

is uniformly continuous from $\mathcal{S}_0(\mathbb{R}^n)$ *to* $\mathcal{S}_0(\mathbb{R}^n)$.

(iv) *There are functions* v, $w \in \mathcal{S}_0(\mathbb{R}^n)$ *and some* $F \in \Sigma_m(\mathbb{H}^n)$ *such that*

$$K = \mathcal{M}_v Op[F] \mathcal{W}_w.$$

(v) *There are operators* A, $B \in CZO$ *such that*

$$K = (-\Delta)^{-m/2} A = B(-\Delta)^{-m/2}.$$

We denote by Σ_m the set of all operators described by the above theorem. Clearly $\Sigma_0 = CZO$. The name smoothing operators is justified since as condition (v) shows, the operators in $\Sigma_m(\mathbb{R}^n)$ have the same smoothing properties as $(-\Delta)^{-m/2}$.

Proof. All assertions apart from the last statement may be proved in the same way as before and we leave the details to the reader. To show (v) note once more that $(-\Delta)^{m/2}$ has matrix elements

$$\left\langle g_{b,a}, (-\Delta)^{m/2} h_{b',a'} \right\rangle = a'^{-m} \frac{1}{a'^n} (\mathcal{W}_g(-\Delta)^{m/2} h) \left(\frac{b-b'}{a'}, \frac{a}{a'} \right).$$

Since $(-\Delta)^{m/2} h \in \mathcal{S}_0(\mathbb{R}^n)$ this implies that $(-\Delta)^{m/2} \in \Sigma_m$. Now, since on $\mathcal{S}_0(\mathbb{R}^n)$ we have $(-\Delta)^{m/2}(-\Delta)^{-m/2} = \mathbb{1}$, we have on $\mathcal{S}_0(\mathbb{R}^n)$

$$K = (-\Delta)^{m/2}[(-\Delta)^{-m/2} K] = [K(-\Delta)^{-m/2}](-\Delta)^{m/2}.$$

The second respectively first term is of order 0 hence a *CZO*. □

The discussion is completed by noting that the mapping properties of $(-\Delta)^m$ may be neatly expressed in terms of the admissible spaces $B(\mathbb{R}^n)$ we have constructed before. Indeed, let $B(\mathbb{R}^n)$ be an admissible Banach space. Then, denote by $(-\Delta)^{m/2} B(\mathbb{R}^n)$ the set of all distributions $\eta \in \mathcal{S}_0'(\mathbb{R}^n)$ for which $(-\Delta)^{-m/2}\eta \in B(\mathbb{R}^n)$. This is again an admissible Banach space with norm

$$\|s\|_{(-\Delta)^{m/2} B(\mathbb{R}^n)} = \left\| (-\Delta)^{-m/2} s \right\|_{B(\mathbb{R}^n)}.$$

To see this take $g \in \mathcal{S}_0(\mathbb{R}^n)$ a radial function, $g = g(|x|)$ and set $h = (-\Delta)^{-m/2} g$. Then observe that $(-\Delta)^{m/2}$ can be written as

$$(-\Delta)^{m/2} = \mathcal{M}_h a^{-m} \mathcal{W}_g.$$

Now $\mathcal{W}_g$ maps $B(\mathbb{R}^n)$ into $B(\mathbb{H}^n)$ continuously and the multiplication by a^{-m} maps $B(\mathbb{H}^n)$ continuously into $a^{-m} B(\mathbb{H}^n)$

Recall that the upper half-space carries a group structure

$$(b, a) \circ (b', a') = (b + ab', aa'),$$

and that we have found a non-euclidian distance function expressed through $\Delta(b, a)$ that satisfies (see Lemma 5.1)

$$\Delta((b, a) \circ (b', a')) \le \Delta(b, a)\Delta(b', a').$$

We do not distinguish any more in notation between $\Delta((b, a))$ and $\Delta(b, a)$.

Definition 7.8 *A non-vanishing, positive function ρ on $\mathbb{H}^n$ is called sub-multiplicative (from the right) if for some c, $K > 0$ we have*

$$\rho(b + ab', aa') = \rho((b, a) \circ (b', a')) \le c\,\Delta(b', a')^K\,\rho(b, a).$$

Clearly all sub-multiplicative weight functions are tempered as follows from the temperedness of $\Delta(b, a)$ itself. Obvious examples of sub-multiplicative weight functions are Δ^k and a^k. The sum and the product of sub-multiplicative weight functions is again a sub-multiplicative weight function.

Consider now a tempered weight function ρ over the half-space $\mathbb{H}^n$. Then let $\Sigma_\rho(\mathbb{H}^n \times \mathbb{H}^n)$ denote the set of all functions on $\mathbb{H}^n \times \mathbb{H}^n$ for which we have

$$|\sigma(b, a; b', a')| \le \rho(b, a)\frac{1}{a'^n} P\left(\frac{b - b'}{a'}, \frac{a}{a'}\right),$$

with some $P \in \mathcal{S}(\mathbb{H}^n)$. Equivalently we may consider the set of all functions for which

$$|\sigma(b, a; b', a')| \le \rho(b', a')\frac{1}{a'^n} P\left(\frac{b - b'}{a'}, \frac{a}{a'}\right).$$

Indeed, the equivalence follows from the fact that

$$\rho(b, a) = \rho((b', a') \circ (b', a')^{-1} \circ (b, a)) \le c\Delta\left(\frac{b - b'}{a'}, \frac{a}{a'}\right) \rho(b', a'),$$

and

$$\rho(b', a') = \rho((b, a) \circ (b, a)^{-1} \circ (b', a')) \le c\Delta\left(\frac{b' - b}{a}, \frac{a'}{a}\right) \rho(b, a),$$

together with the fact that $P\Delta^K$ is still in $\mathcal{S}(\mathbb{H}^n)$.

Obviously the operators $\mathcal{O}p[s]$ with $s \in \Sigma_\rho(\mathbb{H}^n \times \mathbb{H}^n)$ are continuous in $\mathcal{S}(\mathbb{H}^n)$. Thanks to Corollary (6.3) they are continuous from $B(\mathbb{H}^n)$ to $\rho B(\mathbb{H}^n)$ for every admissible Banach space $B(\mathbb{H}^n)$.

Again, these operators may be pulled back to operators acting on function spaces over $\mathbb{R}^n$. It is the analogue of Theorem 7.7.

Theorem 7.9 *Let $K : \mathcal{S}_0(\mathbb{R}^n) \to \mathcal{S}_0'(\mathbb{R}^n)$ be linear and let ρ be sub-multiplicative. Then the following is equivalent*

(i) *The matrix elements*

$$\sigma(b,a;b',a') = \sigma_{g,h}[K](b,a;b',a') = \langle g_{b,a}, Kh_{b',a'}\rangle\,,$$

with respect to some functions g, $h \in \mathcal{S}_0(\mathbb{R}^n)$ satisfying

$$\int_0^\infty \frac{da}{a}\, |\widehat{g}(ak)|^2 \sim \int_0^\infty \frac{da}{a}\, \left|\widehat{h}(ak)\right|^2 \sim 1,$$

satisfy $\sigma \in \Sigma_\rho(\mathbb{H}^n \times \mathbb{H}^n)$.

(ii) *The matrix elements with respect to all functions g' and $h' \in \mathcal{S}_0(\mathbb{R}^n)$ satisfy $\sigma_{g',h'} \in \Sigma_m(\mathbb{H}^n)$.*

(iii) *The family of dilated, translated and rescaled operators*

$$K_{b,a} = \rho(b,a)^{-1}\, D_a^{-1} T_b^{-1} K T_b D_a, \quad (b,a) \in \mathbb{H}^n,$$

is uniformly continuous from $\mathcal{S}_0(\mathbb{R}^n)$ to $\mathcal{S}_0(\mathbb{R}^n)$.

(iv) *There are functions v, $w \in \mathcal{S}_0(\mathbb{R}^n)$ and some $F \in \Sigma_\rho(\mathbb{H}^n \times \mathbb{H}^n)$ such that*

$$K = \mathcal{M}_v \mathit{Op}[F] \mathcal{W}_w.$$

(v) *For all g, $h \in \mathcal{S}_0(\mathbb{R}^n)$, admissible, there are operators A, $B \in CZO$ such that*

$$K = M_\rho\, A = B\, M_\rho,$$

where M_ρ is the wavelet multiplier $s \mapsto \mathcal{M}_h M[\rho] \mathcal{W}_g s$.

Proof. The proof is essentially the same as in the preceding theorems and we may leave it to the reader. □

Let us denote by $\Sigma(\mathbb{H}^n \times \mathbb{H}^n)$ the union of all $\Sigma_\rho(\mathbb{H}^n \times \mathbb{H}^n)$ for all possible sub-multiplicative weight functions ρ. From property (v) it follows that $\Sigma(\mathbb{H}^n \times \mathbb{H}^n)$ is a graded algebra. More precisely we have

$$\mathit{Op}[\Sigma_\rho(\mathbb{H}^n \times \mathbb{H}^n)] \mathit{Op}[\Sigma_\mu(\mathbb{H}^n \times \mathbb{H}^n)] \subset \mathit{Op}[\Sigma_{\rho\mu}(\mathbb{H}^n \times \mathbb{H}^n)].$$

7.6 The small scale averaging operators

The operators in the previous section have been characterized by the uniformity of their mapping properties at all positions and scales. The mapping was from $\mathcal{S}_0(\mathbb{R}^n)$ to $\mathcal{S}_0(\mathbb{R}^n)$. In the following theorem we consider operators that are mappings $\mathcal{S}_0(\mathbb{R}^n) \to \mathcal{S}(\mathbb{R}^n)$. The modification around the 0 frequency is crucial. We have

Theorem 7.10 *For a linear operator $K : \mathcal{S}_0(\mathbb{R}^n) \to \mathcal{S}(\mathbb{R}^n)$ we have that (i) $\Rightarrow$ (ii) $\Leftrightarrow$ (iii), where*

(i) *The family of dilated and translated operators*

$$K_{b,a} = D_a^{-1} T_b^{-1} K T_b D_a, \quad (b,a) \in \mathbb{H}^n,$$

is uniformly continuous from $\mathcal{S}_0(\mathbb{R}^n)$ to $\mathcal{S}(\mathbb{R}^n)$.

(ii) *The matrix elements*

$$\sigma[K](b,a;b',a') = \sigma_{g,h}[K](b,a;b',a') = \langle g_{b,a}, K h_{b',a'} \rangle ,$$

with respect to some functions g, $h \in \mathcal{S}_0(\mathbb{R}^n)$ satisfying

$$\int_0^\infty \frac{da}{a} |\widehat{g}(ak)|^2 \sim \int_0^\infty \frac{da}{a} \left|\widehat{h}(ak)\right|^2 \sim 1,$$

satisfy

$$|\sigma(b,a;b',a')| \leq \frac{1}{a'} r\left(\frac{b-b'}{a'}, \frac{a}{a'}\right),$$

where r satisfies (compare Lemma 6.2) that for all $m > 0$ there is $c > 0$ and $F \in \mathcal{S}(\mathbb{R}^n)$ such that

$$r(b,a) \leq c_m \frac{a^m}{(1+a)^m} \frac{1}{(1+a)^m} F\left(\frac{b}{1+a}\right).$$

(iii) *The matrix elements with respect to all functions g' and $h' \in \mathcal{S}_0(\mathbb{R}^n)$ satisfy the above estimate.*

Note that it is not true that (ii) implies (i). This comes from the fact that it is not possible to characterize $\mathcal{S}(\mathbb{R}^n)$ in wavelet space. However, we have the implication of Lemma 3.21. We call these operators the small scale averaging operators, for the following reason. If for every point $(b,a) \in \mathbb{H}^n$ we consider the so-called Carleson window $T(b,a) = \{|b'-b|, a' \leq a\}$. Then K is essentially majorized by the following non-linear operator that averages the wavelet coefficients over the window $T(b,a)$.

$$\mathcal{W}_g s(b,a) \mapsto \frac{1}{a^n} \int_{T(b,a)} \frac{db'da'}{a'} |\mathcal{W}_g s(b',a')| .$$

Proof. The proof that (i) implies (ii) is the same as the one of Theorem 7.3. The only modification comes from the implication of Lemma 3.21. The implication (ii) $\Rightarrow$ (iii) follows as usual, since the transition from an admissible wavelet to any other is given by convolution with some highly localized kernel, which leaves invariant the type of localization of σ. □

In view of Lemma 6.2 these operators are continuous in all strongly tempered Banach spaces. We will encounter operators of the form

$$(-\Delta)^{m/2}K, \quad K(-\Delta)^{m/2}.$$

In view of Lemma 6.3 the first is smoothing from $B(\mathbb{R}^n)$ to $a^m B(\mathbb{R}^n)$ provided $B(\mathbb{R}^n)$ is strongly tempered, whereas the second is smoothing from $B(\mathbb{R}^n)$ to $a^m B(\mathbb{R}^n)$ if $a^m B(\mathbb{R}^n)$ is strongly tempered.

8 The algebra of smoothly varying space-scale convolution operators

In this section we will consider a sub-algebra of the algebra $\Sigma(\mathbb{H}^n \times \mathbb{H}^n)$ of the previous section. Whereas in $\Sigma(\mathbb{H}^n \times \mathbb{H}^n)$ the product is obtained via integrals, in this new algebra the product becomes rather explicit up to smoothing operators. We can work in the simpler algebra of convolution operators, $\mathcal{S}(\mathbb{H}^n)$. Although, at a first glance, for general functions u, $v \in \mathcal{S}(\mathbb{H}^n)$, the convolution product $u * v$ is not very explicit, things are different in the case where u and v are in the image of wavelet transforms. Indeed, suppose $u = \mathcal{W}_g s$ and $v = \mathcal{W}_h r$. Then

$$u * v = \mathcal{W}_g s * \mathcal{W}_h r = \mathcal{W}_g \mathcal{M}_s \mathcal{W}_h r = \mathcal{W}_g r', \quad \widehat{r'(\xi)} = \widehat{r}(\xi) \int_0^\infty \frac{da}{a} \overline{\widehat{h}}(a\xi)\widehat{s}(a\xi).$$

The analogy with ordinary pseudo-differential operators goes further. Note that one may think of pseudo-differential operators as position dependent convolutions, that is operators of the following formal form

$$s(y) \mapsto F_y * s(x) \,|_{y=x},$$

with F_x in general a distribution. The algebra of wavelet pseudo-differential operators we want to introduce now is of a similar form. However, we allow the convolution operator to vary arbitrarily in the whole position-scale half-space. More precisely, suppose for each point $(b, a) \in \mathbb{H}^n$ we are given a function $P_{(b,a)} \in \mathcal{S}(\mathbb{R}^n)$. Now look at the following mapping

$$r(b, a) \mapsto (P_{(b,a)} * r)(b, a).$$

As it will turn out under some natural regularity requirements we can establish the analogue of Leibniz rule for these operators. In addition as we will see later the small scale behavior of ordinary pseudo-differential operators is described (up to smoothness) by these operators. The big advantage of the wavelet calculus is that we can read directly from the symbol many of its mapping properties in all admissible Banach spaces $B(\mathbb{R}^n)$.

We come to the details. Let $s(b,a;b',a')$ be a complex-valued function over $\mathbb{H}^n \times \mathbb{H}^n$. We associate with it the following new function

$$T[s](b,a;b',a') = \frac{1}{a'^n}\, s\left(b,a;\frac{b-b'}{a'},\frac{a}{a'}\right),$$

and we associate the following operator with s

$$Cop[s] = Op[T[s]].$$

Explicitly $Cop[s]$ acts (formally) on $r(b,a)$ via (writing $s_{b,a}(\cdot,\cdot) = s(b,a;\cdot,\cdot)$)

$$(Cop[s]r)(b,a) = \int_{\mathbb{H}^n} \frac{db'da'}{a'}\frac{1}{a'^n}\, s\left(b,a;\frac{b-b'}{a'},\frac{a}{a'}\right) r(b',a') = (s_{b,a} * r)(b,a).$$

We shall also need a dual version. In that case consider

$$T'[s](b,a;b',a) = \frac{1}{a'^n}\, s\left(b',a';\frac{b-b'}{a'},\frac{a}{a'}\right),$$

and set

$$Poc[s] = Op[T'[s]].$$

Explicitly we have

$$(Poc[s]r)(b,a) = \int_{\mathbb{H}^n} \frac{db'da'}{a'}\frac{1}{a'^n}\, s\left(b',a';\frac{b-b'}{a'},\frac{a}{a'}\right) r(b',a').$$

Recall that a function $\rho(b,a)$ is called sub-multiplicative if it satisfies (c, $M>0$)

$$\rho((b,a)\circ(b',a')) = \rho((b+ab',aa')) \le c\,\Delta((b',a'))^M \rho(b,a).$$

Here Δ is the non-euclidian distance function of Section 5. To simplify notations, we will not distinguish between $\rho((b,a))$ and $\rho(b,a)$. Recall that all sub-multiplicative functions are tempered

$$\rho(b,a) \le O(1)\,(a+1/a)^K(1+|b|)^K.$$

Examples of such sub-multiplicative functions are precisely $\Delta(b,a)$ or a^γ. Recall that the product and the sum of two sub-multiplicative weight functions is again a sub-multiplicative weight function. Moreover, the set of sub-multiplicative weight functions is stable under the formation of positive powers.

Suppose now that s satisfies the following estimate

$$|s(b,a;b',a')| \le \rho(b,a)F(b',a'),$$

with some functions $F \in \mathcal{S}(\mathbb{H}^n)$. Then, as follows from the discussion before Theorem 7.9 we have the following preliminary result.

Proposition 8.1 *The operators $\mathit{Cop}[s]$ and $\mathit{Poc}[s]$ satisfy*

$$\mathit{Cop}[s], \mathit{Poc}[s] \in \mathit{Op}[\Sigma_\rho(\mathbb{H}^n \times \mathbb{H}^n)].$$

In particular, for suitable ρ they are smoothing.

In order to establish a symbolic calculus we have to restrict ourselves to functions s that satisfy a set of differential inequalities. In turn we get a sub-algebra of $\Sigma(\mathbb{H}^n \times \mathbb{H}^n)$ in which the computation of products of operators is reduced to the simpler computation of convolutions of highly localized kernels. We now define a class of symbols.

Definition 8.2 *Let $\rho_{\alpha,l}$, $l = 0, 1, \ldots$, $\alpha \in \mathbb{N}^n$, be a family of sub-multiplicative weight functions. We then say that $s(b, a; b', a')$ is a symbol of type $\{\rho_{\alpha,l}\}$ if for all l and α there are functions $F_{\alpha,l} \in \mathcal{S}(\mathbb{H}^n)$ such that we have*

$$\left|\partial_a^l \partial_b^\alpha s(b, a; b', a')\right| \leq \rho_{\alpha,l}(b, a)\, F_{\alpha,l}(b', a').$$

We denote by $\Xi_{\{\rho_{\alpha,l}\}}(\mathbb{H}^n \times \mathbb{H}^n)$ the set of all such functions. The union of these sets is called $\Xi(\mathbb{H}^n \times \mathbb{H}^n)$.

Note that for the moment we have not introduced any relation between the weight functions. This is because we want to establish the calculus in most general terms. Later in order to use it in the analysis of partial differential operators we will require that $\rho_{\alpha,l}$ gets "smaller" as l and α increase.

For a symbol s of type $\{\rho_{\alpha,m}\}$ obviously have

$$\mathit{Cop}[s] \in \mathit{Op}[\Sigma_{\rho_{0,0}}(\mathbb{H}^n \times \mathbb{H}^n)].$$

We are now ready to state and prove a series of theorems. They allow us to reduce computations of products of operators of the $\mathit{Cop}[s]$ with $s \in \Xi(\mathbb{H}^n \times \mathbb{H}^n)$ in terms of the simpler convolution algebra $\Gamma_\infty(\mathbb{H}^n)$. In practical applications, we will use a still more simple algebra (see below).

Theorem 8.3 *The set $\Xi(\mathbb{H}^n \times \mathbb{H}^n)$ is a graded algebra. More precisely, let $\{\sigma_{\alpha,m}\}$ and $\{\rho_{\alpha,m}\}$ be two families of weight function as in the previous definition. For $s \in \Xi_{\{\sigma_{\alpha,m}\}}(\mathbb{H}^n \times \mathbb{H}^n)$, and $r \in \Xi_{\{\rho_{\alpha,m}\}}(\mathbb{H}^n \times \mathbb{H}^n)$ there is some $u \in \Xi_{\{\nu_{\alpha,m}\}}(\mathbb{H}^n \times \mathbb{H}^n)$, with*

$$\nu_{\alpha,m} = \sum_{\beta \leq \alpha, l \leq m} \sum_{|\gamma|+k=l} \sigma_{\alpha-\beta, m-l}\rho_{\beta+\gamma, m-l},$$

such that

$$\mathit{Cop}[s]\mathit{Cop}[r] = \mathit{Cop}[u].$$

A second theorem allows the expansion of the product symbol u into simpler convolution operators.

Theorem 8.4 *For all $M \in \mathbb{N}$ we have the following expansion*

$$u(b,a;b',a') = \sum_{|\alpha|+m\leq M} (\alpha! m!)^{-1} a^{|\alpha|+m}\, u_{\alpha,m}(b,a;b',a') + a^{M+1} R_M(b,a;b',a'),$$

with

$$u_{\alpha,m}(b,a;b',a') \\ = ([s(b,a;b'',a'')(-b''/a'')^{\alpha}(1/a''-1)^m] * [\partial_a^m \partial_b^{\alpha} r(b,a;b'',a'')])(b',a').$$

The convolution is in the doubly primed variables. The terms $u_{\alpha,m}$ are in $\Xi_{\{\mu_{\beta,l}\}}(\mathbb{H}^n \times \mathbb{H}^n)$ with

$$\mu_{\beta,l} = \sum_{k\leq l,\gamma\leq\beta} \sigma_{\beta-\gamma,l-k}\,\rho_{k+m,\gamma+\alpha},$$

and the remainders R_M satisfy

$$Cop[R_M] \in Op[\Sigma_{\gamma_M}(\mathbb{H}^n \times \mathbb{H}^n)],$$

with

$$\gamma_M = \sum_{|\alpha|+m=M+1} \sigma_{0,0}\,\rho_{\alpha,m}.$$

In order to multiply more than two operator using these formulas we need to know how to commute multiplication with a^{γ} through different symbols. Let us denote by $M[\sigma]$ the multiplication by σ

$$M[\sigma] : s(b,a) \mapsto \sigma(b,a)\, s(b,a).$$

Then we have.

Lemma 8.5 *We have*

$$M[a^{\gamma}]Cop[\sigma] = Cop[\sigma']\, M[a^{\gamma}], \quad \sigma'(b,a;b',a') = \sigma(b,a;b',a')\, a'^{\gamma}.$$

for any $\gamma \in \mathbb{C}$.

Proof. This follows from the identity

$$a^{\gamma} \frac{1}{a'^n} s\left(b,a;\frac{b-b'}{a'},\frac{a}{a'}\right) = \frac{1}{a'^n} s\left(b,a;\frac{b-b'}{a'},\frac{a}{a'}\right) (a/a')^{\gamma}\, a'^{\gamma}.$$

□

So far the theorems concern computation in $Cop[s]$. The relation between $Cop[s]$ and $Poc[s]$ is that of duality for the skew linear and linear pairing

$$\langle s,r\rangle_{\mathbb{H}^n} = \int_{\mathbb{H}^n} \frac{dbda}{a}\, \overline{s}(b,a)\, r(b,a), \quad (s,r)_{\mathbb{H}^n} = \int_{\mathbb{H}^n} \frac{dbda}{a}\, s(b,a)\, r(b,a).$$

Lemma 8.6 *We have for s, $r \in \mathcal{S}(\mathbb{H}^n)$ and $u \in \Xi(\mathbb{H}^n \times \mathbb{H}^n)$*

$$\langle s, Cop[u]r\rangle_{\mathbb{H}^n} = \langle Cop[u]^* s, r\rangle_{\mathbb{H}^n} = \langle Poc[u]s, r\rangle_{\mathbb{H}^n},$$

and

$$(s, Cop[u]^t r)_{\mathbb{H}^n} = (Cop[u]^t s, r)_{\mathbb{H}^n} = (Poc[u]s, r)_{\mathbb{H}^n}.$$

Proof. Direct computation involving only exchange of integration. □

These operators can be treated by the previous calculus.

Theorem 8.7 *Let s be of type $\{\sigma_{\alpha,m}\}$. Then there is u of type $\{\mu_{\alpha,m}\}$ with*

$$\mu_{\alpha,m} = \sum_{|\beta|+l=m} \sigma_{\alpha+\beta,l},$$

such that

$$Poc[s] = Cop[u].$$

Explicitly we have $u(b,a;b',a') = s(b - b'a/a', a/a'; b', a')$.

Note that for the convolution operator $A[P]; s \mapsto P * s$ we have

$$A[P]^* = A[P^*], \quad P^*(b,a) = \overline{P}(-b/a, 1/a)/a^n.$$

The next theorem shows, that for smoothly varying coefficients, this formula holds with "frozen" coefficients.

Theorem 8.8 *Moreover, there are functions $s_{\beta,l}$ of type $\{\mu_{\gamma,k}\}$, $\mu_{\gamma,k} = \sigma_{\gamma+\beta,l+k}$ such that for all $M \in \mathbb{N}$ we have*

$$u(b,a;b',a') = \sum_{|\alpha|+m\le M} (\alpha! m!)^{-1} a^{|\alpha|+m} s_{\alpha,m}(b,a; -b'/a', a/a') + R_M,$$

and $R_M \in \Sigma_{\gamma_M}(\mathbb{H}^n \times \mathbb{H}^n)$ with $\gamma_M = \sum_{|\gamma|+l=M+1} \sigma_{\gamma,l}$. Explicitly we can take

$$s_{\alpha,m}(b,a;b',a') = (\partial_a^m \partial_b^\alpha s)(b,a;b',a') b'^\alpha (a'-1)^m.$$

Since $CZO(\mathbb{H}^n \times \mathbb{H}^n)$ corresponds to those operators, for which the weight functions may be chosen $\rho_{\alpha,m} = 1$, we have that $\Xi(\mathbb{H}^n \times \mathbb{H}^n)$ is stable under $CZO(\mathbb{H}^n \times \mathbb{H}^n)$

$$\begin{aligned} Op[CZO(\mathbb{H}^n \times \mathbb{H}^n)]Cop[\Xi(\mathbb{H}^n \times \mathbb{H}^n)] &\subset Cop[\Xi(\mathbb{H}^n \times \mathbb{H}^n)],\\ Cop[\Xi(\mathbb{H}^n \times \mathbb{H}^n)]Op[CZO(\mathbb{H}^n \times \mathbb{H}^n)] &\subset Cop[\Xi(\mathbb{H}^n \times \mathbb{H}^n)]. \end{aligned}$$

Therefore it makes sense to pull back $Cop[\Xi(\mathbb{H}^n \times \mathbb{H}^n)]$ to an algebra of operators acting on functions over $\mathbb{R}^n$ via the wavelet transform. We denote this algebra by Ξ

and we have $A \in \Xi$ iff there are wavelets g, h, both admissible and $s \in \Xi(\mathbb{H}^n \times \mathbb{H}^n)$ such that

$$A = \mathcal{M}_h Cop[s] \mathcal{W}_g.$$

We now come to the proofs of the statements.

Proof (of Theorem 8.3). We have to show that u defined through

$$\frac{1}{a'^n} u\left(b, a; \frac{b-b'}{a'}, \frac{a}{a'}\right) = \int_{\mathbb{H}^n} \frac{db''da''}{a''} \frac{1}{a''^n} s\left(b, a; \frac{b-b''}{a''}, \frac{a}{a''}\right) \frac{1}{a'^n} r\left(b'', a''; \frac{b''-b'}{a'}, \frac{a''}{a'}\right)$$

satisfies the statements in the theorem. Let $\tilde{u}(b, a; b', a')$ stand for the inverted function

$$\tilde{u}(b, a; b', a') = \frac{1}{a'^n} u\left(b, a; \frac{-b'}{a'}, \frac{1}{a'}\right).$$

Note that $\tilde{\tilde{u}} = u$. We use the analogous notation for $\tilde{s}$ and $\tilde{r}$. In addition we have

$$\partial_a^m \partial_b^\alpha \tilde{u}(b, a; b', a') = \widetilde{\partial_a^m \partial_b^\alpha u}(b, a; b', a'). \tag{8.1}$$

A simple change of variables, $b'' \to ab'' + b$, $a'' \to aa''$ shows that we have

$$\frac{1}{a^n} \tilde{u}\left(b, a; \frac{b'-b}{a}, \frac{a'}{a}\right) = \frac{1}{a^n} \int_{\mathbb{H}^n} \frac{db''da''}{a''} \tilde{s}(b, a; b'', a'') \frac{1}{a''^n} \tilde{r}\left(b + ab'', aa''; \frac{[b'-b]/a - b''}{a''}, \frac{[a'/a]}{a''}\right),$$

or equivalently

$$u(b, a; b', a') = \int_{\mathbb{H}^n} \frac{db''da''}{a''} \frac{1}{a''^n} r\left(b + ab'', aa''; \frac{b'-b''}{a''}, \frac{a'}{a''}\right) s(b, a; b'', a'').$$

Now, for every $m \in \mathbb{N}$

$$\partial_a^m \tilde{r}(b + ab'', aa''; \cdot, \cdot) = \sum_{|\beta|+l=m} a''^{|\beta|+l} (\partial_a^l \partial_b^\beta \tilde{r})(b + ab'', aa''; \cdot, \cdot) b''^\alpha.$$

Thus, by the usual Leibniz rule we have for every multi-index α and $m \in \mathbb{N}$,

$$\begin{aligned}
&\partial_a^m \partial_b^\alpha \tilde{u}(b, a; b', a') \\
&\quad = \int_{\mathbb{H}^n} \frac{db''da''}{a''} \frac{1}{a''^n} \sum_{\beta \le \alpha, l \le m} \binom{\alpha}{\beta} \binom{m}{l} \\
&\quad \sum_{|\gamma|+k=l} a''^k b''^\gamma (\partial_a^k \partial_b^{\beta+\gamma} \tilde{r}) \left(b + ab'', aa''; \frac{b'-b''}{a''}, \frac{a'}{a''}\right) \\
&\quad \times (\partial_a^{m-l} \partial_b^{\alpha-\beta} \tilde{s})(b, a; b'', a'')
\end{aligned}$$

By hypothesis on s and r we may estimate with some function $F \in \mathcal{S}(\mathbb{R}^n)$,

$$|\partial_a^m \partial_b^\alpha \tilde{u}(b,a;b',a')| \le \int_{\mathbb{H}^n} \frac{db''da''}{a''} \sum_{\beta\le\alpha, l\le m} \sum_{|\gamma|+k=l} \rho_{\beta+\gamma,k}(b+ab'',aa'')\sigma_{\alpha-\beta,m-l}(b,a) \frac{1}{a''^n} F\left(\frac{b'-b''}{a''}, \frac{a'}{a''}\right) F(b'',a'').$$

Now writing

$$(b+ab'',aa'') = (b,a)\circ(b'',a''),$$

and using the sub-multiplicativity of the $\rho_{\alpha,m}$ there are c, $K>0$ such that we have for all $\delta\le\alpha$ and $j\le m$

$$\rho_{\delta,j}(a'b''+b',a'a'') \le c\rho_{\delta,j}(b',a')\,(a''+1/a'')^K(1+|b''|)^K.$$

This last term may be absorbed into F and thus with some $G\in\mathcal{S}(\mathbb{H}^n)$ we have

$$|\partial_a^m \partial_b^\alpha \tilde{u}(b,a;b',a')| \le \sum_{\beta\le\alpha, l\le m} \sum_{|\gamma|+k=l} \rho_{\beta+\gamma,k}(b,a)\sigma_{\alpha-\beta,m-l}(b,a)\ (F*G)(b',a').$$

Now $F*G\in\mathcal{S}(\mathbb{H}^n)$ and thus the theorem follows from the identity (8.1). □

Proof (of Theorem 8.4). Let as before $\tilde{u}(b,a;b',a')$ stand for the inverted function

$$\tilde{u}(b,a;b',a') = \frac{1}{a'^n} u\left(b,a;\frac{-b'}{a'},\frac{1}{a'}\right).$$

A simple change of variables, $b''\to ab''+b$, $a''\to aa''$ shows that we have

$$\frac{1}{a^n}\tilde{u}\left(b,a;\frac{b'-b}{a},\frac{a'}{a}\right) = \frac{1}{a^n}\int_{\mathbb{H}^n} \frac{db''da''}{a''}\tilde{s}(b,a;b'',a'')\,\frac{1}{a''^n}\,\tilde{r}\left(b+ab'',aa'';\frac{[b'-b]/a-b''}{a''},\frac{[a'/a]}{a''}\right),$$

or equivalently

$$\tilde{u}(b,a;b',a') = \int_{\mathbb{H}^n} \frac{db''da''}{a''}\frac{1}{a''^n}\tilde{r}\left(b+ab'',aa'';\frac{b'-b''}{a''},\frac{a'}{a''}\right)\tilde{s}(b,a;b'',a'').$$

Using Taylor's theorem we can write

$$\tilde{r}(b+ab'',aa'';b',a') = \sum_{|\alpha|+m\le M} (\alpha!m!)^{-1}a^{|\alpha|+m}(\partial_a^m\partial_b^\alpha\tilde{r})(b,a;b',a')b''^\alpha(a''-1)^m + R'_M(b,a;b',a';b'',a'').$$

Thus

$$\tilde{u}_{\alpha,m}(b,a;b',a')$$
$$= [\partial_a^m \partial_b^\alpha \tilde{r}(b,a;b'',a'')] * [\tilde{s}(b,a;b'',a'')b''^\alpha(a''-1)^m](b',a'),$$

with the convolution understood acting on the doubly primed variables. Now using the formula

$$\tilde{s} * \tilde{r} = \widetilde{r * s},$$

we obtain the terms of the theorem. The remainder can be estimated as follows.

$$R'_M(b,a;b',a';b'',a'')$$
$$= (M+1) \sum_{|\alpha|+m=M+1} \int_0^1 d\lambda (1-\lambda)^M (\alpha! m!)^{-1}$$
$$a^{|\alpha|+m} (\partial_a^m \partial_b^\alpha \tilde{r})(b+\lambda a b'', a(1+\lambda(a''-1)); b', a')\, b''^\alpha (a''-1)^m$$

We write

$$(b+\lambda a b'', a(1+\lambda(a''-1))) = (b,a) \circ (\lambda b'', 1+\lambda(a''-1)).$$

By hypothesis we have with $\rho_{M+1}(b,a) = a^{M+1} \sum_{|\alpha|+m=M+1} \rho_{\alpha,m}(b,a)$,

$$\left| a^{M+1}\, \partial_a^m \partial_b^\alpha \tilde{r}(b,a;b',a') \right| \le \rho_{M+1}(b,a) G(b',a'),$$

with some $G \in \mathcal{S}(\mathbb{H}^n)$. Now ρ_{M+1} is sub-multiplicative and we may write with some $K' > 0$,

$$|R'_M(b,a;b',a';b'',a'')|$$
$$\le |b''|^{M+1} (a''-1)^{M+1} \rho_{M+1}(b,a) G(b',a')$$
$$\int_0^1 d\lambda (1-\lambda)^M \Delta(\lambda b'', 1+\lambda(a''-1))^K$$
$$\le c'(1+|b''|)^{K'} (a''-1/a'')^{K'} \rho_{M+1}(b,a) G(b',a').$$

We have

$$\tilde{R}_M(b,a;b',a') \le \int_{\mathbb{H}^n} \frac{db''da''}{a''} \tilde{s}(b,a;b'',a'') \frac{1}{a''^n} R'_M\left(b,a; \frac{b'-b''}{a''}, \frac{a'}{a''}; b'', a''\right).$$

Now $|\tilde{s}(b,a;b'',a'')| \le \sigma_{0,0}(b,a) F(b'',a'')$. The first factors may thus be absorbed in F, since

$$H = (1+|b''|)^{K'} (a''-1/a'')^{K'} F \in \mathcal{S}(\mathbb{H}^n),$$

and thus

$$R_M(b,a;b',a') \le \sigma_{0,0}(b,a) \rho_{M+1}(b,a) G * H(b',a').$$

We may use again the fact that $G * H \in \mathcal{S}(\mathbb{R}^n)$ to conclude. □

Proof (of Theorem 8.7). We have

$$\frac{1}{a'^n} s\left(b', a'; \frac{b-b'}{a'}, \frac{a}{a'}\right) = \frac{1}{a'^n} u\left(b, a; \frac{b-b'}{a'}, \frac{a}{a'}\right),$$

if and only if

$$u(b, a; b', a') = s(-b'a/a' + b, a/a'; b', a').$$

We have

$$\begin{aligned}\partial_a^m \partial_b^\alpha u(b, a; b', a') \\ = \sum_{|\beta|+l=m} (-b'/a')^\beta a'^{-l} (\partial_a^l \partial_b^{\alpha+\beta} s)(b - ab'/a', a/a'; b', a').\end{aligned}$$

Writing

$$(b - ab'/a', a/a'; b', a') = (b, a) \circ (-b'/a', 1/a'),$$

and using the sub-multiplicativity of $\sigma_{\alpha,m}$ we have

$$|\partial_a^m \partial_b^\alpha u(b, a; b', a')| \leq \sum_{|\beta|+l=m} \sigma_{\alpha+\beta,l}(b, a) F(b', a'),$$

with some $F \in \mathcal{S}(\mathbb{H}^n)$. □

Proof (of Theorem 8.8). To show the expansion we apply Taylor's expansion

$$s(b', a'; b, a) = s(b, a; b', a') + \dots .$$

We leave the details to the reader. □

8.1 The Toeplitz operators

We now want to show that $Cop[\Xi(\mathbb{H}^n \times \mathbb{H}^n)]$ is stable under tempered multipliers

$$M[r] : s \mapsto r \cdot s.$$

Definition 8.9 *Let $\{\rho_{\alpha,m}\}$ be a family of sub-multiplicative weight functions. We say a function $r = r(b, a)$ is of type $\{\rho_{\alpha,m}\}$ if for all multi-indices α and all m there are $c_{\alpha,m} > 0$ such that*

$$|\partial_a^m \partial_b^\alpha r(b, a)| \leq c_{\alpha,m} \rho_{\alpha,m}(b, a).$$

We denote the set of all such multipliers by $\Xi_{\{\rho_{\alpha,m}\}}(\mathbb{H}^n)$ and the union of these sets over all such weight functions by $\Xi(\mathbb{H}^n)$

Proposition 8.10 *The set $\Xi(\mathbb{H}^n)$ is a graded algebra. More precisely for s of type $\sigma_{\alpha,m}$, and r of type $\rho_{\alpha,m}$, $u = sr$ is of type $\mu_{\alpha,m}$ with*

$$\mu_{\alpha,m} = \sum_{\gamma\le\alpha, l\le m} \sigma_{\alpha-\gamma,m-l}\rho_{\gamma,l}.$$

Proof. Leibniz rule of calculus. □

The name Toeplitz operators is justified, since as the next theorem shows, we can compute modulo smoothness e.g. in the algebra generated by the projector $s \mapsto P_g * s$ where P_g is the reproducing kernel associated with the strictly admissible wavelet g and multipliers $s \mapsto \mu s$. More precisely, as it turns out, the algebra $Cop[\Xi(\mathbb{H}^n \times \mathbb{H}^n)]$ is stable under $M[\Xi(\mathbb{H}^n)]$.

$$M[\Xi(\mathbb{H}^n)]Cop[\Xi(\mathbb{H}^n \times \mathbb{H}^n)] \cup Cop[\Xi(\mathbb{H}^n \times \mathbb{H}^n)]M[\Xi(\mathbb{H}^n)] \subset Cop[\Xi(\mathbb{H}^n \times \mathbb{H}^n)].$$

Theorem 8.11 *Let r be of type $\{\rho_{\alpha,m}\}$. Then, let $s \in \Xi(\mathbb{H}^n \times \mathbb{H}^n)$ be of type $\sigma_{\alpha,m}$. It follows that there are u, $u' \in \Xi(\mathbb{H}^n \times \mathbb{H}^n)$ such that*

$$M[r]Cop[s] = Cop[u], \quad Cop[s]M[r] = Cop[u'].$$

We have that u is of type $\{\mu_{\alpha,m}\}$ with

$$\mu_{\alpha,m} = \sum_{\beta\le\alpha, l\le m} \rho_{\beta,l}\sigma_{\alpha-\beta,m-l},$$

and u' is of type $\{\mu'_{\alpha,m}\}$ with

$$\mu'_{\alpha,m} = \sum_{\beta\le\alpha, l\le m} \sum_{|\gamma|+k=l} \rho_{\beta+\gamma,k}\sigma_{\alpha-\beta,m-l}.$$

Moreover, the following expansion holds for the commutator

$$\begin{aligned} &u'(b,a;b',a') \\ &\quad = u(b,a;b',a') + \sum_{1\le|\alpha|+m\le M} (\alpha! m!)^{-1}\, a^{|\alpha|+m} u_{\alpha,m}(b,a;b',a') \\ &\quad\quad + a^{M+1} R_M(b,a;b',a') \end{aligned}$$

with $u_{\alpha,m} \in \Xi(\mathbb{H}^n \times \mathbb{H}^n)$ given by

$$u_{\alpha,m}(b,a;b',a') = s(b,a;b',a')(-b')^\alpha a'^{-|\alpha|-m}(a'-1)^m\, \partial_a^m \partial_b^\alpha r(b,a),$$

and the remainder $R_M(b,a;b',a')$ satisfies $Cop[R_M] \in Op[\Sigma_\nu(\mathbb{H}^n \times \mathbb{H}^n)]$ with $\nu = \sigma_0 \sum_{|\alpha|+m=M+1} \rho_{\alpha,m}$.

Proof. We have

$$u(b,a;b',a') = r(b,a)\,s(b,a;b',a'),$$

and therefore the Leibniz rule shows that $u \in \Xi(\mathbb{H}^n \times \mathbb{H}^n)$ and that it satisfies the estimates of the theorem. On the other hand

$$\frac{1}{a'^n} u'\left(b,a;\frac{b-b'}{a'},\frac{a}{a'}\right) = \frac{1}{a'^n} s\left(b,a;\frac{b-b'}{a'},\frac{a}{a'}\right) r(b',a'),$$

or what is the same

$$u'(b,a;b',a') = s(b,a;b',a')r(b-b'a/a',a/a').$$

Now

$$\begin{aligned}&\partial_a^m \partial_b^\alpha u'(b,a;b',a')\\ &\quad= \sum_{\beta\le\alpha,l\le m} \binom{m}{l}\binom{\alpha}{\beta}\\ &\qquad \sum_{|\gamma|+k=l} a'^l b'^\gamma (\partial_a^k \partial_b^{\beta+\gamma} r)(b-b'a/a',a/a')\,\partial_a^{m-l}\partial_b^{\alpha-\beta} s(b,a;b',a').\end{aligned}$$

using the estimates on the partial derivatives of r and s it follows that $u' \in \Xi(\mathbb{H}^n \times \mathbb{H}^n)$ and that it satisfies the estimates of the theorem.

To prove the commutator formula we use Taylor's theorem to write

$$\begin{aligned}&r(b-b'a/a',a/a')\\ &\quad= \sum_{|\alpha|+m\le M} a^{|\alpha|+m}(\alpha!m!)^{-1}(-b')^\alpha a'^{-|\alpha|-m}(a'-1)^m \partial_a^m\partial_b^\alpha r(b,a)\\ &\qquad +(M+1)\int_0^1 d\lambda(1-\lambda)^M \sum_{|\alpha|+m=M+1} a^{|\alpha|+m}(\alpha!m!)^{-1}(-b')^\alpha a'^{-|\alpha|-m}(a'-1)^m\\ &\qquad \partial_a^m\partial_b^\alpha r(b-\lambda b'a/a',a+\lambda a(1/a'-1)).\end{aligned}$$

The first terms are the expansion terms of the theorem. To estimate the remainder we write

$$(b-\lambda b'a/a',a+\lambda a(1/a'-1)) = (b,a)\circ(-\lambda b'/a',1+\lambda(1/a'-1)).$$

Thanks to sub-multiplicativity we may estimate with ν_M as given in the theorem and some c, $K>0$,

$$|\partial_a^m\partial_b^\alpha r(b-\lambda b'a/a',a+\lambda a(1/a'-1))| \le c\nu_M(b,a)(-\lambda b'/a',1+\lambda(1/a'-1))^K.$$

Using once more the estimate

$$\int_0^1 d\lambda\Delta(\lambda b'',1+\lambda[a''-1])^K \le c'\Delta(b'',a'')^{K'},$$

the theorem follows. □

8.2 A simplification

Until now we have only established formulas on a rather algebraic level. In particular, we have made no hypothesis on the growth behavior of the terms in the various expansions we have established. To do analysis one might require that at least in some regions in the parameter space the higher order terms are smaller. In the following we present a global class of symbols, global in the sense that the higher oder terms are small in the limit $(a+1/a)(1+|b|) \to 0$. But in principle depending on the region in the parameter space that is of interest one can develop a calculus suited for that region of $\mathbb{H}^n$ which is important.

For $\delta > 0$ and $\rho \in \mathbb{R}$ consider the set of all functions $s(b,a;b',a')$ that satisfy

$$|\partial_a^m \partial_b^\alpha s(b,a;b',a')| \le a^{-|\alpha|-m} \Delta(b,a)^{\rho-\delta(|\alpha|+m)}\, F_{\alpha,m}(b',a'),$$

with some $F_{\alpha,m} \in \mathcal{S}(\mathbb{H}^n)$.[1] The set of all such symbols will be denoted by $\Omega_\rho^\delta(\mathbb{H}^n \times \mathbb{H}^n)$. The number ρ is the smoothness variable and δ is the degree of smoothness that is gained by taking derivatives. Note that we have

$$Cop[\Omega_\rho^\delta(\mathbb{H}^n \times \mathbb{H}^n)] \subset Op[\Sigma_{\Delta^\rho}(\mathbb{H}^n \times \mathbb{H}^n)].$$

Definition 8.12 *Let $u_k \in \Omega_{\rho-k}^\delta(\mathbb{H}^n \times \mathbb{H}^n)$, $k = 0,1,\ldots$. We then write*

$$u(b,a;b',a') \sim \sum_{k\ge 0} a^k\, u_k(b,a;b',a'),$$

if for all $M \in \mathbb{N}$ there is $R_M \in \Sigma_{\Delta^{\rho-(M+1)\delta}}(\mathbb{H}^n \times \mathbb{H}^n)$ such that

$$u(b,a;b',a') = \sum_{k=0}^{M} a^k\, u_k(b,a;b',a') \quad + R_M(b,a;b',a').$$

We denote by $\Gamma_\rho^\delta(\mathbb{H}^n \times \mathbb{H}^n)$ the set of all such u.

As a consequence of the calculus we have developed so far we have

- (Smoothing property) For $u \in \Gamma_\rho^\delta(\mathbb{H}^n \times \mathbb{H}^n)$ with

  $$u(b,a;b',a') \sim \sum_{k\ge 0} a^k\, u_k(b,a;b',a'),$$

  we have that each term is majorized by a multiplier $\Delta(b,a)^{\rho-k\delta}$

  $$M[a^k]Cop[u_k] \in Op[\Sigma_{\Delta^{\rho-\delta k}}(\mathbb{H}^n \times \mathbb{H}^n)].$$

  In particular, for $k > \rho/\delta$ the terms are smoothing.

[1] One could use a different sub-multiplicative weight function instead of $\Delta(b,a)$. We have taken this one because in that case the higher order terms are compact with respect to lower order terms (see Theorem 4.14). So, if one is interesten in translation invariant estimates of smoothing terms one would rather take a as weight function.

- (Algebraically closed) For $s \in \Gamma^{\delta}_{\sigma}(\mathbb{H}^n \times \mathbb{H}^n)$ and $r \in \Gamma^{\delta}_{\rho}(\mathbb{H}^n \times \mathbb{H}^n)$ there is $u \in \Gamma^{\delta}_{\sigma+\rho}(\mathbb{H}^n \times \mathbb{H}^n)$ such that

$$Cop[s]Cop[r] = Cop[u].$$

Or in short we have the following graduation

$$Cop[\Gamma^{\delta}_{\sigma}(\mathbb{H}^n \times \mathbb{H}^n)]Cop[\Gamma^{\delta}_{\rho}(\mathbb{H}^n \times \mathbb{H}^n)] \subset Cop[\Gamma^{\delta}_{\sigma+\rho}(\mathbb{H}^n \times \mathbb{H}^n)].$$

- (Symbolic calculus) For

$$s(b,a;b',a') \sim \sum_{k\geq 0} a^k\, u_k(b,a;b',a'), \quad r(b,a;b',a') \sim \sum_{k'\geq 0} a^{k'}\, u_{k'}(b,a;b',a'),$$

and u defined through $Cop[s]Cop[r] = Cop[u]$ we have the following explicit expansion

$$u(b,a;b',a') \sim \sum_{k\geq 0} a^k\, u_k(b,a;b',a'),$$

with

$$u_{k''}(b,a;b',a') = \sum_{l+k+k'=k''} \sum_{|\alpha|+q=l}$$
$$[s_k(b,a;b'',a'')(-b''/a'')^{\alpha}(1/a''-1)^q a''^{-k'}] * [\partial_a^m \partial_b^{\alpha} r_{k'}(b,a;b'',a'')](b',a').$$

In particular, we have

$$Cop[s]Cop[r] = Cop[s(b,a;\cdot,\cdot) * u(b,a;\cdot,\cdot)] + Cop[R_1],$$

with some $R_1 \in \Sigma_{\Delta^{\rho-\delta}}(\mathbb{H}^n \times \mathbb{H}^n)$.

- (Adjoint) For $u \in \Gamma^{\delta}_{\rho}(\mathbb{H}^n \times \mathbb{H}^n)$ the adjoint u^* defined through

$$Cop[u]^* = Cop[u^*],$$

satisfies $u^* \in \Gamma^{\delta}_{\rho}(\mathbb{H}^n \times \mathbb{H}^n)$. In addition it can be expanded as

$$u^*(b,a;b',a') \sim \sum_{k\geq 0} a^k u^*_k(b,a;b',a'),$$

with

$$u^*_k(b,a;b',a') = \sum_{|\alpha|+m=k} \frac{1}{a'^n}(\partial_a^m \partial_b^{\alpha} \overline{u}_{k-|\alpha|-m})(b,a;-b'/a',1/a')(-b'/a')^{\alpha}(1/a'-1)^m.$$

8.3 The algebra of local wavelet changing operators

In this section we want to analyze operators of the form

$$\mathcal{M}_h Cop[u] \mathcal{W}_g,$$

with some admissible wavelets g, $h \in \mathcal{S}_0(\mathbb{R}^n)$ and some symbol u. As it turns out, the computations will be rather explicit in that case. The simplification occurs from the fact that all these operators are parametrized by the space of mappings $\mathbb{H}^n \to C^\infty(S^{n-1})$, where S^{n-1} is the $n-1$-dimensional sphere. This is roughly speaking due to the fact that the set of translation and dilation invariant $A \in CZO$ is parametrized by $C^\infty(S^{n-1})$ via $A : \widehat{s}(\xi) \mapsto m(\xi/|\xi|)\,\widehat{s}(\xi)$, $m \in C^\infty(S^{n-1})$. The generalization we are considering, is that this *CZO* depends in a smooth way on the scale and the position we are looking at. Therefore it is a generalization of the algebra of position dependent singular integral operators that have been considered by A.P.Calderón and A. Zygmund in [CZ57]. More precisely, we want to consider operators K that map a wavelet $g_{b,a}(x) = g([x-b]/a)/a^n$ at position b and scale a into a wavelet $w_{b,a}$ at about the same scale and same position but with $w = w(b,a)$ so that K deforms locally the phase space. Symbolically we would like to have

$$K g_{b,a}(x) = \frac{1}{a^n} w\left(b, a; \frac{x-b}{a}\right).$$

However, in this precise form it is impossible, because of the correlations between the wavelets, as expressed, e.g., throgh the reprducing kernel equation. Because of this, the above identity will only hold modulo smoothnes. To formalize this idea, suppose we have a mapping

$$w : \mathbb{H}^n \to \mathcal{S}_0(\mathbb{R}^n), \quad (b,a) \mapsto w(b,a;x).$$

We then associate with w and some fixed analyzing wavelet g the following operator acting on functions over $\mathbb{R}^n$

$$op[w,g]s(x) = \int_{\mathbb{H}^n} \frac{dbda}{a} \frac{1}{a^n} w\left(b, a; \frac{x-b}{a}\right) \mathcal{W}_g s(b,a).$$

Thus we reconstruct with a wavelet that changes with scales and positions. We also write $op[w]$ if the analyzing wavelet g is fixed. We could also change at scale and position the analyzing wavelet and reconstruct with some fixed g. This idea gives rise to a second type of operators defined through

$$\begin{aligned} po[h,w]s(x) &= \int_{\mathbb{H}^n} \frac{dbda}{a} \frac{1}{a^n} h\left(\frac{x-b}{a}\right) \mathcal{W}[w(b,a),s](b,a) \\ &= \int_{\mathbb{H}^n} \frac{dbda}{a} \frac{1}{a^n} h\left(\frac{x-b}{a}\right) \int_{\mathbb{R}^n} dx' \frac{1}{a^n} \overline{w}\left(b, a; \frac{x'-b}{a}\right) s(x'). \end{aligned}$$

The basic calculus can now be related to the calculus of smoothly varying convolution operators if we use functions w that are in one of the following classes of functions.

Definition 8.13 *Let $\{\rho_{\alpha,m}\}$ be a family of sub multiplicative weight functions. We say a function $w : \mathbb{H}^n \to \mathcal{S}_0(\mathbb{R}^n)$ is of type $\{\rho_{\alpha,m}\}$ if for each pair α, m there is a bounded subset $\Gamma_{\alpha,m} \subset \mathcal{S}_0(\mathbb{R}^n)$, such that for all $(b,a) \in \mathbb{H}^n$ we have*

$$(\rho_{\alpha,m}(b,a))^{-1}\partial_a^m \partial_b^\alpha w(b,a;\cdot) \in \Gamma_{\alpha,m}.$$

In this case we have that $op[w,g]$ and $po[h,w]$ are continuous mappings $\mathcal{S}_0(\mathbb{R}^n) \to \mathcal{S}_0(\mathbb{R}^n)$. Moreover, the following result allows to use the previous theorems about the calculus with smoothly varying convolution operators.

Theorem 8.14 *Let $w(b,a)$ be of type $\{\rho_{\alpha,m}\}$ and let g, $h \in \mathcal{S}_0(\mathbb{R}^n)$ be a fixed analysis reconstruction pair. Then there is u and $u' \in \Xi(\mathbb{H}^n \times \mathbb{H}^n)$ such that*

$$op[w,g] = \mathcal{M}_h Cop[u]\,\mathcal{W}_g, \quad po[h,w] = \mathcal{M}_h Poc[u']\,\mathcal{W}_g.$$

Explicitly we have

$$u(b,a;b',a') = \mathcal{W}[g, w(b,a;\cdot)](b',a'), \quad u'(b,a;b',a') = \mathcal{W}[w(b,a;\cdot),h](b',a').$$

The symbol u and u' are of type $\{\rho_{\alpha,l}\}$. Vice versa, given $u \in \Xi(\mathbb{H}^n \times \mathbb{H}^n)$ of type $\{\rho_{\alpha,m}\}$. Then

$$\mathcal{M}_h Cop[u]\mathcal{W}_g = po[h,w] = op[w',g],$$

with

$$\begin{aligned} w(b',a';x) &= \int_{\mathbb{H}^n} \frac{dbda}{a}\frac{1}{a^n}h\left(\frac{x-b}{a}\right)\frac{1}{a'^n}u\left(b,a;\frac{b-b'}{a'},\frac{a}{a'}\right), \\ w'(b,a;x') &= \int_{\mathbb{H}^n} \frac{db'da'}{a'}\frac{1}{a'^n}u\left(b,a;\frac{b-b'}{a'},\frac{a}{a'}\right)\frac{1}{a'^n}\bar{g}\left(\frac{x'-b'}{a'}\right), \end{aligned}$$

and

$$\mathcal{M}_h Poc[u]\mathcal{W}_g = op[w,g] = po[h,w'],$$

with

$$\begin{aligned} w(b',a';x) &= \int_{\mathbb{H}^n} \frac{dbda}{a}\frac{1}{a^n}h\left(\frac{x-b}{a}\right)\frac{1}{a'^n}u\left(b',a';\frac{b-b'}{a'},\frac{a}{a'}\right), \\ w'(b,a;x') &= \int_{\mathbb{H}^n} \frac{db'da'}{a'}\frac{1}{a'^n}u\left(b',a';\frac{b-b'}{a'},\frac{a}{a'}\right)\frac{1}{a'^n}\bar{g}\left(\frac{x'-b'}{a'}\right). \end{aligned}$$

In all cases w and w' are of type $\{\rho_{\alpha,m}\}$.

Proof. A simple exchange of integration will do. □

A further simplification occurs, since the symbols $s(b,a;b',a')$ that appear if we multiply any of the operators $op[w,g]$, $op[w',g]$, $po[h,w]$, $po[h,w']$ are of the following special form

$$s(b,a;b',a') = \mathcal{W}[f(b,a),v(b,a)](b',a'),$$

where $f,\, v : \mathbb{H}^n \to \mathcal{S}_0(\mathbb{R}^n)$. All the terms that appear in the expansions we have encountered so far are in this algebra up to some smothing remainder. Indeed, this follows from the following facts. The standard Leibniz rule gives

$$\begin{aligned}&\partial_a^m \partial_b^\alpha \mathcal{W}[g(b,a), h(b,a)](b',a')\\ &\quad = \sum_{\beta\le\alpha, l\le m} \binom{m}{l}\binom{\alpha}{\beta} \mathcal{W}[\partial_a^\beta \partial_b^\beta g(b,a), \partial_a^{m-l}\partial_b^{\alpha-\beta} h(b',a)](b',a').\end{aligned}$$

Moreover, we have for all multi-indices α and all $m \in \mathbb{N}$

$$\begin{aligned} b^\alpha \mathcal{W}[g,h](b,a) &= \sum_{\beta\le\alpha} \binom{\alpha}{\beta} \mathcal{W}[(-\Delta)^{|\beta|/2}(x^\beta g), (-\Delta)^{-|\beta|/2}(x^{\alpha-\beta} h)](b,a)\\ &= \sum_{\beta\le\alpha} \binom{\alpha}{\beta} a^{|\beta|} \mathcal{W}[x^\beta g, x^{\alpha-\beta} h](b,a),\end{aligned}$$

as follows from

$$b^\alpha \mathcal{W}[g,h](b,a) = \frac{1}{(2\pi)^n} \int d\xi\, \overline{\widehat{g}}(a\xi)\, \widehat{h}(\xi)\, (-i\partial_\xi)^\alpha e^{ib\xi},$$

through integration by parts. And

$$a^m \mathcal{W}[g,h](b,a) = \mathcal{W}[(-\Delta)^{m/2} g, (-\Delta)^{-m/2} h](b,a),$$

which has been established several times.

In particular, we may consider the functions w which satisfy (compare Section 8.2)

$$a^{|\alpha|+m} \Delta(b,a)^{-\rho+\delta(|\alpha|+m)} \partial_a^m \partial_b^\alpha w(b,a;\cdot) \in F_{\alpha,m},$$

with some bounded set $F_{\alpha,m} \subset \mathcal{S}_0(\mathbb{R}^n)$. Call the set of all such functions w, Ω_ρ^δ.

Definition 8.15 *Let $w_k \in \Omega^\delta_{\rho-k}(\mathbb{H}^n \times \mathbb{H}^n)$, $k = 0, 1, \ldots$. We then write*

$$w(b,a;x) \sim \sum_{k\ge 0} a^k\, w_k(b,a;x),$$

if for all $M \in \mathbb{N}$ there is $R_M \in \Sigma_{\Delta^{\rho-\delta(M+1)}}$ such that

$$op[w] = \sum_{k=0}^{M} (-\Delta)^{-k/2}\, op[w_k] \quad + R_M.$$

We denote by Γ_ρ^δ the set of all such u.

We have that $w \in \Gamma^\delta_\rho$ if and only if

$$op[w] = \mathcal{M}_h Cop[u] \mathcal{W}_g,$$

with some $u \in \Gamma^\delta_\rho(\mathbb{H}^n \times \mathbb{H}^n)$ and some admissible g, $h \in \mathcal{S}_0(\mathbb{R}^n)$. Moreover, in this algebra we may compute modulo smoothness and even modulo compactness using the symbolic calculus established before.

For $w \in \Gamma^\delta_\rho$, $w \sim \sum a^k w_k$ we define the principal symbol as the first term, w_0. It is well-defined modulo smoothing terms. The main point is that we can compute modulo smoothness with the principal symbol only. More precisely, for $w \in \Gamma^\delta_\rho$ and $w' \in \Gamma^\delta_{\rho'}$ with principal symbols w_0 and w'_0 we have (up to compact/smoothing corrections)

$$\begin{aligned} op[w,g]op[w',g] &= op[w_0,g]op[w'_0,g] + \cdots = op[w'',g], \\ w''(b,a;\cdot) &= \mathcal{M}[w_0(b,a;\cdot), \mathcal{W}_g w'_0(b,a;\cdot)] + \ldots . \end{aligned}$$

This corresponds to "freezing" the coefficients. Indeed, the following formula holds for $g_1, g_2, \ldots, g_M \in \mathcal{S}(\mathbb{R}^n)$:

$$\mathcal{M}_{g_M} \mathcal{W}_{g_{M-1}} \mathcal{M}_{g_{M-2}} \ldots \mathcal{W}_{g_1} = \mathcal{M}_h \mathcal{W}_{g_1},$$

where

$$h = \mathcal{M}_{g_M} \mathcal{W}_{g_{M-1}} \mathcal{M}_{g_{M-2}} \ldots \mathcal{W}_{g_3} g_2, \quad \widehat{h}(\xi) = \widehat{g}_2(\xi) \prod_{j=2}^{M} \int_0^\infty \frac{da}{a} \overline{\widehat{g}_{2j-1}(a\xi)} \widehat{g}_{2j}(a\xi).$$

9 Relation with some classical operators

We now show that the algebras of operators we have studied above allow the small scale analysis of pseudo-differential operators. The setting will be as follows. Given an operator $A : \mathcal{S}(\mathbb{R}^n) \to \mathcal{S}'(\mathbb{R}^n)$ we consider the small scale part

$$(1 - \phi(D))A(1 - \phi(D)),$$

where $\phi \in \mathcal{S}(\mathbb{R}^n)$ is some function whose Fourier transform is $= 1$ for $|\xi| \leq 1$ and that decreases rapidly such that $\widehat{\phi}$ is compactly supported by some larger ball. Clearly, $\phi(D)$ is the pseudo-differential operator associated with ϕ. This implies for us that if we take as analyzing wavelets some function with compactly supported Fourier transform then we need to consider only the matrix elements

$$\langle g_{g,a}, A g_{b',a'} \rangle ,$$

with a, $a' \leq 1$. The discussion that follows is not exhaustive. In particular, we shall not introduce the $T(1)$ theorem of G. David and J.L.Journé, [DJ84] that allows the full understanding of the realtion between *CZO* and classes of exotic or forbidden symbols. We refer the reader to the above mentioned paper.

9.1 The pseudo-differential operators

Let $s(x,\xi)$ be a complex-valued function in some suitable symbol class over $\mathbb{R}^n \times \mathbb{R}^n$. As usual we associate with it the pseudo- differential operator

$$\psi[s] : r(x) \mapsto \frac{1}{(2\pi)^n} \int_{\mathbb{R}^n} d\xi\, s(x,\xi) e^{i\xi x} \widehat{r}(\xi).$$

The following two theorems explain the relation between standard pseudo-differential operators on the algebras of Calderón Zygmund operators we have constructed above. As we have seen before, translation invariant CZO can be characterized through their multiplier. In the case of more general CZO this is not possible any more. Instead we have

Theorem 9.1 *Suppose $s(x,\xi)$ satisfies*

$$\left| \partial_x^\alpha \partial_\xi^\beta s(x,\xi) \right| \leq c_{\alpha,\beta} \left|\xi\right|^{-|\beta|},$$

for all $x \in \mathbb{R}^n$ and all $|\xi| \geq 1$. Then there is a sequence of $K_k \in$ CZO such that for every $M \in \mathbb{N}$, there is some small scale averaging operator R_M (see Theorem 7.10*) such that we can write*

$$(1-\phi(D))\psi[s](1-\phi(D)) = \sum_{k=0}^{M} (-\Delta)^{-k/2} K_k + R_M (-\Delta)^{-(M+1)/2}.$$

Thus at small scale at least, a pseudo-differential operator of order 0 behaves like a *CZO*. Recall that by Theorem 6.3 the remainder is continuous between $B(\mathbb{R}^n)$ and $a^M B(\mathbb{R}^n)$ for all strongly tempered Banach spaces (condition (4.1)).

Proof. As we may, we choose a strongly admissible wavelet g such that $\widehat{g}$ is compactly supported. We have

$$\begin{aligned}
\sigma(b,a;b',a') &= \langle g_{b,a}, \psi[s] g_{b',a'} \rangle \\
&= \int_{\mathbb{R}^n} dx \frac{1}{a^n} \overline{g}\left(\frac{x-b}{a}\right) \frac{1}{(2\pi)^n} \int_{\mathbb{R}^n} d\xi\, s(x,\xi)\, e^{i\xi(x-b)} \widehat{g}(a\xi)
\end{aligned}$$

A simple change of variables leads us to

$$\sigma(b,a;b',a') = \int_{\mathbb{R}^n} dx \frac{a'^n}{a^n} \overline{g}\left(\frac{x-[b-b']/a'}{[a/a']}\right) \frac{1}{(2\pi)^n} \int_{\mathbb{R}^n} d\xi\, s(b'+a'x, \xi/a')\, e^{i\xi x} \widehat{g}(\xi).$$

Now Taylor's theorem allows us to write

$$\begin{aligned}
s(b'+a'x,\xi/a') &= \sum_{|\alpha|\leq M} a'^{|\alpha|} (\alpha!)^{-1} (\partial_x^\alpha s)(b',\xi/a') x^\alpha \\
&+ M \int_0^1 d\lambda (1-\lambda)^{M+1} \sum_{|\alpha|=M+1} (\partial_x^\alpha s)(b'+\lambda a'x, \xi/a') a'^{|\alpha|} (\alpha!)^{-1} x^\alpha.
\end{aligned}$$

We have chosen $\widehat{g}$ compactly supported by some annulus $A = \{1/2 \le |\xi| \le 2\}$. By hypothesis on s we therefore have that the functions $(\partial_x^\alpha s)(b', \xi/a')\widehat{g}$ are contained in some bounded set of $C_c^\infty(A)$. It follows that their Fourier transforms are in some bounded set Ω of $\mathcal{S}_0(\mathbb{R}^n)$. Now $x^\alpha \Omega$ is again a bounded set of $\mathcal{S}_0(\mathbb{R}^n)$ and thus we have thanks to Theorem 3.18 that there are symbols σ_α, and functions $F_\alpha \in \mathcal{S}(\mathbb{H}^n)$ such that

$$\sigma(b, a; b', a') = \sum_{|\alpha| \le M} a'^{|\alpha|} \sigma_\alpha(b, a; b', a') + R(b, a; b', a'),$$

with

$$|\sigma_\alpha(b, a; b', a')| \le \frac{1}{a'^n} F_\alpha \left(\frac{b - b'}{a'}, \frac{a}{a'} \right).$$

The remainder R may be estimated as follows. By hypothesis on $s(x, \xi)$

$$f_{\lambda, b', a'}(x) = \int d\xi e^{i\xi x} (\partial_x^\alpha s)(b' + \lambda a' x, \xi/a')\widehat{g}(\xi), \quad 0 \le \lambda \le 1, 0 < a \le 1, b' \in \mathbb{R}^n,$$

is contained in some bounded set Γ of $\mathcal{S}(\mathbb{R}^n)$. Now $x^\alpha \Gamma$ is again a bounded set of $\mathcal{S}(\mathbb{R}^n)$. We may invoke Theorem 3.21 and 7.10 to conclude. □

In case that the symbol satisfies the stronger condition

$$s(x, \lambda\xi) = s(x, \xi), \quad \lambda > 0,$$

we have the following corollary

Corollary 9.2 *In that event the following expansion holds modulo smoothing*

$$(1 - \phi(D))\psi[s](1 - \phi(D)) \sim \sum_\alpha (\alpha!)^{-1} \mathcal{M}_g \mathit{Poc}[u_\alpha] \mathcal{W}_g,$$

with the smoothly varying convolution kernels given by

$$u_\alpha(b, a; b', a') = u_\alpha(b; b', a') = a'^{|\alpha|}\, \mathcal{W}[g, x^\alpha F_{\alpha,b} g](b', a'),$$

with $F_{\alpha,b}$ being the translation invariant CZO given by the following Fourier multiplier

$$F_{\alpha,b} : \widehat{r}(\xi) \mapsto (\partial_x^\alpha s)(b, \xi)\, \widehat{r}(\xi).$$

Recall the relation between the calculus *Cop* and *Poc*, (see Theorem 8.7). Note, in particular, that the multiplication operator $s \mapsto m \cdot s$ reads in wavelet space

$$\mathcal{W}_g s(b, a) \mapsto m(b) \mathcal{W}_g s(b, a) + R,$$

where R is one order smoother than s.

The theorem shows that at small scale, the classical pseudo- differential operators are in the class of local wavelet changing operators. In particular, this shows that a symbol of order 0 is continuous in all tempered Banach spaces $B(\mathbb{R}^n)$ at least locally. On the other hand, as the next theorem shows, all *CZO* satisfy the so-called forbidden estimates (compare the discussion in [CM78]).

Theorem 9.3 *Let $A \in CZO$. Then*

$$A = \psi[s],$$

with s satisfying the estimates

$$\left|\partial_x^\alpha \partial_\xi^\beta s(x,\xi)\right| \leq c_{\alpha,\beta} \left|\xi\right|^{|\alpha|-|\beta|}.$$

Proof. Let $K(b,a;b',a')$ the symbol of A; then we have with some admissible g

$$A = \mathcal{M}_g Op[K] \mathcal{W}_g.$$

Now define $h(b',a';x)$ through

$$\frac{1}{a^n} h\left(b,a;\frac{x-b}{a}\right) = \int_{\mathbb{H}^n} \frac{db'da'}{a'} K(b,a;b',a') \frac{1}{a'^n} g\left(\frac{x-b'}{a'}\right).$$

We then obtain by direct computation involving a change of integration

$$As(x) = \int_{\mathbb{H}^n} \frac{db'da'}{a'} \frac{1}{a'^n} g\left(\frac{x-b'}{a'}\right) \mathcal{W}[h(b,a;\cdot),s](b,a) = po[g,h]s(x).$$

The mapping $(b,a) \to h(b,a;\cdot)$ has as range a bounded set in $\mathcal{S}_0(\mathbb{R}^n)$. From the well known rule

$$\psi[s]e^{ix\xi} = e^{ix\xi} s(x,\xi),$$

it follows that

$$s(x,\xi) = \int_{\mathbb{H}^n} \frac{dbda}{a} \frac{1}{a^n} g\left(\frac{x-b}{a}\right) \widehat{h}_{(b,a)}(a\xi) e^{-i(x-b)\xi},$$

with $h_{(b,a)} = h(b,a;\cdot)$. Now, by the Leibniz rule we have that

$$\begin{aligned}
&\partial_x^\alpha \partial_\xi^\beta s(x,\xi) \\
&= |\xi|^{|\alpha|-|\beta|} \sum_{\gamma\leq\alpha} \sum_{\delta\leq\beta} \binom{\alpha}{\gamma} \binom{\beta}{\delta} \int_{\mathbb{H}^n} \frac{dbda}{a} \frac{1}{a^n} (\partial_x^{\alpha-\gamma} g) \left(\frac{x-b}{a}\right) (-i)^{|\delta|} \left(\frac{x-b}{a}\right)^\delta \times \\
&\quad \times (\partial_\xi^{\beta-\delta} \widehat{h}_{(b,a)})(a\xi)(-ia\xi)^\gamma (|\xi|\, a)^{|\beta|-|\alpha|} e^{-i(x-b)\xi}.
\end{aligned}$$

Now, since $h_{(b,a)}$ is in some bounded set in $\mathcal{S}_0(\mathbb{R}^n)$ there is some function f with

$$f(\lambda) \leq O(\lambda + 1/\lambda)^{-M},$$

for all M such that

$$\left|\partial_\xi^{\beta-\delta} \widehat{h}_{(b,a)}(\xi))\right| \leq f(|\xi|).$$

Moreover, the functions $x^\delta \partial^{\alpha-\gamma} g$ are in $\mathcal{S}(\mathbb{R}^n)$, and thus the integral may be estimated by

$$\leq c \sum_{\gamma \leq \alpha} \sum_{\delta \leq \beta} \|x^\delta \partial^{\alpha-\gamma} g\|_{L^1(\mathbb{R}^n)} \int_0^\infty \frac{da}{a} (a\,|\xi|)^{|\gamma|+|\beta|-|\alpha|} f(a\,|\xi|) \leq c' < \infty.$$

In the last integral we have used the scaling invariance of the measure da/a. This shows the theorem. □

10 Elliptic regularity

We now want to apply the classes introduced above to a problem of elliptic regularity. We first introduce ellipticity in the case of our algebras of wavelet pseudo-differential operators. We then show as an application that some directional elliptic regularity holds for the microlocal classes we have introduced before.

10.1 Elliptic operators

The basic principles are easily explained. Consider the Laplace operator Δ. In wavelet space we have that

$$\mathcal{W}_g \Delta s = a^{-2} \mathcal{W}_{\Delta g} s.$$

Now, together with g, the wavelet Δg is admissible. This is precisely because Δ is elliptic. Therefore, we may conclude from the regularity of Δs to localization properties of $\mathcal{W}_g \Delta s$, which implies localization properties of $\mathcal{W}_{\Delta g} s$. Now, since Δg is admissible, this implies a certain regularity of s. We have shown in fact the following proposition.

Proposition 10.1 *Let $s \in \mathcal{S}_0'(\mathbb{R}^n)$ with $B(\mathbb{R}^n)$ some admissible Banach space, satisfy*

$$\Delta s \in B(\mathbb{R}^n).$$

Then it follows that $s \in a^2 B(\mathbb{R}^n)$.

Based on this example, we define

Definition 10.2 *Let $\Omega \subset \mathbb{H}^n$ satisfy*

$$\forall k > 0 \quad \Omega \cap \{\Delta(b,a) \geq k\} \neq \emptyset.$$

A symbol $w \in \Gamma_\rho^\delta$ is called elliptic in Ω, if there is some $k \in \mathbb{N}$ and some closed set $\Gamma \subset \mathcal{S}_0(\mathbb{R}^n)$ consisting of admissible vectors, such that for all $(b,a) \in \Omega \cap \{\Delta(b,a) \geq k\}$ we have

$$\Delta(b,a)^{-\rho} w(b,a;\cdot) \in \Gamma.$$

As usual, ellipticity can be read from the principal symbol.

Proposition 10.3 *w is elliptic in Ω, iff there is some principal symbol w_0 which is elliptic in Ω.*

Proof. Apply the calculus. □

Therefore at least at small and large scale and large position, we obtain for $op[w, g]$ the mapping property of the Laplacian operator that it maps admissible wavelets to admissible wavelets[2].

The next proposition allows us to define local ellipticity for operators.

Proposition 10.4 *Suppose g, $h \in \mathcal{S}_0(\mathrm{I\!R}^n)$ are admissible and let $\Omega \supset \Sigma$ be two subsets of $\mathrm{I\!H}^n$ such that Ω^c, the complement of Ω, is well separated from Σ (Definition 5.2). Then, let $w \in \Gamma_\rho^\delta$ be elliptic in Ω. If we have*

$$op[w, g] = op[w', h],$$

then w' is elliptic in Σ.

Proof. Use the calculus together with the proof of Lemma 5.7. We leave the details to the reader. □

In view of this proposition we can define

Definition 10.5 *An operator $K : \mathcal{S}_0(\mathrm{I\!R}^n) \to \mathcal{S}_0'(\mathrm{I\!R}^n)$ is called elliptic in $\Sigma \subset \mathrm{I\!H}^n$, if there is some admissible $g \in \mathcal{S}_0(\mathrm{I\!R}^n)$, some $\Omega \supset \Sigma$ such that Ω^c is well separated from Σ and some $w \in \Gamma_\rho^\delta$ which is elliptic in Ω, and which satisfies*

$$K = op[w, g].$$

As usual, ellipticity implies stable invertibility, which is explained in the following result. As above Ω will also indicate the multiplication by the characteristic function of Ω.

Theorem 10.6 *If $K = op[w]$ with $w \in \Gamma_\rho^\delta$ is elliptic in Ω, then there is F, $G \in \Gamma_{-\rho}^\delta$ such that the following four operators are in $op[\Gamma_{\rho-\delta}^\delta]$:*

$$\mathcal{M}_h \Omega \mathcal{W}_g (KF - 1\!\!1),\ \mathcal{M}_h \Omega \mathcal{W}_g (FG - 1\!\!1),\ (KF - 1\!\!1) \mathcal{M}_h \Omega \mathcal{W}_g,\ (FK - 1\!\!1) \mathcal{M}_h \Omega \mathcal{W}_g,$$

for all g, $h \in \mathcal{S}_0(\mathrm{I\!R}^n)$.

Proof. Use the calculus together with the proof of Lemma 5.7. We leave the details to the reader. □

[2]Note that if we had defined symbol classes with respect to say a instead of $\Delta(b, a)$ as weight functions, ellipticity would only be a small scale concept.

10.2 Directional elliptic regularity in domains with cusps

Recall the definition of the microlocal classes in Section 5.4 and, in particular, the directional classes of Section 5.5. We say an open domain $\Omega \subset \mathbb{R}^n$ satisfies the cusp condition of degree $\delta > 0$ at $x \in \partial\Omega$ in direction ξ, $\xi \in \mathbb{R}^n \setminus \{0\}$ if there is some $c > 0$ such that

$$\{y : |y-x| \le c\,|(y-x)\xi|^\delta, |y-x| \le c\} \subset \Omega.$$

Note that for these domains no non-tangential appproach in the sense of NTA domaines considered in [JK82] is possible. The next theorem however, shows that directional elliptic regularity applies to these classes.

Theorem 10.7 *Let Ω satisfy a cusp condition at 0 of type δ in direction ξ. Suppose that η is a tempered distribution that satisfies inside Ω*

$$\Delta\eta = f,$$

for some distribution f supported by $\overline{\Omega}$. If f is of type (ξ, γ, α), with $\gamma > \delta$, it follows that η is of type $(\xi, \gamma, \alpha + 2)$.

This theorem is a special case of a more general theorem. Let $B(\mathbb{R}^n)$ be a local regularity space of the type we have considered before with $B(\mathbb{H}^n)$ being the associated space of functions over $\mathbb{H}^n$.

For a set $\Omega \subset \mathbb{R}^n$ we consider the set

$$I(\Omega) = \bigcup_{b\in\Omega} K(b),$$

where $K(b) \subset \mathbb{H}^n$ is the cone of opening angle 1 with vertex in b

$$K(b) = \{(\beta, \alpha) : |\beta - b| \le \alpha\}.$$

We call this set the influence region of Ω in the upper half-space.

The general theorem can now be stated as follows.

Theorem 10.8 *Let $\Omega \subset \mathbb{R}^n$ be open. Suppose that $\Xi \subset \mathbb{H}^n$ is well separated from the influence region $I(\Omega)$ of Ω. Let $E \in op[\Gamma^\delta_\rho]$ be elliptic in Ξ. Suppose that $\Xi' \subset \Xi$ such that Ξ' and Ξ^c are well separated. Suppose that η is a tempered distribution that satisfies inside Ω*

$$E\eta = f,$$

for some distribution $f \in S'(\mathbb{R}^n)$ supported by $\overline{\Omega}$. If now f satisfies

$$\mathcal{M}_g \Xi \mathcal{W}_g f \in B(\mathbb{R}^n),$$

with some admissible wavelet $g \in \mathcal{S}_0(\mathbb{R}^n)$, then it follows that

$$\mathcal{M}_g \Xi' \mathcal{W}_g \eta \in (\rho^{-1} B)(\mathbb{R}^n).$$

Proof. In the case where η is known globally, the proof follows from Theorem 10.6. The theorem is therefore proved if we can show that any distribution u that coincides with f inside Ω satisfies again

$$\mathcal{M}_g \Xi \mathcal{W}_g E u \in B(\mathbb{R}^n).$$

But this follows from the next lemma, which justifies the name influence region for Ω.

Lemma 10.9 *Let $\Xi \subset \mathbb{H}^n$ be well separated from the influence region of Ω. Then, for all $\rho \in \mathcal{S}'(\mathbb{R}^n)$ with support in Ω we have that*

$$\mathcal{M}_g \Xi \mathcal{W}_g \rho \in \mathcal{S}_0(\mathbb{R}^n).$$

Proof. By hypothesis there is an $\epsilon > 0$ such that for $(b,a) \in \Gamma_\epsilon(\Xi)$ there is an n-dimensional Euclidian ball of radius $> \Delta((b,a))^\epsilon$ around b that is contained in the complement of the influence region of Ω. Denote by ϕ a C^∞ function which is identically 1 on the complement of the unit ball of radius 1 and which is supported on the complement of a slightly smaller ball. Then denote by $\phi_{b,a}$ the family of translates and dilates of ϕ. Therefore if ρ is supported by Ω, it follows that

$$(b,a) \in \Xi \Rightarrow \mathcal{W}_g \rho(b,a) = \mathcal{W}_g(\phi_{b,\Delta((b,a))^\epsilon} \rho)(b,a) = \left\langle g_{b,a} \phi_{b,\Delta((b,a))^\epsilon}, \rho \right\rangle .$$

But now for every $\epsilon > 0$ we have that $g_{b,a}\phi_{b,\Delta((b,a))^\epsilon}$ tends to 0 in $\mathcal{S}_0(\mathbb{R}^n)$ as $\Delta((b,a)) \to \infty$ in such a way that for all semi-norms and all $M > 0$ we have

$$\left\| g_{b,a} \phi_{b,\Delta((b,a))^\epsilon} \right\|_{l,m} \leq c_{l,m,M} \Delta((b,a))^{-M}.$$

This proves the lemma. □

The theorem is proved. □

11 The diffeomorphism in wavelet space

First some notations. Let $\phi : \mathbb{R}^n \to \mathbb{R}^n$ be a diffeomorphism. In components it reads

$$\phi : (x_j) \mapsto (\phi^k((x_j))).$$

Our standing assumption in order to avoid unnecessary difficulties is that this map and its inverse are uniformly smooth, that is all derivatives of all orders of all its components are bounded functions over $\mathbb{R}^n$. This hypothesis may be relaxed. For every $x \in \mathbb{R}^n$ we denote by $\partial\phi_x$ the differential map

$$\partial\phi_x : \mathbb{R}^n \to \mathbb{R}^n, \quad x_k \mapsto \sum_{j=1}^n (\partial_j \phi^k(x)) x_j .$$

We sometimes also write $\partial[\phi, x]$. As usual the Jacobian is defined as

$$J[\phi](x) = J(x) = |\det \partial\phi_x| \,.$$

Consider now the pull-back operator associated with ϕ

$$\Phi[\phi] = \Phi : s(x) \mapsto J(\phi^{-1}(x))^{-1} s(\phi^{-1}(x)).$$

It maps functions localized around y to functions localized around $\phi(y)$

$$\Phi[\phi]\delta_y = \delta_{\phi(y)}.$$

It preserves the L^1-norm. Its adjoint reads

$$\Phi^* : s(x) \mapsto s(\phi(x)),$$

and we have for $s \in L^p(\mathbb{R}^n)$, $r \in L^q(\mathbb{R}^n)$, $1/p + 1/q = 1$,

$$\int dx \Phi^* s(x)\, r(x) = \int dx s(x) \Phi r(x).$$

For $x \in \mathbb{R}^n$ we denote by $\partial\Phi_x$ the following linear map

$$\partial\Phi[\phi, x] = \partial\Phi_x : s(y) \mapsto (J(x))^{-1} s([\partial\phi_x]^{-1}(y)).$$

We have

$$\partial\Phi[\phi, x] = \Phi[\partial\phi_x].$$

It describes the action of the tangent map on functions. Its adjoint reads

$$(\partial\Phi_x)^* : s(y) \mapsto s(\partial\phi_x y).$$

In analogy we define the following operator acting on functions over $\mathbb{H}^n$

$$\Psi = \Psi[\phi] : \; s(b,a) \mapsto J(\phi^{-1}(x))^{-1}\, s(\phi^{-1}(b), a) = (\Phi s(\cdot, a))(b).$$

The next theorem describes the diffeomorphism in wavelet space.

Lemma 11.1 *There is a sequence of symbols $F_k(b, a; b', a')$, $k = 1, 2, \ldots$ satisfying the estimates of Lemma* 6.1 *such that for all $K \in \mathbb{N}$ we may write for a, $a' \leq 1$*

$$\begin{aligned} \sigma_{g,g}\left[\Phi[\phi]\right](b, a; b', a') &= \frac{1}{a'^n} P\left(b'; \frac{b - \phi(b')}{a'}, \frac{a}{a'}\right) \\ &+ \sum_{k=1}^{K-1} a'^k F_k(b, a; \phi(b'), a') + a'^K R_K(b, a; \phi(b'), a') \end{aligned}$$

where $P(b'; b, a) = \mathcal{W}[g, \partial\Phi_{b'} g](b, a)$. The remainder R_K satisfies the estimates of Lemma 6.2.

Remark. If g is such that its Fourier transform is supported by a corona $\{c^{-1} \leq |\xi| \leq c\}$, it follows that the principal term vanishes whenever a/a' is too large or too small. Indeed, all wavelets $\partial\Phi_{b'}g$ are supported by a corona and there is one that contains them all. Hence all the functions $P(b';\cdot,\cdot)$ are supported by some strip $a + 1/a \leq c$. Therefore the principal part of $\Phi[\phi]$ leaves invariant the spaces of functions whose wavelet transforms with respect to g are supported by a strip. This remark will play a role in some demonstrations.

In view of this theorem we have that in wavelet space the diffeomorphism reads at small scale ($a < 1$)

$$\Phi[\phi] : \mathcal{W}_g s \mapsto \mathcal{P}oc[P]\, \Psi[\phi]\mathcal{W}_g s + R_1 s$$

where R_1 is smoothing from $B(\mathbb{H}^n)$ to $aB(\mathbb{H}^n)$ for any admissible, deformable Banach space $B(\mathbb{H}^n)$ (see Definition 4.8). Here we have used $\mathcal{P}oc$ for $P(b, a; b', a') = P(b; b', a')$ in the obvious way. Upon taking transpose, we see that we also have

$$\Phi[\phi] : \mathcal{W}_g s \mapsto \Psi[\phi]\mathcal{C}op[P']\mathcal{W}_g s + R_2 s$$

with $P'(b; b', a') = P(b, -b'/a', 1/a')$. But now as can be seen from the lemma the remainder R_2 is smoothing only in stronlgy tempered spaces.

Proof. First observe that $\partial\Phi_b$ is the small scale behavior of Φ at $b \in \mathbb{R}^n$ in the following precise sense. Let as usual $D_a : s(x) \mapsto s(x/a)/a^n$ denote the dilation operator and let T_b denote the translation operator $s(x) \mapsto s(x - b)$. We then have

$$D_a^{-1}T_{\phi(b)}^{-1}\Phi[\phi]T_b D_a g(x) = \frac{1}{J(\phi^{-1}(\phi(b) + ax))}\, g\left(\frac{\phi^{-1}(ax + \phi(b)) - b}{a}\right),$$

and thus it follows that

$$\Phi_{b,a}[\phi] = D_a^{-1}T_{\phi(b)}^{-1}\Phi[\phi]T_b D_a \to \partial\Phi_b, \quad (a \to 0).$$

More precisely, let us introduce, for each b the following operators $A_k(b)$ via ($s \in \mathcal{S}(\mathbb{R}^n)$)

$$(A_k(b)s)(x) = (\partial\Phi_b)^{-1} \lim_{a\to 0} \frac{d^k}{da^k}\Phi_{b,a}[\phi]s(x).$$

Repeated application of the chain rule shows that the $A_k(b)$ are in the algebra of partial differential operators generated by x^α and ∂_x^β

$$A_k(b) = \sum_{|\alpha|+|\beta|\leq M_k} c_{k,\alpha,\beta}(b)x^\alpha\partial_x^\beta.$$

Moreover, since the diffeomorphism is uniformly smooth we have that all the $c_{k,\alpha,\beta}(b)$ are bounded functions. It follows that $\{A_k(b)g : b \in \mathbb{R}^n\}$ is a bounded subset of

$\mathcal{S}_0(\mathbb{R}^n)$. Now Taylor's theorem takes the form

$$\Phi_{b,a}[\phi]g = \partial\Phi_b g + \partial\Phi_b \sum_{k=1}^{K-1} a^k A_k(b)g + R(b,a), \tag{11.1}$$

and the remainder $R = R_K$ satisfy

$$a^{-K}R(b,a) \in \Sigma,$$

for some bounded set $\Sigma \subset \mathcal{S}(\mathbb{R}^n)$. We may write using the fact that $(T_\beta)^* = T_\beta^{-1}$ and $(D_\alpha)^* = \alpha^{-n}D_\alpha^{-1}$

$$\begin{aligned}\sigma_{g,g}[\Phi[\phi]](b,a;b',a') &= \langle g_{b,a}, \Phi g_{b',a'}\rangle \\ &= \langle g_{b,a}, \Phi T_{b'} D_{a'} g\rangle \\ &= a'^{-n}\left\langle D_{a'}^{-1}T_{\phi(b')}^{-1} g_{b,a}, \Phi_{b',a'} g\right\rangle \\ &= a'^{-n}\left\langle g_{[b-\phi(b')]/a',a/a'}, \Phi_{b',a'} g\right\rangle .\end{aligned}$$

We now may insert expression (11.1). The principal term is the one stated in the theorem. The terms F_k are localized since they are contained in the set of wavelet transforms of a bounded set of functions in $\mathcal{S}_0(\mathbb{R}^n)$. Finally, by Lemma 3.21 the remainder satisfies the estimate of Lemma 6.2. □

In exactly the same way we can show the next lemma, the proof of which is left to the reader.

Lemma 11.2 *There is a sequence of symbols $F_k(b,a;b',a')$, $k = 1,2,\ldots$ satisfying the estimates of Lemma 6.1 such that for all $K \in \mathbb{N}$ we may write for a, $a' \le 1$*

$$\begin{aligned}\sigma_{g,g}\left[\Phi[\phi]M[\mu]\right](b,a;b',a') &= \mu(b')\,\frac{1}{a'^n}P\left(b';\frac{b-\phi(b')}{a'},\frac{a}{a'}\right) \\ &+ \sum_{k=1}^{K-1} a'^k F_k(b,a;\phi(b'),a') + a'^K R_K(b,a;\phi(b'),a')\end{aligned}$$

where $P(b';b,a) = \mathcal{W}[g,\partial\Phi_{b'}g](b,a)$. The remainder R_K are satisfying the estimates of Lemma 6.2.

11.1 The locally linear approximation

We even may go one step further and consider the diffeomorphism at small scale and locally.

Lemma 11.3 *For a, a', $|b'-\beta| \le 1$, $\beta \in \mathbb{R}^n$, we have*

$$\frac{1}{a'^n}P\left(b';\frac{b-\phi(b')}{a'},\frac{a}{a'}\right) = \frac{1}{a'^n}P\left(\beta;\frac{b-\phi(b')}{a'},\frac{a}{a'}\right) + |b'-\beta|\,R(b,a;\phi(b'),a'),$$

with $P(b'; b, a) = \mathcal{W}[g, \partial\Phi_{b'} g](b, a)$. *The remainder* R *satisfies the estimates of Lemma* 6.1.

Note however that in general, we may not replace $\phi(b)$ in the kernel by some linear approximation.

Proof. For $g \in \mathcal{S}_0(\mathbb{R}^n)$, there is a bounded set $\Sigma \subset \mathcal{S}_0(\mathbb{R}^n)$, such that for $|b' - \beta| \leq 1$ we have

$$|b' - \beta|^{-1} (\partial\Phi_{b'} - \partial\Phi_{\beta}) g \subset \Sigma.$$

Therefore, upon taking a wavelet transform, the lemma follows. □

11.2 Transformation behavior of the directional micro-local classes

As an immediate application of the previous discussion we have that the following theorem holds.

Theorem 11.4 *If* η *is at* x *of type* (ξ, γ, α), *then* $\Phi[\phi]\eta$ *is at* $\phi(x)$ *of type* (ξ', γ, α) *with* $\xi' = \partial\phi_x \xi$.

Proof. We may use the local linear approximation to conclude. □

12 The essential spectrum of some Fourier integral operators

As an application of the previous section, we want to establish a formula for the essential spectral radius of some Fourier integral operators, acting in Banach spaces of periodic functions. We limit our discussion to the discussion of the so called transfer operator. However all arguments extend easily to more general operators.

12.1 Definition of the transfer operator

Let ϕ_k, $k = 1, \ldots, K$, be a finite family of diffeomorphisms of $\mathbb{R}^n$. Let further smooth, periodic functions $\mu_k : \mathbb{T}^n \to \mathbb{C}$ be given and let us look at the following operator

$$\mathcal{L} = \sum_{k=1}^{K} \Phi[\phi_k] M[\mu_k] : \quad s(x) \mapsto \sum_{k=1}^{K} \mu_k(\phi_k^{-1}(x)) J[\phi_k](\phi_k^{-1}(x))^{-1} s(\phi_k^{-1}(x)).$$

We then suppose further that the ϕ_k and the the μ_k are such that $\mathcal{L}$ defines a mapping of $C^\infty(\mathbb{T}^n)$ into itself in that it maps $[0,1]^n$-periodic functions into periodic functions with the same period. Then for every $m \in \mathbb{N}$ we have that $\mathcal{L}^m$ has a similar form giving rise to a definition of multipliers $\mu_{m,k}$, and diffeomorphisms $\phi_{m,k}$, $k = 1, \dots, K^m$ in the obvious way:

$$\mathcal{L}^m = \sum_{k=1}^{K^m} \Phi[\phi_{m,k}] M[\mu_{m,k}].$$

We fix a strictly admissible wavelet g. Later we will impose some more conditions on it. Let $\sigma[\mathcal{L}]$ be the symbol of the transfer operator, with respect to g, and let $\mathcal{L}^\sharp$ be the associated operator acting on functions over $\mathbb{H}^n$

$$\mathcal{L}^\sharp = Op\big[\sigma[\mathcal{L}]\big].$$

As we have seen in Lemma 11.2, the main term of $\mathcal{L}^\sharp$ at small scale is of the form $\mathcal{L}_\partial = Op[L]$ with

$$L(b,a;b',a') = \sum_{k=1}^{K} L_k(b,a;b',a') = \sum_{k=1}^{K} \mu_k(b') \frac{1}{a'^n} P_k\left(b'; \frac{b-\phi_k(b')}{a'}, \frac{a}{a'}\right), \tag{12.1}$$

with $P_k(b';b,a) = \mathcal{W}[g, \partial\Phi[\phi_k, b']g](b,a)$.

Proposition 12.1 *Let $B(\mathbb{H}^n)$ be an admissible, deformable Banach space. Then the operator $\mathcal{L}$ is acting continuously on $B(\mathbb{T}^n)$.*

Proof. We write

$$\mathcal{L}^\sharp = \mathcal{L}_\partial + \mathcal{R},$$

Then, upon choosing K large enough such that $a^K B(\mathbb{H}^n)$ is strongly tempered it follows from Lemma 11.2 that $\mathcal{R}$ maps $B(\mathbb{H}^n)$ continuously to $a^K B(\mathbb{H}^n)$ since $a^K B(\mathbb{H}^n)$ satisfies again condition 4.4. The principal term $\mathcal{L}_\partial$ is continuous in $B(\mathbb{H}^n)$ and the proof is finished. □

Note that the principal part $\mathcal{L}_\partial$ is acting on $B(\mathbb{M}^n)$ and on $\Pi B(\mathbb{M}^n)$, where Π is the extension to $B(\mathbb{H}^n)$ of the orthogonal projection onto the image of $L^2(\mathbb{R}^n)$ under $\mathcal{W}_g$; that is, $\Pi s = P_g * s$, $P_g = \mathcal{W}_g g$. This follows from the fact that

$$\Pi\mathcal{L}_\partial = \Pi Op[L] = \mathcal{L}_\partial,$$

since the for fixed b', a' the kernel $L(\cdot,\cdot;b',a')$ is in the image of the wavelet transform. In the same way we may consider the iterates $\mathcal{L}^m$ to define $(\mathcal{L}^m)_\partial$. Indeed, all powers of $\mathcal{L}$ can be written as

$$\mathcal{L}^m = \sum_{k=1}^{K^m} \Phi[\phi_{m,k}] M[\mu_{m,k}].$$

This serves as a definition for the $\phi_{m,k}$ and the $\mu_{m,k}$. The small part $(\mathcal{L}^m)_\partial$ of $\mathcal{L}^m$ is now defined by using the $\mu_{m,k}$ and $\phi_{m,k}$. Note that we have to distinguish between $(\mathcal{L}^m)_\partial$ and $(\mathcal{L}_\partial)^m$.

12.2 A formula for the essential spectral radius

In order to state the formula we need some additional notation. Let $B = B(\mathbb{R}^n)$ be a Banach space of functions over $\mathbb{R}^n$. For any $K \in \mathbb{N}$, consider a sequence $s_k \in B(\mathbb{R}^n)$, $k = 1, \ldots, K$ and look at the following limit

$$F((s_k)) = \lim_{\lambda\to\infty} \sup \left\{ \left\| \sum s_k(\cdot - \beta_k) \right\|_{B(\mathbb{R}^n)} : |\beta_k - \beta_{k'}| > \lambda \ (k \neq k') \right\}.$$

That is we look at the behavior of the norm as we shift different peaces independently away from each other. For completely regular spaces, the simple limit actually exists. More than that, there are functions $\gamma_K : \mathbb{R}_+^K \to \mathbb{R}_+$, defining a norm on $\mathbb{C}^K$ via $\|(x_k)\| = \gamma_K((|x_k|))$ such that

$$F((s_k)) = \gamma_K((\|s_k\|)).$$

We make $B^K = B \times B \cdots \times B$ (K factors) a Banach space with norm

$$\|(s_k)\|_{B^K} = F((s_k)) = \gamma_K((\|s_k\|)).$$

Often, the γ_K may be computed explicitly. For instance in the case of $L^p(\mathbb{R}^n)$, $1 \leq p \leq \infty$ we clearly have

$$F((s_k)) = \left(\sum \|s_k\|^p_{L^p(\mathbb{R}^n)} \right)^{1/p},$$

with the obvious modification for $p = \infty$.

Now consider a family of linear operators $A_k : B \to B$, $k = 1, \ldots, K$. We then define an operator via

$$\bigoplus_{k-1}^{K} A_k : B^K \to B, \quad (s_k) \mapsto \sum A_k s_k. \tag{12.1}$$

We shall need its norm which explicitly reads

$$\left\| \bigoplus_{k=1}^{K} A_k \right\| = \sup \left\{ \left\| \sum_{k=1}^{K} A_k s_k \right\|_{B(\mathbb{R}^n)} : s_k \in B(\mathbb{R}^n), \gamma_K((\|s_k\|)) = 1 \right\}.$$

With these notations we are now ready to state the main theorem.

Theorem 12.2 *Let $B(\mathbb{R}^n)$ be a completely regular Banach space and let $B(\mathbb{T}^n)$ be its associated space of periodic functions. Then the essential spectral radius of the transfer operator can be computed as follows in terms of the norms of the linearized maps $\mu_{m,k}(\phi_{m,k}^{-1}(x))\partial\Phi[\phi_{m,k}, \phi_{m,k}^{-1}(x)]$, $k = 1, \ldots, K^m$.*

$$\rho_{ess}(\mathcal{L}) = \lim_{m\to\infty} \left(\sup_{x\in\mathbb{T}^n} \left\| \bigoplus_{k=1}^{K^m} \mu_{m,k}(\phi_{m,k}^{-1}(x))\partial\Phi[\phi_{m,k}, \phi_{m,k}^{-1}(x)] \right\|_{B^{K^m}(\mathbb{R}^n)\to B(\mathbb{R}^n)} \right)^{1/m}.$$

This is therefore a sharpening of a result of Ruelle who has given estimates of the essential spectral radius in terms of the local expansion rate in [Rue89]. In the one dimensional case we thus reobtain the results of [BJLI95]. See also [CK95]. In the case of the spaces $F^{p,q}_\gamma$ this formula is valid. However we have the following simpler estimate.

Theorem 12.3 *We have that $\mathcal{L}$ is acting on $F^{p,q}_\gamma(\mathbb{T}^n)$, $1 \le p \le \infty$, $1 \le q < \infty$ or $p = q = \infty$, and we may estimate*

$$\rho_{ess}(\mathcal{L}) \le \limsup_{m\to\infty} \sup_{x\in\mathbb{T}^n}$$
$$\left(\sum_{k=1}^{K^m} \left|\mu_{m,k}(\phi^{-1}_{m,k}(x))\right|^v \left\|\partial\Phi[\phi_{m,k},\phi^{-1}_{m,k}(x)]\right\|^v_{F^{p,q}_\gamma(\mathbb{R}^n)\to F^{p,q}_\gamma(\mathbb{R}^n)}\right)^{1/vm},$$

with v, the dual exponent of p, $v^{-1} + p^{-1} = 1$.

12.3 An abstract formula for ρ_{ess}

Let B be an infinite dimensional Banach space and $\mathcal{F}$ an operator acting continuously on B. We denote by B' the dual space consisting of linear functional v, $s \mapsto (v \mid s)$, and by $\mathcal{F}^t$, the transpose operator defined through $(v \mid \mathcal{F}s) = (\mathcal{F}^t v \mid s)$. The following formula for the essential spectral radius of $\mathcal{F}$ has been given in ([Nus70])

$$\rho_{ess}(\mathcal{F}) = \lim_{m\to\infty} (\inf\{\|\mathcal{F}^m - \mathcal{K}\| \,:\, \mathcal{K} \text{ is compact}\})^{1/m}.$$

In [Fri95] the formula has been modified to

$$\rho_{ess}(\mathcal{F}) = \lim_{m\to\infty} (\inf\{\|\mathcal{F}^m - \mathcal{K}\| \,:\, \mathcal{K} \text{ is of finite rank}\})^{1/m}.$$

Recall that the operators of finite rank can be written as

$$s \mapsto \sum_\alpha v_\alpha(s) u_\alpha,$$

with a finite collection of $v_\alpha \in B'$ and $u_\alpha \in B$. We now want to show a formula that describes the essential spectral radius in terms of the "infinitesimal (=small scale)" behavior of $\mathcal{F}$. What we precisely mean by that becomes more explicit when we take a concrete realization of the family of operators Π_j below. For the moment we state the formula in an abstract setting.

Theorem 12.4 *Suppose there is a family of bounded operators $\Pi_j : B \to B$, $j = 1$, $2, \ldots,$ subject to the following conditions*
(i) $\exists\, c > 0 \,:\, \forall j,\ \|\Pi_j\| \le c$.
(ii) $(\mathbb{1} - \Pi_j)$ *is compact.*

Suppose that B satisfies
(iii) $B_0 = \{r \in B : \lim_{j\to\infty} \Pi_j r = 0\}$ *is dense in B.*
Then $B_0 = B$ and we have the following formula

$$\rho_{ess}(\mathcal{F}) = \lim_{m\to\infty} (\limsup_{j\to\infty} \|\Pi_j \mathcal{F}^m\|)^{1/m} = \lim_{m\to\infty} (\liminf_{j\to\infty} \|\Pi_j \mathcal{F}^m\|)^{1/m}.$$

If instead we have
(iii′) $B'_0 = \{r \in B' : \lim_{j\to\infty} \Pi_j^t r = 0\}$ *is dense in B'.*
Then $B'_0 = B'$ and we have the following formula

$$\rho_{ess}(\mathcal{F}) = \lim_{m\to\infty} (\limsup_{j\to\infty} \|\mathcal{F}^m \Pi_j\|)^{1/m} = \lim_{m\to\infty} (\liminf_{j\to\infty} \|\mathcal{F}^m \Pi_j\|)^{1/m}.$$

If furthermore the operators Π_j are hierarchical in the sense that they satisfy
(iv) $\Pi_j \Pi_{j+1} = \Pi_{j+1}\Pi_j = \Pi_{j+1}$,
and if the dynamic acts in scales as expressed through
(v) $\exists k \in \mathbb{N} : \lim_{j\to\infty} \|(\mathbb{1} - \Pi_j)\mathcal{F}\Pi_{j+k}\| = \lim_{j\to\infty} \|\Pi_{j+k}\mathcal{F}(\mathbb{1} - \Pi_j)\| = 0.$
Then, if either (iii) *or* (iii′) *holds true, any of the above formulas may be used to compute $\rho_{ess}(\mathcal{F})$.*

Remark: In the second part of the theorem, if Π_j satisfies that $c \le 1$, then the interior limit actually exists. Indeed, for any bounded $\mathcal{A}$ by (ii) we have

$$\|\mathcal{A}\Pi_j\| = \|\mathcal{A}\Pi_{j-1}\Pi_j\| \le c\|\mathcal{A}\Pi_{j-1}\| \le \|\mathcal{A}\Pi_{j-1}\|.$$

Therefore the sequence $\|\mathcal{A}\Pi_j\|$ is monotonically decreasing. The same argument applies to $\|\Pi_j\mathcal{A}\|$.

If one thinks of Π_j as projection operators on the small scales of a multi-resolution analysis on a compact space, then the formula shows, that it is the small scale or infinitesimal behavior of $\mathcal{F}$ that determines the essential spectrum.

Proof. We proceed in eight elementary steps. First suppose that the density property (iii′) holds. The case (iii) is completely analoguous and we leave it to the reader. We also consider only the case lim sup and leave the lim inf part to the reader.

Step 1: We have

$$\inf\{\|\mathcal{F} - \mathcal{K}\| : \mathcal{K} \text{ compact}\} \le \|\mathcal{F}\Pi_j\|.$$

Indeed, we may write

$$\mathcal{F} - \mathcal{F}\Pi_j = \mathcal{F}(\mathbb{1} - \Pi_j).$$

The right hand side is a compact operator, since it contains a compact factor by hypothesis (ii) and since $\mathcal{F}$ is bounded.

Step 2: We have $B'_0 = B'$. Indeed, suppose $B'_0 \ni s_k \to s \in B'$. Then for every $\epsilon > 0$, we find K such that $k \ge K$ implies $\|s_k - s\| \le \epsilon$. It follows that

$$\left\|\Pi_j^t s\right\| \le \left\|\Pi_j^t s_k\right\| + \left\|\Pi_j^t(s - s_k)\right\| \le \left\|\Pi_j^t s_k\right\| + c\epsilon \to c\epsilon \quad (j \to \infty).$$

Since ϵ was arbitrary, the statement follows.

Let Ξ be the set of finite rank operators

$$s \mapsto \sum_\alpha v_\alpha(s) u_\alpha,$$

where the sum is finite with $v_\alpha \in B'$ and $u_\alpha \in B$.

Step 3: We have for all $\mathcal{K} \in \Xi$

$$\lim_{j\to\infty} \|\mathcal{K}\Pi_j\| = 0.$$

Indeed, we have for all s with $\|s\| = 1$

$$\|\mathcal{K}\Pi_j s\| = \left\| \sum_\alpha (v_\alpha \mid \Pi_j s) u_\alpha \right\| \le \sum_\alpha \left\|\Pi_j^t v_\alpha\right\|_{B'} \|u_\alpha\|_B .$$

This tends to 0 as $j \to \infty$, since by step 2 $B_0' = B'$, and since the sum contains only finitely many terms.

Step 4: We have for all $\mathcal{K} \in \Xi$

$$\|\mathcal{F} - \mathcal{K}\| \ge \frac{1}{c} \limsup_{j\to\infty} \|\mathcal{F}\Pi_j\| .$$

Here $\infty > c \ge 1$ is given by condition (i). Indeed, we have, since $\|\Pi_j\| \le c$

$$\|(\mathcal{F} - \mathcal{K})\Pi_j\| \le c \|\mathcal{F} - \mathcal{K}\| ,$$

for each j. Now

$$\|(\mathcal{F} - \mathcal{K})\Pi_j\| \ge \|\mathcal{F}\Pi_j\| - \|\mathcal{K}\Pi_j\| .$$

We may take the limit superior $j \to \infty$ and obtain the stated estimate.

Step 5: We have, upon replacing $\mathcal{F}$ by $\mathcal{F}^m$, for all m

$$c^{-1/m} \left(\limsup_{j\to\infty} \|\mathcal{F}^m\Pi_j\|\right)^{1/m} \le \left(\inf\{\|\mathcal{F}^m - \mathcal{K}\| : \mathcal{K} \text{ finite rank}\}\right)^{1/m},$$

and

$$\left(\inf\{\|\mathcal{F}^m - \mathcal{K}\| : \mathcal{K} \text{ compact}\}\right)^{1/m} \le \left(\limsup_{j\to\infty} \|\mathcal{F}^m\Pi_j\|\right)^{1/m}.$$

We now may go to the limit $m \to \infty$ to conclude for the first set of formulas. The last three steps prove the second part.

Step 6: Note that if $\mathcal{A}$ satisfies at (v) with some k it also satisfies the same with $k+1$ (and hence for all $k' > k$). Indeed, we may write

$$\|(\mathbb{1} - \Pi_j)\mathcal{A}\Pi_{j+k+1}\| = \|(\mathbb{1} - \Pi_j)\mathcal{A}\Pi_{j+k}\Pi_{j+k+1}\| \le c \|(\mathbb{1} - \Pi_j)\mathcal{A}\Pi_{j+k}\| .$$

The last expression tends to 0 by hypothesis on $\mathcal{A}$. The analogue argument applies for the second limit.

Step 7: The set of all bounded operators satisfying (v) forms an algebra. Indeed if $\mathcal{G}$ satisfies as $\mathcal{F}$ condition (v) then their sum and scalar multiples obviously satisfy the same and we are left to check the product. By step 6 we may suppose that $\mathcal{F}$ and $\mathcal{G}$ satisfy (v) with the same k. Then we have thanks to (iv),

$$\|(\mathbb{1}-\Pi_j)\mathcal{F}\mathcal{G}\Pi_{j+2k+1}\| = \|(\mathbb{1}-\Pi_j)\mathcal{F}[\Pi_{j+k} + (\mathbb{1}-\Pi_{j+k})(\mathbb{1}-\Pi_{j+k+1})]\mathcal{G}\Pi_{j+2k+1}\|\,.$$

Using the triangular inequality and (i), the last expression may be majorized by

$$\leq c\,\|\mathcal{G}\|\,\|(\mathbb{1}-\Pi_j)\mathcal{F}\Pi_{j+k}\| + (1+c)^2\,\|\mathcal{F}\|\,\|(\mathbb{1}-\Pi_{j+k+1})\mathcal{G}\Pi_{j+2k+1}\|\,.$$

By hypothesis, both expressions tend to 0 as $j \to \infty$.

Step 8: Conclusion. For any $\mathcal{A}$, bounded in B satisfying (v) and thus in particular for $\mathcal{A} = \mathcal{F}^m$ we have

$$\|\mathcal{A}\Pi_{j+k}\| \leq \|\Pi_j\mathcal{A}\Pi_{j+k}\| + \|(\mathbb{1}-\Pi_j)\mathcal{A}\Pi_{j+k}\|\,.$$

In the limit $j \to \infty$ we obtain

$$\limsup_{j\to\infty} \|\mathcal{A}\Pi_{j+k}\| \leq \limsup_{j\to\infty} \|\Pi_j\mathcal{A}\|\,\|\Pi_{j+k}\| \leq c\,\limsup_{j\to\infty} \|\Pi_j\mathcal{A}\|\,.$$

In the same way we obtain

$$\limsup_{j\to\infty} \|\Pi_{j+k}\mathcal{A}\| \leq c \limsup_{j\to\infty} \|\mathcal{A}\Pi_j\|\,.$$

The constant c does not depend on $\mathcal{A}$ and thus the stated equality of all the limits follows. □

As the proof shows we have the following corollary

Corollary 12.5 *Suppose* Π_j *satisfies* (i), (ii) *and* (iv) *and suppose that* $\mathcal{F}$ *satisfies at* (v) *then the following four limits are equal*

$$\limsup_{m\to\infty}(\inf)(\limsup_{j\to\infty}(\inf)\,\|\mathcal{F}^m\Pi_j\|)^{1/m}.$$

We shall need a second theorem for the essential spectral radius. This is because the as it will turn out, the space $\Lambda^\alpha(\mathbb{R}^n)$ does not satisfy the density properties (iii) nor (iii′). However we will be able to use the following theorem that is similar to a theorem of Keller proved in [Kel84].

Theorem 12.6 *Suppose that* Π_j *satisfies conditions* (i) *and* (ii) *of the previous theorem. The following condition is supposed to hold for* $\mathcal{F}$ *and all its iterates* $\mathcal{F}^m$. *We suppose that there is a double sequence of infinite dimensional closed spaces*

Step 3. $\mathcal{L}_\partial$ satisfies condition (iv) of theorem 12.4 holds and hence the first pair of formulas holds. We have to show that we have

$$\lim_{j\to\infty} \|(\mathbb{1} - \Pi_{j'})\mathcal{L}_\partial \Pi_j\| = \lim_{j\to\infty} \|\Pi_j \mathcal{L}_\partial (\mathbb{1} - \Pi_{j'})\| = 0,$$

whenever $j - j' \to \infty$. We even have

$$\mathcal{L}_\partial \Pi_j = \Pi_{j-k} \mathcal{L}_\partial \Pi_j, \quad \Pi_j \mathcal{L}_\partial = \Pi_j \mathcal{L}_\partial \Pi_{j+k},$$

for some k large enough. This follows from the fact that all kernels in $\mathcal{L}_\partial$ are supported by a strip in the half-space and that Π_j acts like the reproducing kernel in $\{a \le 4^{-j-2}\}$, whereas in $\{a \ge 4^{-j+2}\}$ it acts like the zero operator. Now we may write

$$\|(\mathbb{1} - \Pi_{j'})\mathcal{L}_\partial \Pi_j\| \le \|(\mathbb{1} - \Pi_{j'})\Pi_{j-k}\| \, \|\mathcal{L}_\partial \Pi_j\|,$$

and

$$\|\Pi_j \mathcal{L}_\partial (\mathbb{1} - \Pi_{j'})\| \le \|\Pi_j \mathcal{L}_\partial\| \, \|\Pi_{j+k}(\mathbb{1} - \Pi_{j'})\|.$$

Both expressions are zero for $j - j'$ large enough.

This shows that the second pair of equations holds.

Step 4. The difference $(\mathcal{L}_\partial)^m - (\mathcal{L}^m)_\partial$ is smoothing from $B(\mathbb{H}^n)$ to $aB(\mathbb{H}^n)$ and from $a^{-1}B(\mathbb{H}^n)$ to $B(\mathbb{H}^n)$. Indeed, we have

$$\mathcal{L}^\sharp = \mathcal{L}_\partial + \mathcal{R}_1,$$

and

$$(\mathcal{L}^\sharp)^m = (\mathcal{L}^m)_\partial + \mathcal{R}_2,$$

where $\mathcal{R}_i$, $i = 1, 2$ are smoothing from $a^k B(\mathbb{H}^n)$ to $a^{k+1}B(\mathbb{H}^n)$ for all $k \in \mathbb{Z}$. Therefore upon taking the m-th power and taking the difference, the statement follows.

Step 5. Hence $\|((\mathcal{L}_\partial)^m - (\mathcal{L}^m)_\partial)\Pi_j\|$ and $\|\Pi_j((\mathcal{L}_\partial)^m - (\mathcal{L}^m)_\partial)\|$ tend to zero as $j \to \infty$. Indeed, the mapping Π_j from $aB(\mathbb{H}^n)$ to $B(\mathbb{H}^n)$ has a norm that is $\le c4^{-j}$. Indeed, $\Pi_j s(b, a) = 0$ for $a > 4^{-j+2}$ and thus

$$\|\Pi_j s\|_{B(\mathbb{H}^n)} = \left\|aa^{-1}\Pi_j s\right\|_{B(\mathbb{H}^n)} \le 4^{-j+2} \left\|a^{-1}\Pi_j s\right\|_{B(\mathbb{H}^n)} = 4^{-j+2} \|s\|_{aB(\mathbb{H}^n)}.$$

The same estimate shows that Π_j as mapping from $B(\mathbb{H}^n)$ to $a^{-1}B(\mathbb{H}^n)$ has norm $\le c4^{-j}$. Therefore,

$$\lim_{j\to\infty} \left\|\Pi_j\left((\mathcal{L}_\partial)^m - (\mathcal{L}^m)_\partial\right)\right\| = 0 = \lim_{j\to\infty} \left\|(((\mathcal{L}_\partial)^m - (\mathcal{L}^m)_\partial)\Pi_j\right\|.$$

This shows the theorem. □

12.5 A more explicit formula for $\rho_{ess}(\mathcal{L})$

We now can actually use the previous theorem to compute the essential spectral radius in terms of the tangential maps. We use the notations introduced in the previous section. In particular recall the definition of γ_K (see (4.5)) and the notations $F((s_k))$ (see 4.7) and $\bigoplus A_k$ (see (12.1)). For the convenience of the reader we recall the theorem we want to prove.

Theorem 12.8 *Let $B(\mathbb{R}^n)$ be a completely regular Banach space. Let $B(\mathbb{T}^n)$ be the induced Banach space over $\mathbb{T}^n$. Then the transfer operator acts continuously on $B(\mathbb{T}^n)$. Its essential spectral radius can be computed as follows in terms of the norms of the linearized maps $\mu_{m,k}(\phi_{m,k}^{-1}x)\partial\Phi[\phi_{m,k},x]$, $k = 1,\ldots,K^m$ acting on $B(\mathbb{R}^n)$.*

$$\rho_{ess}(\mathcal{L}) = \lim_{m\to\infty} \sup_{x\in\mathbb{T}^n} \left\| \bigoplus_{k=1}^{K^m} \mu_{m,k}(\phi_{m,k}^{-1}x)\partial\Phi[\phi_{m,k},\phi_{m,k}^{-1}x] \right\|^{1/m} .$$

Note that $\mathcal{L}$ is acting on periodic functions, whereas the norm on the right hand side is computed by the action of $\partial\Phi[\phi_{m,k}]$ on functions over $\mathbb{R}^n$. This formula is easy to understand. It states, that the small scale norm of $\mathcal{L}$ and all its derivatives is produced by the strongest local deformation measures by the norm $\|\partial\Phi[\phi_{m,k},x]\|$ of the tangential map at x acting in $B(\mathbb{R}^n)$.

Proof. The main idea of the proof is as follows. Consider a small neighborhood of $x \in \mathbb{R}^n$. The points that map to x are $\phi_k^{-1}(x)$. At small scales the dynamic can be described by the tangential map. Since the local regions can be varied independently and since by homogeneity we can localize around the $\phi_k^{-1}(x)$ we make appear the norm of the sum of the tangential maps as stated in the theorem.

For the sake of simplicity we will prove the theorem under the hypothesis that $B(\mathbb{H}^n)$ satisfies the stronger condition $(CR2)$ instead of $(CR2')$. The modifications necessary to prove the theorem under the slightly weaker conditions on $B(\mathbb{H}^n)$ are straight forward, and we leave the details to the reader.

The central argument to prove the lower bound is given by the next two lemmas.

Lemma 12.9 *For $s = \mathcal{W}_g r$, $r \in \mathcal{S}_0(\mathbb{R}^n)$, $\|s\| = 1$ consider the family of dilated and translated functions $s_{\beta,\alpha} = \mathcal{W}_g T_\beta D_\alpha r = T_\beta^\sharp D_\alpha^\sharp s$. We then have for $\mathcal{F} = \Pi Op[F]\Pi$, with F given by (compare 12.1)*

$$F(b,a;b',a') = \mu(b')\frac{1}{a'^n} D\left(b'; \frac{b-\phi(b')}{a'}, \frac{a}{a'}\right),$$

with $D(b';b,a) = \mathcal{W}[g,\partial\Phi[\phi,b']g](b,a)$ that

$$\lim_{\alpha\to 0} \|\mathcal{F}s_{\beta,\alpha}\| \,/\, \|s_{\beta,\alpha}\| = |\mu(\beta)|\, \|\partial\Phi[\phi,\beta]r\|_{B(\mathbb{R}^n)} .$$

Proof. By homogeneity of the norm of $B(\mathbb{H}^n)$ we may suppose that $\beta = 0$ and that $\phi(0) = 0$. Suppose $\|s\| = 1$ and let $s_\alpha = s_{0,\alpha} = D^\sharp_\alpha s$. We write $\partial\phi$ for $\partial\phi_0$. We then may write, using again the homogeneity

$$\|\mathcal{F}s\| = a^\gamma \left\|(D^\sharp_\alpha)^{-1}\mathcal{F}D^\sharp_\alpha s\right\|_{B(\mathbb{H}^n)}.$$

Now $\mathcal{F}$ may be factorized as follows

$$\mathcal{F} = A[P]Poc[D']\Psi[\phi]M[\mu] = Poc[D']\Psi[\phi]M[\mu],$$

with $D'(x; b, a) = D(\phi^{-1}(x); b, a)$. We now have three elementary steps.

Step 1. The family of operators $\mathcal{F}_\alpha = (D^\sharp_\alpha)^{-1}\mathcal{F}D^\sharp_\alpha$, $0 < \alpha \leq 1$ is uniformly continuous from $\mathcal{S}(\mathbb{H}^n)$ to $\mathcal{S}(\mathbb{H}^n)$ (it maps highly localized function into functions having the same localization). Clearly it is enough to check this for each factor since the symbols satisfying these estimates constitute an algebra (= the algebra of highly regular Calderón Zygmund operators). For this let us denote for any operator $\mathcal{G}$ by $\mathcal{G}_\alpha$ the rescaled operator $(D^\sharp_\alpha)^{-1}\mathcal{G}D^\sharp_\alpha$. We may write

$$\mathcal{F}_\alpha = Poc[D']_\alpha\Psi[\phi]_\alpha M[\mu]_\alpha.$$

Suppose that $\mathcal{G} = Op[G]$ with G satisfying with some $v \in \mathcal{S}(\mathbb{H}^n)$ (compare Lemma 6.1)

$$|G(b, a; b', a')| \leq \frac{1}{a'^n} v\left(\frac{b - b'}{a'}, \frac{a}{a'}\right).$$

Then we may find G_α with $\mathcal{G}_\alpha = Op[G_\alpha]$ where the G_α satisfies the same estimate with the same function v. In addition these operators are uniformly continuous in $\mathcal{S}(\mathbb{H}^n)$. Therefore, $Poc[D']_\alpha$ is uniformly continuous in $\mathcal{S}(\mathbb{H}^n)$. Now we have

$$M[\mu]_\alpha = M[\mu_\alpha], \quad \mu_\alpha(b, a) = \mu(\alpha b, a),$$

thereby showing that $M[\mu]_\alpha$ is uniformly continuous in $\mathcal{S}(\mathbb{H}^n)$. Finally we may write

$$\Psi[\phi]_\alpha = \Psi[\phi_\alpha], \quad \phi_\alpha(x) = \alpha\phi(x/\alpha).$$

Now, since $\left|\phi^{-1}(x)\right| \geq c\,|x|$ we have with the same c that $\left|(\phi_\alpha)^{-1}(x)\right| \geq c\,|x|$. Therefore again the localization of $\Psi[\phi]_\alpha s$ may be estimated in terms of the localization of s, uniformly in α.

This shows that there is some function $u \in \mathcal{S}(\mathbb{H}^n)$ such that

$$|\mathcal{F}_\alpha s(b, a)| \leq u(b, a).$$

Step 2. We have point-wise with $W(b, a) = D'(0; b, a)$

$$\mathcal{F}_\alpha s(b, a) \to \mu(0)A[W]\Psi[\partial\phi]s(b, a) \quad (\alpha \to 0).$$

Indeed, we have point-wise $\Psi[\phi]_\alpha M[\mu]_\alpha s \to \mu(0)\Psi[\partial\phi]s$. Now $\Psi[\phi]_\alpha M[\mu]_\alpha s$ is uniformly localized by step 1. Therefore, by standard Lebesgue dominated convergence we have point-wise

$$Poc[D']_\alpha \Psi[\phi]_\alpha M[\mu]_\alpha s(b,a) \to \mu(0)A[W]\Psi[\partial\phi]s(b,a).$$

Step 3. The family $\mathcal{F}_\alpha s(b,a)$ is uniformly continuous on compact sets. Indeed, thanks to corollary 4.12 uniform localization of $v_\alpha = Poc[D']_\alpha \Psi[\phi]_\alpha M[\mu]_\alpha s$ implies uniform localization of all partial derivatives of $\mathcal{F}_\alpha s(b,a) = v_\alpha(b,a) = P * v_\alpha(b,a)$. We may evoke Lemma 4.5 to conclude for the convergence of the norms. □

In exactly the same way we can show, the next lemma.

Lemma 12.10 *For K functions $s_k = \mathcal{W}_g r_k$, $r_k \in \mathcal{S}_0(\mathbb{R}^n)$, and diffeomorphisms ϕ_k, operators $\mathcal{F}_k$ associated with them as in Lemma 12.9 and pair-wise distinct points β_k with $\phi_k\beta_k = \beta$ we have*

$$\lim_{\alpha\to 0}\left\|\sum_{k=1}^{K}\mathcal{F}_k s_{k;\beta_k,\alpha}\right\| / \left\|\sum s_{k;\beta_k,\alpha}\right\| = \left\|\sum_{k=1}^{K}\mu_k(\beta_k)\partial\Phi[\phi_k,\beta_k]r_k\right\| / F((s_k)).$$

Proof. We only have to note that thanks to formula 4.7 we have

$$\lim_{\alpha\to 0}\alpha^{-\gamma}\left\|\sum s_{k;\beta_k,\alpha}\right\| = \lim_{\alpha\to 0}\left\|\sum s_{k;\beta_k/\alpha,1}\right\| = F((s_k)).$$

where the exponent γ is given by hypothesis 4.2. □

Until now we have considered functions over $\mathbb{R}^n$. However if we introduce the periodization operator $\mathcal{P}$ via ($s \in \mathcal{S}(\mathbb{H}^n)$)

$$\mathcal{P}s = \sum_{k\in\mathbb{Z}^n} s(\cdot + k, \cdot),$$

then $\mathcal{P}s \in \mathcal{S}(\mathbb{M}^n)$. Then if we set $s^\circ_{\beta,\alpha} = \mathcal{P}s_{\beta,\alpha}$ we have again that (supposing $\|s\| = 1$)

$$\lim_{\alpha\to 0}\left\|\mathcal{F}s^\circ_{\beta,\alpha}\right\| / \left\|s_{\beta,\alpha}\right\| = |\mu(\beta)|\,\|\partial\Phi[\phi_k,\beta]r\|_{B(\mathbb{R}^n)}\,.$$

We now come back to the transfer operator $\mathcal{L}_\partial = \sum_{k=1}^K \mathcal{L}_k$, $\mathcal{L}_k = Op[L_k]$ (compare 12.1). In the same way as before we may prove the next lemma. We may leave the details to the reader.

Lemma 12.11 *Consider the operator $\mathcal{L}_\partial = \sum_{k=1}^K \mathcal{L}_k$ acting on $\mathcal{S}(\mathbb{M}^n)$. Let $x \in \mathbb{T}^n$ and let $\beta_k = \phi_k^{-1}(x) \in \mathbb{T}^n$ be the preimages of the diffeomorphism ϕ_k. Then for $s_k \in \Pi\mathcal{S}(\mathbb{H}^n)$, $s_k = \mathcal{W}_g r_k$ consider the functions $s_\alpha = \sum_{k=1}^K \mathcal{P}s_{k;\beta_k,\alpha}$. Then we have*

$$\lim_{\alpha\to 0}\|\mathcal{L}_\partial s_\alpha\|_{B(\mathbb{M}^n)} / \|s_\alpha\|_{B(\mathbb{M}^n)} = \left\|\left(\bigoplus_{k=1}^K A_k\right)(r_k)\right\|_{B(\mathbb{R}^n)} / F((s_k)),$$

where $A_k = \mu(\beta_k)\partial\Phi[\phi_k,\beta_k]$.

We shall show now that the lower bound of the theorem holds. Pick a sequence $x_m \in \mathbb{T}^n$ for which

$$\left\| \bigoplus_{k=1}^{K} \mu_k(\phi_k^{-1}(x_m)) \partial\Phi[\phi_k, \phi_k^{-1}(x_m)] \right\| \\ \to \sup_{x\in\mathbb{T}^n} \left\| \bigoplus_{k=1}^{K} \mu_k(\phi_k^{-1}(x)) \partial\Phi[\phi_k, \phi_k^{-1}(x)] \right\|, \quad (m \to \infty).$$

Now, for each m, pick a sequence of K-touples of functions $s_{k,l,m} \in \Pi\mathcal{S}(\mathbb{H}^n)$, $s_{k,l,m} = \mathcal{W}_g r_{k,l,m}$, $k = 1, \ldots, K$, $l = 1, 2, \ldots$, such that

$$F((s_{k,l,m})_{k=1,\ldots,K}) = 1,$$

and

$$\left\| \sum_{k=1}^{K} \mathcal{L}_k s_{k,l,m} \right\|_{B(\mathbb{H}^n)} \to \left\| \bigoplus_{k=1}^{K} \mu_k(\phi_k^{-1}(x_m)) \partial\Phi[\phi_k, \phi_k^{-1}(x_m)] \right\|, \quad (m \to \infty).$$

Such a sequence exists by density. In addition we may suppose that each $s_{k,l,m}$ is supported in some (different) strip $\{1/c \le a \le c\}$. Pick a sequence $\alpha_p > 0$, $\alpha_p \to 0$ as $p \to \infty$. Consider the dilated and periodized functions

$$s^\circ_{k,l,m,p} = \sum_{q\in\mathbb{Z}^n} \alpha_p^\gamma s_{k,l,m}([\cdot - \phi_k^{-1}(x_m) + q]/\alpha_p, \cdot/\alpha_p) = \alpha_p^{n+\gamma} \mathcal{P} T^\sharp_{\phi_k^{-1}(x_m)} D^\sharp_{\alpha_p} s_{k,l,m},$$

and set $u^\circ_{l,m,p} = \sum_{k=1}^{K} s^\circ_{k,l,m,p}$ Then $u^\circ_{l,m,p} \in \mathcal{S}(\mathbb{M}^n)$ and

$$\left\| u^\circ_{l,m,p} \right\|_{B(\mathbb{M}^n)} \to 1, \quad (l \to \infty),$$

and

$$\left\| \mathcal{L}_\partial u^\circ_{l,m,p} \right\|_{B(\mathbb{M}^n)} \to \left\| \bigoplus_{k=1}^{K} \mu_k(\phi_k^{-1}(x_m)) \partial\Phi[\phi_k, \phi_k^{-1}(x_m)] \right\|, \quad (p \to \infty).$$

Moreover for l large enough we have $u^\circ_{l,m,p} \in \Pi_j \mathcal{S}(\mathbb{M}^n)$. Therefore we can find a diagonal sequence $u^\circ_j \in \Pi_j B(\mathbb{M}^n)$ for which

$$\left\| u^\circ_j \right\|_{B(\mathbb{M}^n)} \to 1,$$

and

$$\left\| \mathcal{L}_\partial u^\circ_j \right\|_{B(\mathbb{M}^n)} \to \sup_{x\in\mathbb{T}^n} \left\| \bigoplus_{k=1}^{K} \mu_k((\phi_k^{-1}(x)) \partial\Phi[\phi_k, \phi_k^{-1}(x)] \right\|.$$

This shows that

$$\|\mathcal{L}_\partial\| \geq \sup_{x\in\mathbf{T}^n} \left\| \bigoplus_{k=1}^{K} \mu_k(\phi_k^{-1}(x))\partial\Phi[\phi_k, \phi_k^{-1}(x)] \right\|.$$

The same argument applies to $(\mathcal{L}^m)_\partial$ which shows the lower bound of the essential spectral radius.

We now prove the upper bound for $\mathcal{L}_\partial$. We prove that

$$\|\mathcal{L}_\partial \Pi_j\|_{\Pi B(\mathbb{M}^n)\to B(\mathbb{M}^n)} \leq c \sup_{x\in\mathbf{T}^n} \left\| \bigoplus_{k=1}^{K} \mu_k((\phi_k^{-1}(x))\partial\Phi[\phi_k, \phi_k^{-1}(x)] \right\|,$$

with a c not depending on $\mathcal{L}$. The theorem follows by looking at $(\mathcal{L}^m)_\partial$.

We need some notation. For each $l \in \mathbb{N}$ consider the grid $\Gamma_l = 2^{-l}\mathbb{Z}^n \cap \mathbb{T}^n$. Γ_l is an abelian group in the obvious way (quotient of $\mathbb{Z}^n$ by $2^l\mathbb{Z}^n$). Here we have again identified the torus with $[0,1)^n$. Consider the hypercubes of equal size centered localizationese grid points. They cover $\mathbb{T}^n$ and they intersect only on sets of measure zero. Call these cubes Q_σ, $\sigma \in \Gamma_l$ and their centers x_σ. Note that they depend on l but we do not write this dependency explicitly for notational simplicity. Now consider the set P_σ, of pre images

$$P_\sigma = \bigcup_{k=1}^{K} \phi_k^{-1} Q_\sigma.$$

(we work modulo $[0,1)^n$). The sets P_σ have pair-wise disjoint interiors. Now fix $j \in \mathbb{N}$. We define sets in $\mathbb{M}^n$ as follows. The set of points with positions in $b \in P_\sigma$ and scales $0 < a \leq 2^{-j}$ shall be denoted by $\tilde{P}_\sigma$ and analogously for $\tilde{Q}_\sigma$. Note that these sets depend on j and l but we do again not write it for the sake of notational simplicity. In the same vain we denote the associated projection multipliers given by the characteristic functions, with the same symbols. Thus e.g. $\tilde{P}_\sigma$ stands for the multiplier with the characteristic function of the set $\tilde{P}_\sigma$.

Step 1: For σ, $\sigma' \in \Lambda_l$ that are not next or next nearest neighbors, we have

$$\lim_{j\to\infty} \left\| \tilde{Q}_{\sigma'} \mathcal{L}_\partial \tilde{P}_\sigma \right\| = 0.$$

This follows easily from the locality property of the kernel of $\mathcal{L}_\partial$. Therefore, we need only consider the following diagonalized operator

$$\sum_{\sigma\in\Gamma_l} \sum_{\rho\in\{-1,0,+1\}^n} \tilde{Q}_{\sigma+\rho} \mathcal{L}_\partial \tilde{P}_\sigma,$$

Note that every $\sigma \in \Gamma_l$ can be written in a unique way as $\lambda + \rho$ with $\lambda \in \{0,1\}^n$ and $\rho \in \Gamma_{l-1}$.

Step 2: We have

$$\left\|\sum_{\sigma\in\Gamma_l}\sum_{\rho\in\{-1,0,+1\}^n}\tilde{Q}_{\sigma+\rho}\mathcal{L}\partial\tilde{P}_\sigma\right\| \le 6^n \max_{\delta\in\{0,1\}^n}\max_{\rho\in\{-1,0,+1\}^n}\max_{\sigma\in\Gamma_{l-1}}\left\|\tilde{Q}_{\sigma+\rho+\alpha}\mathcal{L}\partial\tilde{P}_{\sigma+\delta}\right\|$$

$$\le c \max_{\delta\in\{0,1\}^n}\max_{\sigma\in\Gamma_{l-1}}\left\|\mathcal{L}\partial\tilde{P}_{\sigma+\delta}\right\|$$

Indeed, we may write

$$\sum_{\sigma\in\Gamma_l}\sum_{\rho\in\{-1,0,+1\}^n}\tilde{Q}_{\sigma+\rho}\mathcal{L}\partial\tilde{P}_\sigma = \sum_{\delta\in\{0,1\}^n}\sum_{\rho\in\{-1,0,+1\}^n}\sum_{\sigma\in\Gamma_{l-1}}\tilde{Q}_{\sigma+\rho+\delta}\mathcal{L}\partial\tilde{P}_{\sigma+\delta}.$$

Therefore we have, since the first sums are over 2^n3^n terms

$$\left\|\sum_{\sigma\in\Gamma_l}\sum_{\rho\in\{-1,0,+1\}^n}\tilde{Q}_{\sigma+\rho}\mathcal{L}\partial\tilde{P}_\sigma\right\| \le 6^n \max_{\delta\in\{0,1\}^n}\max_{\rho\in\{-1,0,+1\}^n}\left\|\sum_{\sigma\in\Gamma_{l-1}}\tilde{Q}_{\sigma+\rho+\delta}\mathcal{L}\partial\tilde{P}_{\sigma+\delta}\right\|.$$

Now for fixed l but j small enough, the sets $\tilde{P}_{\sigma+\delta}$ have pair-wise disjoint bases for fixed δ. The same holds for the sets $\tilde{Q}_{\sigma+\rho+\delta}$ for fixed δ, ρ. We now may invoke Lemma 4.7 to conclude.

The next lemma allows us then to conclude.

Lemma 12.12 *Let ϕ_k, $k = 1,\ldots,K$ be uniformly smooth diffeomorphisms. For $\beta \in \mathbb{R}^n$, define β_k through $\phi_k(\beta_k) = \beta$. Let $\Pi_{k,\epsilon}$ denote the multiplier with the characteristic function of $\{|b-\beta_k| \le \epsilon, 0 < a \le \epsilon\}$ and set $\Pi_\epsilon = \sum_k \Pi_{k,\epsilon}$. Let $\mathcal{F} = Op[L]$ with L given by (compare 12.1)*

$$L(b,a;b',a') = \sum_{k=1}^{K} L_k(b,a;b',a') = \sum_{k=1}^{K}\mu_k(b')\frac{1}{a'^n}D_k\left(b';\frac{b-\phi_k(b')}{a'},\frac{a}{a'}\right),$$

with $D_k(b';b,a) = \mathcal{W}[g,\partial\Phi[\phi_k,b']g](b,a)$. It follows that

$$\|\mathcal{F}\Pi_\epsilon\| \le \left\|\bigoplus_k Op[\Lambda_k]\right\| + \rho(\epsilon),$$

where $\rho(\epsilon)\to 0$ as $\epsilon \to 0$ and ρ may be chosen independent of the β. The symbols Λ_k are given by

$$\Lambda_k(b,a;b',a') = \frac{\mu_k(\beta_k)}{a'^n}D_k\left(\beta_k;\frac{b-\beta-\partial[\phi_k,\beta_k](b'-\beta_k)}{a'},\frac{a}{a'}\right).$$

Proof. Around each β_k we may use Lemma 11.3 to write

$$L_k(b,a;b',a') = \frac{\mu_k(\beta_k)}{a'^n} D_k\left(\beta_k; \frac{b-\phi_k(b')}{a'}, \frac{a}{a'}\right) + |b-\beta_k|\, R_k(b,a;\phi_k(b'),a'). \tag{12.1}$$

The remainder $r_k = |b-\beta_k|\, R_k$ is such that

$$\|Op[r_k]\Pi_{k,\epsilon}\| \le \epsilon\, \|Op[r_k]\|\| \le c\epsilon.$$

Here c may be chosen uniformly for all β_k. We therefore need only consider the first term in 12.1 which we call T_k. We have the following factorization into a convolution operator with $P_k(b,a) = \mu_k(\beta_k)D_k(\beta_k;b,a)$ and the diffeomorphism acting on the b variable only

$$Op[T_k] = P_k * \Psi[\phi_k] = A[P_k]\Psi[\phi_k].$$

If we write $\delta\phi_k$ for the linearization of ϕ_k around β_k, then we have

$$Op[T_k]\Pi_{k,\epsilon} = A[P_k]\Psi[\delta\phi_k]\Psi[(\delta\phi_k)^{-1}\phi_k]\Pi_{k,\epsilon}.$$

Now the mapping

$$\Gamma_\epsilon : B(\mathbb{H}^n)^K \to B(\mathbb{H}^n)^K, \quad (s_k) \mapsto (\Psi[(\delta\phi_k)^{-1}\phi_k]\Pi_{k,\epsilon}s_k),$$

satisfies by Lemma 4.6

$$\|\Gamma_\epsilon\| \le 1 + \rho'(\epsilon),$$

where $\rho'(\epsilon) \to 0$ as $\epsilon \to 0$ independent of β. If we denote by Ξ the operator $(s_k) \mapsto \sum_k s_k$, and by Λ the operator with components Λ_k, then we have

$$\left\| \bigoplus_{k=1}^{K} Op[T_k]\Pi_{k,\epsilon} \right\| = \|\Xi\Lambda\Gamma_\epsilon\| \le \|\Xi\Lambda\|\, \|\Gamma_\epsilon\|,$$

and the proof is finished. □

This shows the theorem. □

12.6 A simpler upper bound

Let $\gamma_K : \mathbb{R}_+^K \to \mathbb{R}_+$ denote the function for which $F((s_k)) = \gamma_K((\|s_k\|))$. As we have seen, it defines a norm on $\mathbb{R}^K$ via $|x| = \gamma_K((|x_k|))$. Denote by γ'_K the dual norm, defined through ($x = (x_k)$)

$$\gamma'_K((x_k)) = \sup_{\gamma_K(y)=1} |(x \mid y)|\,.$$

We then have clearly under the same hypothesis as in theorem 12.8

$$\rho_{ess}(\mathcal{L}) \le \limsup_{m\to\infty} \sup_{x\in\mathbb{T}^n} \gamma'_{K^m}\left(\left(\left|\mu_{m,k}(\phi_{m,k}^{-1}(x))\right| \left\|\partial\Phi[\phi_{m,k},\phi_{k,m}^{-1}x]\right)\right\|\right)\right)^{1/m}.$$

In particular this proves theorem 12.3.

12.7 The case $\Lambda_\gamma(\mathbb{T}^n)$

The theorem that we have proved in the previous section applies a priori to $\lambda_\gamma(\mathbb{T}^n)$ but not to $\Lambda_\gamma(\mathbb{T}^n)$ because in the latter space $\mathcal{S}_0(\mathbb{T}^n)$ is not dense. Therefore in particular condition (iii) of theorem 12.4 does not hold. However we have

Theorem 12.13 *The formula of theorem* 12.8 *holds for* $\Lambda_\gamma(\mathbb{T}^n)$, $\gamma \in \mathbb{R}$.

Note that in the case of Λ_γ we have that

$$\gamma((\|s_k\|)) = \max_k \|s_k\| .$$

Proof. The upper bound holds without any change as the proof of theorem 12.4 and the proof of theorem 12.8 show. To prove the lower bound we invoke theorem 12.6. We therefore have to construct the spaces V_j^m. We will do this first for the family of operators $(\mathcal{L}^m)_\partial$ and we will argue why the same spaces work for $(\mathcal{L}_\partial)^m$, which differs only by a small scale smoothing operator. Denote for fixed x and m the linearized diffeomorphism in wavelet space by

$$\partial\mathcal{L}_k = \partial\mathcal{L}_k(x,m) = \mu_{m,k}(\phi_{m,k}^{-1}(x))\partial\Phi[\phi_{m,k},\phi_{m,k}^{-1}(x)]^\sharp. \quad k=1,\ldots,K^m.$$

That is, $\partial\mathcal{L}_k = Op[D_k]$ with

$$D_k(b,a;b',a') = \mu_{m,k}(\phi_{m,k}^{-1}(x))\frac{1}{a'^n}P\left(\frac{b-\delta\phi_{k,m}(b')}{a'},\frac{a}{a'}\right),$$

where $\delta\phi_{k,m}$ is the linearization of ϕ around $\phi_{m,k}^{-1}(x)$ and

$$P(b,a) = \mathcal{W}[g,\partial\Phi[\phi_{m,k},\phi_{m,k}^{-1}(x)]g](b,a).$$

Step 1. There is a constant $c>0$ such that for all m and all x there is a sequence of vectors $s_{k,l} \in \Pi\mathcal{S}(\mathbb{H}^n)$, each being supported by some strip $\{1/c_l \le a \le c_l\}$ such that

$$\max_k \|s_{k,l}\| = 1,$$

and that

$$\lim_{l\to\infty}\left\|\sum_{k=1}^{K^m}\partial\mathcal{L}_k s_{k,l}\right\| \to c\left\|\bigoplus_{k=1}^{K^m}\partial\mathcal{L}_k\right\|.$$

Since $\mathcal{S}(\mathbb{H}^n)$ is not dense in $\Lambda_\gamma(\mathbb{H}^n)$ we need an extra argument. However, there clearly is a sequence $r_{k,l} \in \Pi\Lambda_\gamma(\mathbb{H}^n)$, $l=1,2,\ldots$, $k=1,2,\ldots,K^m$ with $\max_k \|r_{k,l}\| = 1$ such that

$$\left\|\sum_{k=1}^{K^m}\partial\mathcal{L}_k r_{k,l}\right\| \to \left\|\bigoplus_{k=1}^{K^m}\partial\mathcal{L}_k\right\|, \quad (l\to\infty).$$

Let P_w be the multiplier with the characteristic function of $\{a + 1/a + |b| \le w\}$ and let $E_w = \Pi P_w \Pi$. We now set

$$v_{k,l,w} = E_w r_{k,l}.$$

Clearly $v_{k,l,w} \in \Pi\mathcal{S}(\mathbb{H}^n)$, and moreover, it is supported by some strip in the halfspace thanks to the localization of the reproducing kernel. We have $\|E_w\| \le c_1$ and thus

$$\|v_{k,l,w}\| \le c_1 \|r_{k,l}\|.$$

Again by the locality of the reproducing kernel we have for all functions $f \in \Pi\Lambda_\gamma(\mathbb{H}^n)$ that point-wise

$$\lim_{w\to\infty} E_w f(b,a) = f(b,a).$$

Now, since $\|f\| = \sup a^{-\gamma} f$ we have

$$\liminf_{w\to\infty} \|v_{k,l,w}\| = \liminf_{w\to\infty} \|E_w r_{k,l}\| \ge \|r_{k,l}\|. \tag{12.1}$$

For the same reason, since the kernel of $\partial\mathcal{L}_k$ are localized too we have

$$\liminf_{w\to\infty} \left\| \sum_{k=1}^{K^m} \partial\mathcal{L}_k v_{k,l,w} \right\| \ge \left\| \sum_{k=1}^{K^m} \partial\mathcal{L}_k r_{k,l} \right\|.$$

We may extract a diagonal sequence $s_{k,p}$ from $v_{k,l,w}$, normalize it in such a way that $\max_k \|s_{k,p}\| = 1$ and that

$$\lim_{p\to\infty} \left\| \sum_{k=1}^{K^m} \partial\mathcal{L}_k s_{k,p} \right\| = c \left\| \bigoplus_{k=1}^{K^m} \mu_{m,k}(\phi_{m,k}(x)) \partial\Phi[\phi_{m,k}, \phi_{m,k}^{-1}(x)] \right\|.$$

Step 2. There is a sequence of functions $s_l \in \Pi\mathcal{S}(\mathbb{H}^n)$, such that the following holds.

- Their wavelet transforms have disjoint supports $\Xi_l \subset \mathbb{H}^n$ such that the ratios $a/a' + a'/a$ with $a \in \Xi_l$ and $a' \in \Xi_{l'}$ tend to ∞ as $l, l' \to \infty$ with $l \neq l'$.
- we have $\|s_l\| = 1$
- The following limit holds with c as above

$$\lim_{l\to\infty} \|(\mathcal{L}^m)_\partial s_l\| \to c \sup_{x\in\mathbb{T}^n} \left\| \bigoplus_{k=1}^{K^m} \mu_{m,k}(\phi_{m,k}^{-1}(x)) \partial\Phi[\phi_{m,k}, \phi_{m,k}^{-1}(x)] \right\|.$$

Indeed, by homogeneity of the norm, we may dilate the $s_{k,l}$ in such a way, that the supports for different l satisfy the above requirement. Then, periodizing a suitable sequence of superpositions as before in the proof of theorem 12.8 will do the job.

Step 3. Construction of the spaces V_j^m. Let us define V_j^m to be the closure in $\Lambda_\gamma(\mathbb{M}^n)$ of the space spanned by those s_l above, for which l is large enough (= that are living at small enough scale) so that $\Pi_j s_l = s_l$. The vectors in V_j^m have supports in a collection of disjoined strips Ξ_l whose relative scale ratio tends to infinity at small scale. If we decompose $r \in V_j^m$ into components r_l, each having support in one of these strips

$$r \in V \Rightarrow r = \sum r_k, \quad \text{with } \operatorname{supp} r_k \subset \Xi_k,$$

then we have by lacunarity

$$\|r\| = \sup_k \|r_k\|\,. \tag{12.2}$$

Now, the transfer operator $(\mathcal{L}^m)_\partial$ respects the band structure in wavelet space. More precisely, it maps the space of functions with support in Ξ_l to a space of functions supported by the strip

$$\Xi_l' = \{a : \exists a' \in \Xi_l, c_2^{-1} a' \le a \le c_2 a'\},$$

with c_2 independent of l. Thus for l large enough the bands Ξ_l' have disjoint supports. Therefore we have for $r \in V_j^m$, j large enough

$$\|(\mathcal{L}^m)_\partial r\| = \left\| \sum_k (\mathcal{L}^m)_\partial r_k \right\| = \sup_k \|(\mathcal{L}^m)_\partial r_k\|\,. \tag{12.3}$$

Now, by construction of V_j^m we have for any sequence $r_l \in V_j^m$, $\|r_l\| = 1$, with support in Ξ_l that

$$\lim_{l\to\infty} \|(\mathcal{L}^m)_\partial r_l\| \to c \sup_{x\in\mathbb{T}^n} \left\| \bigoplus_{k=1}^{K^m} \mu_{m,k}(\phi_{m,k}^{-1}(x)) \partial\Phi[\phi_{m,k}, \phi_{m,k}^{-1}(x)] \right\| .$$

Therefore in particular for j large enough we have for all $r_l \in V_j^m$, $\|r_l\| = 1$, with support in Ξ_l that

$$\|(\mathcal{L}^m)_\partial r_l\| \ge (c/2) \sup_{x\in\mathbb{T}^n} \left\| \bigoplus_{k=1}^{K^m} \mu_{m,k}(\phi_{m,k}^{-1}(x)) \partial\Phi[\phi_{m,k}, \phi_{m,k}^{-1}(x)] \right\| . \tag{12.4}$$

Now suppose we are given a sequence $u_j \in V_j^m$ with $\|u_j\| = 1$. We decompose it into lacunary components $u_j = \sum_k r_{j,k}$ with $r_{j,k}$ supported by Ξ_k. It follows from 12.2 that for all j there is some l such that

$$\|r_{j,l}\| \ge 1/2.$$

Therefore, by 12.3 and 12.4 we have for j large enough

$$\|(\mathcal{L}^m)_\partial u_j\| \ge \|(\mathcal{L}^m)_\partial r_{j,l}\| \ge (1/4) c \sup_{x\in\mathbb{T}^n} \left\| \bigoplus_{k=1}^{K^m} \mu_{m,k}(\phi_{m,k}^{-1}(x)) \partial\Phi[\phi_{m,k}, \phi_{m,k}^{-1}(x)] \right\| .$$

Step 4. Conclusion. We have constructed a space V_j^m for $(\mathcal{L}^m)_\partial$ and not for $(\mathcal{L}_\partial)^m$ as we should in order to use theorem 12.6. However we may use once more step 4 of the proof of theorem 12.7 to conclude. □

13 Appendix

A.1 On the modulus of continuity

In this section we shall discuss how the function spaces we have defined in some previous section relate to the the function spaces that are defined in terms of local regularity in the classical sense. The results of this section have found applicatins in the analysis of fractals (see e.g. [AGH88] and [HT91]). For similar results see also [Jaf89]. Particular instances of the regularities we can describe through the order of magnitude of the wavelet coefficients are the Hölder classes and other regularity classes defined in terms of the order of magnitude of the modulus of continuity $\Omega[s]$ respectively $\Omega[s,x]$ defined as

$$\Omega[s,x](t) = \sup_{|y|\leq t} |s(x+y) - s(x)| \, .$$

The point-wise Hölder classes are then defined via

$$\begin{aligned} s \in \Lambda^\alpha(x) \quad &\Longleftrightarrow \quad \Omega[s,x](t) \leq O(t^\alpha) \ (t \to 0) \\ s \in \lambda^\alpha(x) \quad &\Longleftrightarrow \quad \Omega[s,x](t) \leq o(t^\alpha) \ (t \to 0) \end{aligned}$$

We shall consider the following natural generalizations. We say a function $\rho : \mathbb{R}_+ \to \mathbb{R}_+$ is sub-multiplicative if for some $c > 0$ we have

$$\rho(aa') \leq c\rho(a)\rho(a'), \quad a, a' > 0.$$

Obvious examples are $\rho(t) = t^\gamma$ or $\rho(t) = (1 + |\log t|)^\beta t^\gamma$. Natural regularity classes are therefore given by the vector space of continuous functions that satisfy

$$\begin{aligned} \Omega[s,x_0](t) \quad &\leq O(\rho(t)) \ (t \to 0), \\ \Omega[s,x_0](t) \quad &\leq o(\rho(t)) \ (t \to 0). \end{aligned}$$

A slight generalization consists in considering the approximation by local polynomials. In order to measure the regularity of a function s in a point $x_0 \in \mathbb{R}^n$ we estimate how well it can be approximated locally by a polynomial P_m of degree m in n variables. That is we suppose that we may write

$$s(x_0 + x) = P_m(x) + s_{loc}(x),$$

where the remainder becomes small compared to the principal term P_m; that is $s_{loc}(x) = o(|x|^m)$ as $x \to 0$.

In general the faster $s_{loc}(x)$ goes to 0 together with x the higher is the regularity of s at x_0. It therefore is natural to consider the regularity classes defined by.

$$\begin{aligned} |s_{loc}(x)| &\le O(\rho(|x|))\ (x \to 0) \\ |s_{loc}(x)| &\le o(\rho(|x|))\ (x \to 0) \end{aligned}$$

In particular, for $m = 0$, the constant polynomial, we reobtain the modulus of continuity at x_0.

For the sake of completeness we shall prove the next theorem for more general wavelets.

Theorem A.1 *Let ρ be a sub-multiplicative weight function and let s satisfy*

$$|s(x_0 + x) - P_m(x)| \le c\rho(|x|), \quad (x \in \mathbb{R}^n),$$

with some polynomial P_m of degree m. Suppose the analyzing wavelet satisfies

$$\begin{aligned} &\text{(i)} \quad g \in L^1(\mathbb{R}^n),\ \rho \cdot g \in L^1(\mathbb{R}^n) \\ &\text{(ii)} \quad \int dx\, g(x)\, x^\alpha = 0, \quad |\alpha| \le m. \end{aligned}$$

Then

$$|\mathcal{W}_g s(t_0 + b, a)| \le c_1(\rho(|b|) + \rho(a)),$$

with some c_1 depending only on g and c. If s satisfies in addition at

$$s(x_0 + x) = P_m(x) + o(\rho(|x|)), \quad (x \to 0),$$

then we have

$$\mathcal{W}_g s(x_0 + b, a) = o(\rho(|b|) + \rho(a)), \quad (b, a \to 0).$$

Proof. Since under the conditions stated in the theorem the wavelet does not see the polynomial part, we only need to estimate $\mathcal{W}_g r$, where $r(x) = \rho(|x|)$. We first consider the O-part.

By symmetry and monotony we have For $t,\ t' > 0$ either $2|t| \ge |t| + |t'|$ or $2|t'| \ge |t| + |t'|$. Therefore by monotony of r we have

$$r(t + t') \le r(2|t|) + r(2|t'|) \le c\,(r(t) + r(t')).$$

Therefore we may estimate

$$\begin{aligned} |\mathcal{W}_g r(b, a)| &= \left| \int dx\, \frac{1}{a^n}\, \overline{g}\left(\frac{x - b}{a}\right) r(x) \right| = \left| \int dx\, \overline{g}(x)\, r(ax + b) \right| \\ &\le \int dx\, |g(x)|\, \left(r(a\,|x|) + r(|b|)\right) \\ &\le O(1)\ \|r\, g\|_{L^1(\mathbb{R}^n)}\ r(a) + O(1)\ \|g\|_{L^1(\mathbb{R}^n)}\ r(|b|) \end{aligned}$$

and the first part of the theorem is proved.

The case o follows easily, since $o(\rho)$ stands by definition for a function such that for all $\epsilon > 0$ we have for t small enough

$$o(\rho(t)) \leq \epsilon\rho(t),$$

and the previous result may be applied. □

Note that we have proved the following uniform estimate

$$\exists c \ : \ \forall x \ \ \Omega[s,x](t) \leq ct^{\alpha} \Rightarrow |\mathcal{W}_g s(b,a)| \leq c'a^{\alpha},$$

with some c' depending only on c and g but not on b.

A.2 Inverse theorems for global regularity

Here we shall prove some inverse theorems concerning the proof of uniform global regularity through the wavelet coefficients. The general setting will be as follows. Suppose we have a position-scale representation of a distribution s (in the sense of distributions)

$$s(x) = \int_0^{\infty} \frac{da}{a} \int db\, r(b,a)\, \frac{1}{a^n}\, h\left(\frac{x-b}{a}\right),$$

with some polynomial bounded scale-position coefficients

$$|r(b,a)| \leq c\,(1+|b|)^K\,(a+1/a)^K.$$

We then may split s into two parts

$$\begin{aligned} s(x) &= \int_0^{\infty} \frac{da}{a} \int db\, r(b,a)\, \frac{1}{a^n}\, h\left(\frac{x-b}{a}\right) \\ &= \left\{\int_0^1 \frac{da}{a} + \int_1^{\infty} \frac{da}{a}\right\} \int db\, r(b,a)\, \frac{1}{a^n}\, h\left(\frac{x-b}{a}\right) \\ &= s_{\text{small}}(x) + s_{\text{large}}(x). \end{aligned}$$

Now the large-scale reconstruction s_{large} is a smooth function. Therefore the local behavior of s is only determined by the small scale behavior of r. In the following theorem we show, that uniform Hölder regularity and similar ones may be completely characterized by the decay of the modulus of the scale- space coefficients as the scale gets small.

Theorem A.2 *Let ρ be a non-negative, monoton sub-multiplicative function that satisfies for some $m \in \mathbb{N}_0$*

(i) $\displaystyle\int_0^1 \frac{dt}{t^{1+m}} \rho(t) < \infty$ *and* $\rho(t) = o(t^m)$ $(t \to 0)$

(ii) $\displaystyle\int_1^{\infty} \frac{dt}{t^{2+m}}\, \rho(t) < \infty$

Let r be some scale-space coefficients of some function s. Suppose $r(\cdot, a) = 0$, for $a > 1$. Then, if

$$|r(b,a)| \leq c\,\rho(a),$$

it follows that $\partial^\alpha s$ exists for $|\alpha| \leq m$ and that its derivative satisfies ($|\alpha| = m$)

$$\Omega[\partial^\alpha s, x](t) \leq c'\rho(t)/t^m.$$

with some constant c' not depending on x. If the space scale coefficients satisfy in addition uniformly in b

$$|r(b,a)| = o(\rho(a)) \quad (a \to 0),$$

we have uniformly in x for $|\alpha| = m$

$$\Omega[\partial^\alpha s, x](t) \leq o\big(\rho(t)/t^n.$$

In both cases the reconstruction wavelet h is supposed to be compactly supported having $m+1$ continuous derivatives.

First some comments. Condition (i) ensures a certain decay at small scales. Condition (ii) ensures that this decay is not too fast. So in some sense ρ has to stay away from polynomial behavior.

Indeed, because of the sub-multiplicativity of ρ we have $\rho(t) \geq c/\rho(1/t)$. Now by (ii) and monotony of ρ, we have $t^{1+n}\rho(t) \to 0$ as $t \to \infty$. Therefore, we have $t^{-n-1}\rho(t) \to \infty$ as $t \to 0$, or, what is the same

$$t^{1+n} = o\big(\rho(t)\big), \quad (t \to 0).$$

Thus in some sense condition (i) and (ii) ensure that the local fluctuation is "bounded away" from the polynomial behavior, which would be $\sim t^n$ and $\sim t^{n+1}$.

Proof. Thanks to the formula (verified by integration by parts)

$$\mathcal{M}[h, \partial^\alpha s](b,a) = (-1)^{|\alpha|} a^{-|\alpha|}\, \mathcal{M}[\partial^\alpha h, s](b,a),$$

we may limit ourselves to $m = 0$. By translation invariance it is enough to analyze s around 0. By an over-all re-scaling we may suppose that the support of h is contained in the ball of radius $1/2$ centered at the origin. By hypothesis on s we may write point-wise ($0 < |x| \leq 1$)

$$\begin{aligned} s(x) &= \left\{ \int_0^{|x|} \frac{da}{a} + \int_{|x|}^1 \frac{da}{a} \right\} \int db\, \frac{1}{a^n} h\left(\frac{x-b}{a}\right) r\,(b,a) \\ &= X_1(x) + X_2(x) \end{aligned}$$

We now prove the O part.

X_1. Using the decrease of the wavelet coefficients $r = O(\rho(a))$ at small scales we may write

$$\begin{aligned} |X_2| &\leq \int_0^{|x|} \frac{da}{a} \int db \left| \frac{1}{a^n} h\left(\frac{x-b}{a}\right) \right| |r(b,a)| \\ &= O(1)\ \|h\|_{L^1(\mathbb{R}^n)} \int_0^{|x|} \frac{da}{a} \rho(a) \\ &= O(\rho(|x|)). \end{aligned}$$

In the last equation we have used the fact, that because of the sub-multiplicativity of ρ we have

$$\int_0^t \frac{da}{a} \rho(a) = \int_0^1 \frac{da}{a} \rho(a\,t) \leq O(\rho(t)) \int_0^1 \frac{da}{a} \rho(a) = O(\rho(t)). \tag{A.1}$$

X_2. For later reference we estimate this term for general multi-index α. Here now we use the fact that h is $m+1$ times differentiable. We thus may write

$$h_{b,a}(x) - h_{b,a}(0) = \xi a^{-1} (\partial_{x/|x|} h)_{b,a}.$$

Here ξ is some intermediate point on the line between 0 and x and $\partial_{x/|x|} h$ is the derivative of h in direction x. In particular, we have

$$\int db\, |h_{b,a}(x) - h_{b,a}(0)| \leq c_2\, |x|\, /a,$$

with some $c_2 > 0$. Thus we may write

$$\begin{aligned} |X_2(x) - X_2(0)| &\leq \int_{|x|}^1 \frac{da}{a} \int db\ |h_{b,a}(x) - h_{b,a}(0)|\, |r(b,a)| \\ &\leq O(|x|) \int_{|x|}^1 \frac{da}{a} a^{-1} \rho(a) \\ &\leq O(\rho(|x|)). \end{aligned}$$

In the last equation we have again used the sub-multiplicativity of ρ via

$$\begin{aligned} t \int_t^1 \frac{da}{a^2} \rho(a) &= \int_1^{1/t} \frac{da}{a^2} \rho(a\,t) \\ &\leq O(\rho(t)) \int_1^{1/t} \frac{da}{a^2} \rho(a) \\ &= O(\rho(t)), \end{aligned} \tag{A.2}$$

since by hypothesis on ρ the integral is convergent.

To prove the o part of the lemma, we remark that it is enough to replace the estimates by their o version. Suppose therefore that u is a non-negative function that satisfies

$$u(t) = o\big(\rho(t)\big) \quad (t \to 0).$$

Then, for every $\epsilon > 0$, we can find an η, $0 < \eta < 1$, such that $|u(t)| < \epsilon\rho(t)$ whenever $0 \le t \le \eta$. Therefore, for t small enough

$$\int_0^t \frac{da}{a}\, u(a) \le \epsilon \int_0^t \frac{da}{a}\, \rho(a) \le \epsilon\, \rho(t),$$

showing because ϵ was arbitrarily chosen the o analog of (A.1).

For the analog of (A.2) we obtain for fixed ϵ and t small enough with the same η as previously

$$t \int_t^1 \frac{da}{a^2}\, u(a) = t \left\{ \int_t^\eta \int_\eta^1 \right\} \frac{da}{a^2}\, u(a) = \sigma_1(t) + \sigma_2(t).$$

Now by sub-multiplicativity of ρ we have as before

$$\sigma_2(t) \le O(t) \int_\eta^1 \frac{da}{a^2}\, u(a) \le O(\rho(t) \int_{\eta/t}^{1/t} \frac{da}{a^2}\, \rho(a).$$

By condition (ii) the integral tends to 0 as $t \to \infty$ and thus $\sigma_2(t) = o\big(\rho(t)\big)$.

By hypothesis (ii) on ρ we have for t small enough as in

$$\sigma_1(t) \le t \int_t^\eta \frac{da}{a^2}\, u(a) \le \epsilon\, t \int_t^1 \frac{da}{a^2}\, \rho(a) \le \epsilon\, O\big(\rho(t)\big).$$

Since ϵ was arbitrary the theorem follows. □

A.3 The class of Zygmund

For the sake of simplicity we consider only the one-dimensional case. Until now we had to exclude the case $\Lambda^\alpha(\mathbb{R}^n)$ with $\alpha \in \mathbb{N}$. Indeed, the hypothesis of Theorem A.2 does not apply. This no accident. Consider the "discrete" Laplacian in n dimensions

$$\Delta_\epsilon : s(x) \mapsto 2ns(x) - \sum_{k=1}^n s(x_1, \ldots, x_k + \epsilon, \ldots, x_n) + s(x_1, \ldots, x_k - \epsilon, \ldots, x_n).$$

For $\xi \in SO(n)$, let U_ξ be the rotation acting on functions in $\mathbb{R}^n$.

Definition A.3 *A function s is in the class of Zygmund, written $\Lambda^*(\mathbb{R}^n)$, if it is continuous and if it satisfies*

$$|\Delta_\epsilon U_\xi s(x)| \le O(\epsilon), \quad (\epsilon \to 0),$$

uniformly in $x \in \mathbb{R}^n$ and $\xi \in SO(n)$. We say s is in $\lambda^*(\mathbb{R}^n)$ if the same estimate holds with O replaced by o.

The class of Zygmund can be characterized in wavelet space.

Theorem A.4 *Let s be a bounded function that is locally in the class of Zygmund $\Lambda^*(\mathbb{R}^n)$. Then for all $g \in \mathcal{S}_0(\mathbb{R}^n)$ we have*

$$\mathcal{W}_g s(b,a) = O(a) \qquad (a \to 0).$$

Vice versa, if s can be written as a scale-space integral with compactly supported, twice differentiable reconstruction wavelet and coefficients r satisfying this estimate, then s is in the clcass of Zygmund. The analogue holds for $\lambda^*(\mathbb{R}^n)$.

Proof. First suppose that g is radial and admissible

$$g(x) = g(|x|).$$

Then we have by symmetry and $\int dx g(x) = 0$ as usual

$$\begin{aligned} \mathcal{W}_g s(b,a) &= \int dx \, \frac{1}{a^n} \bar{g}\left(\frac{x}{a}\right) (U_\xi s)(\pm x + b) \\ &= \frac{1}{2n} \int dx \, \frac{1}{a^n} \bar{g}\left(\frac{x}{a}\right) \Delta_{|x|} U_{\xi(x)} s(b) \end{aligned}$$

Here $\xi(x)$ is some rotation of x into the first coordinat direction. Using the estimate from the definition and the fact that s is bounded we may estimate

$$\left|\Delta_{|x|} U_{\xi(x)s(b)}\right| \le O(|x|),$$

and the first part of the theorem follows for radial g.

Suppose that h is not symmetric. The wavelet transform with respect to h is obtained from the one with respect to g by an the action of the cross kernel. This function is localized enough and thus it does not change the decrees of the wavelet coefficients at small scale.

We now prove the converse. Again we have supposed that the reconstruction wavelet h has compact support. By a cross-kernel argument we may assume that h is radial. By translation and rotation invariance it is enough to estimate

$$\Delta_\epsilon s(0).$$

To each term of $\Delta_\epsilon s(0)$ corresponds an integral over some influence cone of h. The integrals over the region $0 < a \le \epsilon$ may be estimated as previously. In the only new term, X_3 say, we may write

$$\Delta_\epsilon h_{b,a}(x) = (\Delta_{\epsilon/a} h)_{b,a}(x).$$

Therefore, using the regularity of h we have

$$\int dx \, |\Delta_\epsilon h_{b,a}(x)| \le c\,\epsilon^2/a^2.$$

We therefore end up with

$$X_3 = \int_\epsilon^1 \frac{da}{a} \int db\, r(b,a) \Delta_\epsilon h_{b,a}(x),$$

and thus

$$|X_3| \le c\epsilon^2 \int_\epsilon^1 \frac{da}{a} a^{-1} \le O(\epsilon),$$

as was to be shown.

The proof of the o part is similar. □

By what we have shown so far this implies via the characterization of the Hölder spaces through wavelet transforms the following for the regularity of the Zygmund class. We have

$$\Lambda^1(\mathbb{R}^n) \Rightarrow \Lambda^*(\mathbb{R}^n) \Rightarrow \Lambda^\alpha(\mathbb{R}^n),$$

for $0 < \alpha < 1$. But as follows easily from the proof of theorem we even have a little more precise information: a function in $\Lambda^*(\mathbb{R}^n)$ satisfies

$$|s(x_0 + t) - s(x_0)| = O(|x| \log |x|) \quad (x \to 0).$$

Indeed, the logarithmic correction comes from the term X_2 as can be easily seen.

A.4 Inverse theorems for point-wise regularity

In this section we are concerned with the point-wise inverse of Theorem A.2. As we have seen, a local Hölder regularity implies a local decrees of the wavelet coefficients. On the other hand as we just have proved an overall decrease of the scale-space coefficients proves a global regularity of the reconstructed function. However, to prove local regularity through scale-space coefficients, we must suppose some stronger conditions.

Theorem A.5 *Let s have a scale-space-representation r that satisfies small scale $a < 1$ for some for some $\gamma > 0$*

$$\begin{array}{ll} \text{(i)} & r(b,a) = O\left(a^\gamma\right) \qquad \textit{uniformly in } b, \\ \text{(ii)} & r\left(x_0 + b, a\right) = O(a^\alpha) + O(b^\alpha \log^{-1} b) \qquad (b, a \to 0) \end{array}$$

At large scale we suppose that r is rapidly decreasing. Then s has at x_0 a local regularity exponent α in the sense that

$$|s(x_0 + x) - P_m(x)| = O\left(|x|^\alpha\right) \qquad (x \to 0),$$

with some polynomial of order n, $\alpha - 1 < n < \alpha$.

Again this theorem holds for sub-multiplicative remainders in general.

Theorem A.6 *Let ρ be a non-negative, sub-multiplicative function that satisfies for some $n \in \mathbb{N}_0$*

$$\text{(i)} \quad \int_0^1 \frac{dt}{t^{n+1}} \rho(t) < \infty, \ \textit{and} \ \rho(t) = o(t^m) \ (t \to 0)$$

$$\text{(ii)} \quad \int_1^\infty \frac{dt}{t^{n+2}} \rho(t) < \infty.$$

Let s have scale-space coefficients r supported by $a \leq 1$ that satisfy with some $\gamma > 0$

$$\begin{array}{lll} \text{(iii)} & r(b,a) = O\left(a^\gamma\right) & \textit{uniformly in } b, \\ \text{(iv)} & r(x_0 + b, a) = O\big(\rho(a)\big) + O\big(\rho(b)/\log\rho(b)\big) & (b, a \to 0). \end{array}$$

Then there is a polynomial P_m of degree m such that

$$s(x_0 + x) = P_m(x) + O\big(\rho(|x|)\big), \qquad (x \to 0).$$

If we have in addition

$$\text{(iv}'\text{)} \ r\left(x_0 + b, a\right) = o\big(\rho(a)\big) + o\big(\rho(b)/\log\rho(b)\big) \quad (b, a \to 0),$$

then

$$s(x_0 + x) = P_m(x) + o\big(\rho(|x|)\big), \qquad (x \to 0).$$

The reconstruction wavelet should be $n+1$ times continuously differentiable and compactly supported.

Proof. We may suppose $\gamma < 1$. Then with the help of the function $\eta(t) = \rho(t)^{2/\gamma}$ we may split the integral into several parts.

$$\begin{aligned} s(x) &= \int_0^{\eta(|x|)} \frac{da}{a} \int db \, \frac{1}{a^n} h\left(\frac{x-b}{a}\right) r(b,a) \\ &+ \int_{\eta(|x|)}^{|x|} \frac{da}{a} \int db \, \frac{1}{a^n} h\left(\frac{x-b}{a}\right) r(b,a) \\ &+ \int_{|x|}^1 \frac{da}{a} \int db \, \frac{1}{a^n} h\left(\frac{x-b}{a}\right) r(b,a) = X_1 + X_2 + X_3 \end{aligned}$$

The contributions of these terms will be estimated independently.

X_1. Using the global Hölder regularity of s we may estimate $r = O\left(a^\gamma\right)$ which leads us to

$$\begin{aligned} |X_1| &\leq \int_0^{\eta(|x|)} \frac{da}{a} \int db \left| \frac{1}{a^n} h\left(\frac{x-b}{a}\right) \right| |r(b,a)| \\ &= O(1) \, \|h\|_{L^1(\mathbb{R}^n)} \int_0^{\eta(|x|)} \frac{da}{a} a^\gamma \end{aligned}$$

By the choice of η we obtain $X_1 = O\big(\rho(|x|)\big)$.

X_2. Since h is compactly supported we have under the integral $a \le O(|x|)$ and $b \le O(|x|)$. Therefore we may estimate

$$|r(b,a)| = O\big(\rho(a)\big) + O\big(\rho(t)\,\log^{-1}\rho(t)\big).$$

The contribution of the fist term may be estimated as the term X_1 in the proof of theorem to be of order $O\left(\rho(t)\right)$, and thus we end up with

$$\begin{aligned}
|X_2| &\le \int_{\eta(|x|)}^{t} \frac{da}{a} \int db \left| \frac{1}{a^n} h\left(\frac{x-b}{a}\right)\right| |r\,(b,a)| \\
&= O(1)\,\frac{\rho(|x|)}{|\log\rho(|x|)|} \int_{\eta(|x|)}^{|x|} \frac{da}{a} \int db \left| \frac{1}{a^n} h\left(\frac{x-b}{a}\right)\right| + O\left(\rho(|x|)\right) \\
&= \|h\|_{L^1(\mathbb{R}^n)}\, O(1)\, \frac{\rho(|x|)}{|\log\rho(|x|)|} \log\left(\frac{|x|}{\eta(|x|)}\right) + O\left(\rho(|x|)\right) \\
&= O\left(\rho(|x|)\right)
\end{aligned}$$

by the choice of η and, since by hypothesis (ii) on ρ we have $\log(|x|) = O(\log(\rho(t))$.

X_3. This term will produce the local polynomial. Using Taylor's theorem we see that we can write

$$h(x-b) = \sum_{|\alpha|\le m} c_\alpha(b)x^\alpha + w(b,x),$$

with some regular (bounded) coefficients c_α. The remainder satisfies the following estimate valid for $|b| + |x| \le 2$

$$|w(b,x)| \le O(|x|^{m+1}).$$

Now under the integral we always have $|x| \le 2a$ and $|b| \le 2a$. Thus we may write under the integral

$$\begin{aligned}
h(x/a - b/a) &= \sum_{|\alpha|\le m} c_\alpha(b/a)x^\alpha a^{-|\alpha|} + w(b/a, x/a) \\
&= \sum_{|\alpha|\le m} c_\alpha(b/a)x^\alpha a^{-|\alpha|} + O(|x|^{m+1}/a^{m+1}).
\end{aligned}$$

The integral over the first term yields a polynomial in x, since all integrals converge absolutely thanks to the decay of the wavelet coefficients. The remaining term reads (after integrating over b)

$$|x|^{n+1} \int_{|x|}^{1} \frac{da}{a} a^{-m-1} \rho(a).$$

This term may be treated in the same way as before, and the "O"-part of the theorem is done.

The "o" part follows by considerations as before. □

Corollary A.7 *If r satisfies with $\alpha \in (0,1)$*

$$\begin{aligned} |r(b,a)| &\leq O(a^{\gamma}) \quad \text{with some } \gamma > 0 \\ r(x_0 + b, a) &\leq O\big(b^{\alpha} + a^{\alpha}\big) \qquad (b, a \to 0), \end{aligned}$$

then $|s(x_0 + x) - s(x_0)| = O\big(\,|x|^{\alpha} \log(|x|)\big)$, *as* $x \to 0$.

Proof. The logarithm is exactly the contribution of the second term in the previous demonstration. □

References

[AAGM95] S. T. Ali, J. P. Antoine, J. P. Gazeau, and U. A. Mueller. Coherent states and their gneeralizations: A mathematical overview. *Rev. Math. Phys.*, 7(7):1013–1104, 1995.

[AGH88] A. Arneodo, G. Grasseau, and M. Holschneider. On the wavelet transform of multifractals. *Phys. Rev. Lett.*, 61:2281–2284, 1988.

[BJLI95] V. Baladi, Y. Jiang, and O. E. Lanford III. Transfer Operators Acting on Zygmund Functions. Technical report, ETH Zürich, March 1995.

[CGT89] J. M. Combes, A. Grossmann, and Ph. Tchamitchiam. *Wavelets, Time-Frequency Methods and Phase Space.* Springer-Verlag, Berlin, Heidelberg, New York 1989.

[Chu92] C. K. Chui. *An Introduction to Wavelets.* Academic Press, New York1992.

[CK95] J. Cooper and H. Koch. The spectrum of a hyperbolic evolution operator. *J. Funct. Anal.*, 133(2):301–328, 1995.

[CM78] R. Coifman and Y. Meyer. Au-delà des opérateurs pseudo-différentiels. *Astérisque*, 57, 1978.

[CM90] R. Coifman and Y. Meyer. *Ondelettes et Opèrateurs III.* Hermann, Paris 1990.

[Con95] A. Connes. Geometry from the Spectral Point of View. *Letters in Mathematical Physics*, 34:203–238, 1995.

[CZ57] A. P. Calderón and A. Zygmund. Singular integral operators and differential equations. *Amer. J. Math.*, 79:901–921, 1957.

[Dau92] I. Daubechies. *Ten Lectures on Wavelets.* SIAM, Philadelphia1992.

[DJ84] G. David and J. L. Journé. A boundedness criterion for generalized Calderón-Zygmund operators. *Ann. of Math.*, 120:371–397, 1984.

[DPS95] W. Dahmen, S. Prössdorf, and R. Schneider. Wavelet approximation methods for pseudodifferential operators. *Math. Z.*, 215:583-620, 1994

[FG89a] H. G. Feichtinger and K. Gröchenig. Banach spaces related to integrable group representations and their atomic decompositions I. *J. Funct. Anal.*, 86:107–115, 1989.

[FG89b] H. G. Feichtinger and K. Gröchenig. Banach spaces related to integrable group representations and their atomic decompositions II. *Mh. Math.*, 108:129–148, 1989.

[Fla93] P. Flandrin. *Temps-Fréquence.* Hermès, Paris 1993.

[Fri95] D. Fried. Essential spectrum and Banach algebras. Technical report, Boston University, 1995.

[GM84] A. Grossmann and J. Morlet. Decomposition of Hardy functions into square integrable wavelets of constant shape. *SIAM J. Math. Anal.*, 15:723–736, 1984.

[Hol93a] M. Holschneider. Inverse radon transfoms through inverse wavelet transforms. *Inverse Problems*, 7:853–861, 1993.

[Hol93b] M. Holschneider. Localization properties of wavelet transforms. *Journal of Mathematical Physics*, 34(7):3227–3244, July 1993.

[Hol94] M. Holschneider. Functional calculus using wavelet transforms. *Journal of Mathematical Physics*, 35(7):3745–3752, July 1994.

[Hol95] M. Holschneider. *Wavelets: An Analysis Tool.* Oxford University Press, Oxford 1995.

[Hör82] L. Hörmander. *The Analysis of Linear Partial Differential Operators I.* Springer Verlag, Berlin–Heidelberg–New York–Paris–London–Tokyo–Hong Kong 1982.

[HT91] M. Holschneider and Ph. Tchamitchian. Pointwise regularity of riemanns "nowhere differentiable" function. *Inventiones Mathematicae*, 105:157–175, 1991.

[Jaf89] S. Jaffard. Estimations hölderiennes ponctuelles des fonctions au moyen de leurs coefficients d'ondelettes. *C.R. Acad. Sci., Paris, Sér. I*, 308:79–81, 1989.

[Jaf92] S. Jaffard. Wavelet methods for fast resolution of elliptic problems. *SIAM J. Numer. Anal.*, 29:965–987, 1992.

[JK82] D. Jerison and C. Kenig. Boundary behavior of harmonic functions in nontangentially accessible domains. *Advances in Math*, 146:80–147, 1982.

[Kel84] G. Keller. On the rate of convergence to equilibrium in one-dimensional systems. *Commun. Math. Phys.*, 96:181–193, 1984.

[KMG87] R. Kronland, J. Morlet, and A. Grossmann. Analysis of sound patterns through wavelet transforms. *International Journal of Pattern Recognition and Artificial Intelligence*, 1(2):97–126, 1987.

[Mey90] Y. Meyer. *Ondelettes et Opèrateurs I&II.* Hermann, Paris 1990.

[Mey95] Y. Meyer. Chirps n-dimensionels et analyse 2-microlocale. *Séminair équations aux derivées partielles, Ecole Polytechnique*, 1994/95.

[Nus70] R. D. Nussbaum. The radius of the essential spectrum. *Duke Math. J.*, 37:473–478, 1970.

[Pau84] Th. Paul. Functions analytic on the half-plane as quantum mechanical states. *J. Math. Phys.*, 25:3252–3263, 1984.

[Rue89] D. Ruelle. The thermodynamic formalism for expanding maps. *Comm. Math. Phys.*, 125:239–262, 1989.

[Sar94] G. Saracco. Propagation of transient waves through a stratified fluid medium: Wavelet analysis of a nonasymptotic decomposition of the propagator. *J. Acoust. Soc. Am.*, 95(3):1191–1205, 1994.

[Ste70] E. Stein. *Singular integrals and differentiability properties of functions.* Princeton University Press, Princeton (N.J.) 1970.

[Tch95] Ph. Tchamitchian. Wavelets and Differential Operators. In: I. Daubechies (editor), *Different Perspectives on Wavelets*, volume 47 of *Proceedings of Symposia in Applied Mathematics*, pages 77–88. A.M.S., Providence (R.I.) 1995.

[Tor95] B. Torrésani. *Analyse continue par ondelettes.* CNRS Editions, 1995.

[Tre67] F Treves. *Topological Vector Spaces, Distributions, and Kernels.* Academic Press, San Diego–New York–London 1967.

[Tri84] H. Triebel. *Theory of Function Spaces II.* Birkhäuser Verlag, Basel 1984.

On the Index of Elliptic Operators on a Cone

B. V. Fedosov[1] and B.-W. Schulze

Introduction

We consider a special case of Mellin pseudodifferential operators on a cone $X^\wedge = X \times \overline{\mathbb{R}}_+ / X \times 0$ where the base X is a smooth compact manifold of dimension n without boundary. The operators have the form

$$(Au)(t) = \frac{1}{2\pi i} \int_\Gamma dz \int_0^\infty \left(\frac{t_1}{t}\right)^z a(t,z) u(t_1) \frac{dt_1}{t_1}. \tag{0.1}$$

Here

$$u(t) \in C_0^\infty(\mathbb{R}_+, C^\infty(X,E)) \tag{0.2}$$

which means that $u(t)$ is a function with compact support on $\mathbb{R}_+$ whose values are sections of a vector bundle E over X. The weight line Γ may be any vertical line $\Gamma_\beta = \{\Re z = \beta\}$ in a complex plane, we assume without loss of generality that Γ coincides with the imaginary axis Γ_0.

The operator-valued Mellin symbol $a(t,z)$ satisfies the following conditions:

1.

$$a(t,z) \in C^\infty(\overline{\mathbb{R}}_+, L^\mu_{cl}(X,\Gamma)), \tag{0.3}$$

which means that a is a smooth function in $t \in \overline{\mathbb{R}}_+$ whose values are parameter-dependent pseudodifferential operators of order μ on X with a parameter $z = i\tau,\ \tau \in \mathbb{R}$,

[1]Supported by the Deutsche Forschungsgemeinschaft

2. for $t \in [C, \infty)$ the symbol is independent of t, more precisely

$$a(t, z) = a(\infty, z) \quad t \in [C, \infty),$$

3. for $t \in [0, c]$ with $0 < c < C$ the symbol $a(t, z)$ admits an analytical continuation to some strip $S = \{|\Re z| < \varepsilon\}$ and on each line $\Gamma_\beta = \{\Re z = \beta\}$ is a parameter-dependent operator on X of oder μ, that is satisfies (0.3) uniformly in $|\beta| \leq \varepsilon_1 < \varepsilon$.

We use a notation $ML^\mu(X^\wedge)$ for the set of such operators and $ML_0^\mu(X^\wedge)$ if $a(\infty, z) = 0$. Troughout this paper we assume $\mu \leq 0$. The operators (0.1) are of great importance for the calculus of pseudodifferential operators on manifolds with conical singularities (see e.g. [1]). Here we restrict ourselves to a model case when a singular manifold is a pure cone and the operator may be written globally in a Mellin form. It is well known that $A \in ML^\mu(X^\wedge)$ is a bounded operator in weighted Sobolev spaces $A : H^{s,(n+1)/2} \to H^{s-\mu,(n+1)/2}$.

We will consider elliptic operators of zero order ($\mu = 0$). The operator $A \in ML^0(X^\wedge)$ is called elliptic if its symbol satisfies the following additional conditions:

1. for $t \in \mathbb{R}_+$ $\partial(\approx, F)$ is a parameter-dependent elliptic operator on X with a parameter $z \in \Gamma$,

2. $a(t, z)$ is invertible for for $t \in [0, c]$ and any z in the strip $|\Re z| < \varepsilon$,

3. $a(\infty, z) = 1$ where 1 stands for identity operator.

We prove (section 3) that ellipticity implies Fredholm property in Sobolev spaces and obtain an index formula. A basic observation is that ellipticity conditions imply that the elliptic family $a(t, z)$ paramatrized by $t \in \mathbb{R}_+, F \in \Gamma$ is trivial (that is $a(t, z)$ is invertible) outside a compact set in $\mathbb{R}_+ \times \Gamma$. Thus it defines an index bundle $\operatorname{ind} a \in K_c(\mathbb{R}_+ \times \Gamma)$ where K_c means K-functor with compact supports (see [4]). Its Chern character is represented by a closed differential form with compact support, and we prove the following result

$$\operatorname{ind} A = \int_0^\infty \int_\Gamma \operatorname{ch}(\operatorname{ind} a). \tag{0.4}$$

Another useful form of (0.4) may be obtained in terms of the family $a(t, z)$ and its parametrix $r_0(t, z)$ such that $1 - r_0 a$ and $1 - a r_0$ are trace class operators for any $(t, z) \in \mathbb{R}_+ \times \Gamma$ and $r_0 = a^{-1}$ outside a compact. Then (see e.g. [5]) using the formula for $\operatorname{ch}(\operatorname{ind} a)$ in terms of a and r_0 we may write

$$\operatorname{ind} A = \frac{1}{2\pi i} \int_{\mathbb{R}_+ \times \Gamma} \operatorname{tr}(dr_0 + r_0 da r_0) \wedge da \tag{0.5}$$

and precisely in this form we prove our result.

The proof follows the scheme developed in [6]. It consists in comparing three expressions:

1. analytical index

$$\operatorname{ind} A = \operatorname{tr}(1 - RA) - \operatorname{tr}(1 - AR) \tag{0.6}$$

where R is a parametrix of A up to a trace class operator,

2. algebraical index

$$\operatorname{ind} A = \operatorname{tr}(1 - r \circ a) - \operatorname{tr}(1 - a \circ r) \tag{0.7}$$

where r is a formal complete symbol of R and $\circ$ means a composition of formal complete symbols,

3. topological index given by (0.6).

The most important step is transition from 1 to 2 or, using the terminology of [6], the theorem on a regularized trace of a product. In contrast to [6] we have not only to watch the order but also to gain weight, which is much more difficult. The second transition from 2 to 3 is based on the machinery developed in [5] and requires no new ideas.

Acknowledgments. We are very grateful to Professor Schrohe for numerous clarifying discussions especially on Fredholm property and parametrix construction. The first author would like to thank Professor R. Nest for an invitation to Copenhagen University where the article was completed.

1 Preliminary estimates

We will use the following notations. For a function $a = a(t, z)$ we set

$$a^{(k)} = \frac{\partial^k}{\partial z^k} a(t, z),$$

$$a_{(k)} = D^k a(t, z) = \left(-t\frac{\partial}{\partial t}\right)^k a(t, z).$$

For two functions $a(t, z)$ and $b(t, z)$ set

$$a \circ b|_N = \sum_{k=0}^{N-1} \frac{1}{k!} a^{(k)}(t, z) b_{(k)}(t, z). \tag{1.1}$$

For a function $u(t),\ t \in [0, \infty)$ with values in $C^\infty(X, E)$ we denote by $|u(t)|$ any norm in $C^\infty(X, E)$ at a given t. A similar notation $|a(t, z)|$ for operator-valued functions

$$a(t, z) : C^\infty(X, E) \to C^\infty(X, F)$$

means any operator norm.

Introduce

$$\langle z\rangle^s = (1-z^2)^{s/2} = \exp\left(\frac{s}{2}\ln(1-z^2)\right)$$

for $|\Re z| < \varepsilon < 1$, $s \in \mathbb{C}$, assuming that the branch of $\ln(1-z^2)$ in the strip $|\Re z| < \varepsilon$ is real at $z \in \Gamma_0$. This function is holomorphic in z belonging to the strip $|\Re z| < \varepsilon$ and in s for all $s \in \mathbb{C}$. If s belongs to a horizontal strip $|\Im s| < \delta$, then the inequality holds

$$|\langle z\rangle^s| \leq C(1+|z|^2)^{\Re s/2} \tag{1.2}$$

We also need an operator-valued version of the above order reducing function. Let Δ be the Laplace operator on X. Define an order reducing family

$$R^s(z) = (1-\Delta-z^2)^{s/2} \tag{1.3}$$

for $|\Re z| < \varepsilon < 1$ and $|\Im s| < \delta$. A complex power is understood in the sense of elliptic theory (see [7]). This family is holomorphic in z and s belonging to the mentioned strips.

For $u(t) \in C_0^\infty(\mathbb{R}_+)$ its Mellin transform

$$\widehat{u}(z) = \int_0^\infty t^{z-1}u(t)dt$$

is an entire function in z. In virtue of the identity

$$z^k\widehat{u}(z) = \widehat{u}_{(k)}(z) = \int_0^\infty t^{z-1}D^k u(t)dt \tag{1.4}$$

we see that $\widehat{u}(z)$ is rapidly decreasing on any vertical line, that is

$$|\widehat{u}(z)| \leq C|\,\langle z\rangle^{-k}\,| \tag{1.5}$$

for any $k \in \mathbb{N}$, $|\Re z| < \varepsilon$.

Finally we will consider Mellin transforms of operator-valued symbols $a(t,z) \in ML_0(X^\wedge)$, that is

$$\widehat{a}(\zeta,z) = \int_0^\infty t^{\zeta-1}a(t,z)dt.$$

The integral converges at $\Re\zeta > 0$ so that $\widehat{a}(\zeta,t)$ is a holomorphic function in a half-plane $\Re\zeta > 0$. Similarly to (1.4) we have

$$\widehat{a}(\zeta,t) = \frac{1}{\zeta}\widehat{a}_{(1)}(\zeta,t),$$

where

$$\widehat{a}_{(1)} = \int_0^\infty t^{\zeta-1}Da(t,z)dt$$

is holomorphic in a half-plane $\Re\zeta > -1$ since $Da(t,z)$ vanishes at $t = 0$. Repeating this procedure we may write similarly to (1.4)

$$\widehat{a}(\zeta, z) = \frac{1}{\zeta^k}\widehat{a}_{(k)}(\zeta, z), \tag{1.6}$$

where $\widehat{a}_{(k)}(\zeta, z)$ is holomorphic in $\Re\zeta > -1$. From (1.6) it follows that $\widehat{a}(\zeta, z)$ is rapidly decreasing in ζ on any vertical line Γ_β, so that

$$|\widehat{a}(\zeta, z)| \leq C(z)|\langle\zeta\rangle|^{-k} \tag{1.7}$$

for any $k \in \mathbb{N}$, $\zeta \in \Gamma_\beta$.

For symbols $a(t,z) \in ML_0^\mu(X^\wedge)$ with $\mu \leq 0$ more precise estimates hold

$$||\widehat{a}(\zeta, z)|| \leq C|\langle\zeta\rangle|^{-k}|\langle z\rangle|^\mu. \tag{1.8}$$

Here $||\cdot||$ means the norm of the operator

$$\widehat{a}(\zeta, t) : L^2(X, E) \to L^2(X, E). \tag{1.9}$$

If $\mu < -n/2$ the operator (1.9) belongs to the Hilbert-Schmidt class and its Hilbert-Schmidt norm $||\cdot||_2$ satisfies an equality

$$||\widehat{a}(\zeta, z)||_2 \leq C|\langle\zeta\rangle|^{-k}|\langle z\rangle|^{\mu+n/2}. \tag{1.10}$$

Finally, if $\mu < -n$ then the operator (1.9) belongs to the trace class and the trace norm $||\cdot||_1$ satisfies an estimate

$$||\widehat{a}(\zeta, z)||_1 \leq C|\langle\zeta\rangle|^{-k}|\langle z\rangle|^{\mu+n}. \tag{1.11}$$

In (1.8)–(1.11) $\zeta \in \Gamma_{\beta_1}$, $\beta_1 \neq 0$, $z \in \Gamma_{\beta_2}$, $|\beta_2| < \varepsilon$ and k may be any integer positive number.

The operator (0.1) corresponding to the operator-valued Mellin symbol $a(t,z)$ will be denoted by $\mathrm{Op}(a)$. With these notations the main theorem of this section is as follows.

Theorem 1.1 *Let $a(t,z) \in ML_0^\mu(X^\wedge)$, $b(t,z) \in ML_0^\nu(X^\wedge)$ with $\mu, \nu \leq 0$. Then for N sufficiently large the operator*

$$C_N = \mathrm{Op}(a)\,\mathrm{Op}(b) - \mathrm{Op}(a \circ b|_N)$$

as an operator in the space $H^{0,(n+1)/2}$ belongs to the trace class.

Proof. Introduce a partition of unity $\rho_0(t), \rho_1(t), \rho_2(t)$ such that

$$\operatorname{supp}\rho_0 \in [0, c],\ \operatorname{supp}\rho_1 \in [c/2, c],\ \operatorname{supp}\rho_2 \in [c/2, \infty),$$
$$\operatorname{supp}\rho_0 \cap \operatorname{supp}\rho_2 = \emptyset$$

and

$$\rho_0(t) \equiv 1,\ t \in [0, c/2],\ \rho_2(t) \equiv 1,\ t \in [c, \infty).$$

Then any operator $\mathrm{Op}(a)$ may be represented as a sum

$$\mathrm{Op}(a) = \sum_{k=0}^{2} \mathrm{Op}(\rho_k a).$$

We emphasize that the symbols $\rho_0(t)a(t,z)$ and $\rho_1(t)a(t,s)$ are holomorphic in z belonging to the strip S.

The operator C_N is then a sum

$$\sum_{i,j=0}^{2} \mathrm{Op}(\rho_i a)\mathrm{Op}(\rho_j b) - \mathrm{Op}((\rho_i a) \circ (\rho_j b)|_N) \tag{1.12}$$

and we consider several cases according to the values of i, j.

Case 1 ($i, j \neq 0$).

In this case the supports of symbols $\rho_i a$ and $\rho_j b$ are separated from $t = 0$. The Mellin calculus for such operators on a half– line $t \geq 0$ may be reduced to the usual Fourier calculus of p.d.o. on the whole line $y \in (-\infty, \infty)$ by change of variables $t = e^y,\ z = i\tau$. Indeed, formula (0.1) goes to

$$(Au)(e^y) = \frac{1}{2\pi} \int_{-\infty}^{\infty} d\tau \int_{-\infty}^{\infty} e^{i\tau(y_1 - y)} a(e^y, i\tau) u(e^{y_1}) dy_1,$$

and the $\circ$-product of two symbols becomes

$$a(t,z) \circ b(t,z)|_N = \sum_{k=0}^{N-1} \frac{i^k}{k!} \frac{\partial^k}{\partial \tau^k} a(e^y, i\tau) \frac{\partial^k}{\partial y^k} b(e^y, i\tau)$$

which is the usual composition rule for Fourier p.d.o. The symbols

$$\rho_i(e^y)a(e^y, i\tau),\ \rho_j(e^y)b(e^y, i\tau)$$

have compact support in y, so the theorem follows from the usual Fourier calculus of p.d.o. (see e.g. [6]).

Case 2 ($i = 2, j = 0$).

In this case $(\rho_2 a) \circ (\rho_0 b)|_N = 0$ since the supports of ρ_0 and ρ_2 do not intersect, so we need to prove that the operator

$$\mathrm{Op}(\rho_2 a)\mathrm{Op}(\rho_0 b) = \mathrm{Op}(\rho_2 a)\rho_0 \mathrm{Op}(b)$$

belongs to trace class (the so-colled pseudolocality property). The operator $\mathrm{Op}(b)$ is bounded in $H^{0,(n+1)/2}$ since its order $\nu \leq 0$, and thus it is sufficient to prove that $\mathrm{Op}(\rho_2 a)\rho_0$ belongs to trace class. To this end we represent it as a composition

$$H^{0,(n+1)/2} \xrightarrow{C} H^{s,\gamma} \hookrightarrow H^{0,(n+1)/2} \tag{1.13}$$

with some integer $s > (n+1)/2$ and some $\gamma > (n+1)/2$ and show that both operators in (1.13) are Hilbert-Schmidt ones. The second operator is an embedding.

Lemma 1.2 *For $s > (n+1)/2,\ \delta > 0$ the embedding*

$$H^{s,(n+1)/2+\delta} \hookrightarrow H^{0,(n+1)/2}$$

considered on a subspace of functions $u \in H^{s,(n+1)/2+\delta}$ whose supports belong to the interval $t \in [0,T]$ is a Hilbert-Schmidt operator.

Proof. Let first $u(t) \in C_0^\infty(0,T)$. Then $\widehat{u}(z)T^{-z}$ is an entire function rapidly decreasing in the right half-plane. The norm of $u(t)$ in the Sobolev space $H^{s,(n+1)/2+\delta}$ is equal to an L^2-norm of $R^s(z)\widehat{u}(z)$ on the line $\Gamma_{-\delta}$. Here $R^s(z)$ is an order-reducing family (1.3). The norm of $u(t)$ in $H^{0,(n+1)/2}$ is an L_2-norm of the restriction of $\widehat{u}(z)$ to the line Γ_0. The restrictions of $\widehat{u}(z)$ to $\Gamma_{-\delta}$ and Γ_0 are connected by the Cauchy integral

$$\widehat{u}(\zeta) = \frac{1}{2\pi i}\int_{\Gamma_{-\delta}} \frac{T^{\zeta - z}}{z-\zeta}\widehat{u}(z)dz \tag{1.14}$$

where $\zeta \in \Gamma_0$. Denoting $\widehat{v}(z) = R^s(z)\widehat{u}(z)$ we write (1.14) in the form

$$\widehat{u}(\zeta) = \frac{1}{2\pi i}\int_{\Gamma_{-\delta}} \frac{T^{\zeta - z}}{z-\zeta}R^{-s}(z)\widehat{v}(z)dz.$$

This operator acts between L^2 spaces on lines $\Gamma_{-\delta}$ and Γ_0 and its Hilbert-Schmidt norm is equal to the L^2-norm of its Schwarz kernel

$$K(\zeta,z) = \frac{T^{\zeta - z}}{z-\zeta}R^{-s}(z)$$

which is an operator-valued function. For $s > n/2$ we have an estimate for the Hilbert-Schmidt norm in $L^2(X)$

$$||R^{-s}(z)||_2^2 \le |\,\langle z\rangle\,|^{-2s+n}$$

(cf. (1.10)), so that

$$||K(\zeta,z)||_2^2 \le \frac{T^{2\delta}}{\delta^2 + (\Im\zeta - \Im z)^2}|\,\langle z\rangle\,|^{-2s+n}.$$

Integrating over $\Im\zeta$ and $\Im z$ and using that $2s-n>1$, we obtain that the L^2-norm of $K(\zeta,z)$ is finite, whence the lemma follows. □

To prove that C in (1.13) is a Hilbert-Schmidt operator we use that $\rho_0\rho_2$ vanishes on the diagonal $t = t_1$. So, writing

$$\left(\frac{t_1}{t}\right)^z = (\ln t_1 - \ln t)^{-N}\frac{\partial^N}{\partial z^N}\left(\frac{t_1}{t}\right)^z$$

for any $N \in \mathbb{N}$ and integrating by parts we represent C as an integral operator

$$(Cu)(t) = \frac{1}{2\pi i} \int_\Gamma dz \int_0^\infty \left(\frac{t_1}{t}\right)^z \frac{\rho_2(t)\rho_0(t_1)a^{(N)}(t,z)}{(\ln t - \ln t_1)^N t_1^{1/2}} v(t_1)dt_1$$

where

$$v(t_1) = t_1^{-1/2} u(t_1) \in L^2(\mathbb{R}_+, \mathbb{L}^{\not\models}(\mathbb{X}))$$

if $u \in H^{0,(n+1)/2}$. Inclusion $Cu \in H^{s,\gamma}$ for s integer means that for any $k = 0, 1, \ldots, s$ and for any differential operator P of order $s - k$ on X

$$t^{n/2-\gamma} D^k PCu \in L^2(\mathbb{R}_+, \mathbb{L}^{\not\models}(\mathbb{X})).$$

This operator acting on $v = t^{-1/2}u$ has a Schwarz kernel

$$K(t,t_1) = \int_\Gamma t^{n/2-\gamma} D_t^k \frac{\rho_2(t)\rho_0(t_1)Pa^{(N)}(t,z)}{(\ln t - \ln t_1)^N t_1^{1/2}} \left(\frac{t_1}{t}\right)^z dz \tag{1.15}$$

and we need to estimate the L^2-norm of this kernel, more precisely, we need to prove that

$$\int_0^\infty dt \int_0^\infty dt_1 \|K(t,t_1)\|_2^2 \leq \infty$$

(in fact the integration is over $t \in [c/2, C]$ and over $t_1 \in [0, c]$). We may omit the factor $t^{n/2-\gamma}$ since it is bounded on $[c/2, C]$. Next, since

$$D_t^k \left(\frac{t_1}{t}\right)^z = z^k \left(\frac{t_1}{t}\right)^z$$

and

$$\left|\left(\frac{t_1}{t}\right)^z\right| \leq 1$$

for $z = i\tau$ the Hilbert-Schmidt norm of the integrand may be esimated as

$$C_1 \frac{|z|^k \|Pa^{(N)}(t,z)\|_2}{(1 + |\ln t - \ln t_1|)^N t^{1/2}} \leq \frac{C_2 |\langle z\rangle|^{\mu-N+s+n/2}}{(1 + |\ln t - \ln t_1|)^N t_1^{1/2}}. \tag{1.16}$$

Here we have made use of the fact that $t_1 \neq t$, so that $|\ln t_1 - \ln t| \neq 0$. Moreover,

$$\|D_t^k \rho_2 Pa^{(N)}(t,z)\|_2 \leq C|\langle z\rangle|^{\mu-N+n/2}.$$

For N large enough the integral of (1.16) over z converges and we obtain

$$\|K(t,t_1)\|_2 \leq \frac{C_3}{(1 + |\ln t_1 - \ln t|)^N t_1^{1/2}}.$$

Thus, for L^2-norm of $||K(t,t_1)||$ we obtain an estimate

$$C_4 \int_{c/2}^{C} dt \int_0^c \frac{dt_1}{(1+|\ln t_1 - \ln t|)^N t^1}$$
$$= C_4 \int_{\ln c/2}^{\ln C} e^y dy \int_{-\infty}^{\ln c} \frac{dy_1}{(1+|y-y_1|)^N} < \infty$$

for $N > 1$. This proves case 2.

Case 3 ($i, j = 0, 1$).

This is the most difficult case. Here we will make use that $\rho_i(t)a(t,z)$ and $\rho_j(t)b(t,z)$ are holomorphic functions in the strip S. To simplify notations, we omit the factors $\rho_i(t), \rho_j(t)$ including them into a and b.

For $u \in C_0^\infty(\mathbb{R}_+)$ we have

$$(Bu)(t) = \ \mathrm{Op}(b)u = \frac{1}{2\pi i}\int_{\Gamma_1} t^{-z} b(t,z)\widehat{u}(z)dz$$

where Γ_1 may be any vertical line with $|\Re\Gamma_1| < \varepsilon$. Then for $\Re\zeta > \Re\Gamma_1$ the Mellin transform of Bu has the form

$$\widehat{Bu}(\zeta) = \frac{1}{2\pi i}\int_{\Gamma_1} \widehat{b}(\zeta - z, z)\widehat{u}(z)dz.$$

Similarly for $\widehat{ABu}(w)$ we obtain

$$\widehat{ABu}(w) = (2\pi i)^{-2}\int_{\Gamma_2} d\zeta \int_{\Gamma_1} dz \widehat{a}(w-\zeta,\zeta)\widehat{b}(\zeta - z, z)\widehat{u}(z) \tag{1.17}$$

with $\Re w > \Re\Gamma_2 > \Re\Gamma_1$. The integral converges because of estimates (1.5), (1.8).

Now using the Taylor formula, we write

$$\widehat{a}(w-\zeta,\zeta) = \sum_{k=0}^{N-1} \frac{1}{k!}\widehat{a}^{(k)}(w-\zeta,z)(\zeta-z)^k + T_N\widehat{a}(\zeta - z)^N \tag{1.18}$$

where

$$(T_N\widehat{a})(w-\zeta,z,\zeta) = \int_0^1 \widehat{a}^{(N)}(w-\zeta, z+\theta(\zeta-z))\frac{(1-\theta)^{N-1}}{(N-1)!}d\theta. \tag{1.19}$$

The regular terms in (1.18) after substitution into (1.17) and integration over ζ give

$$\frac{1}{2\pi i}\int_{\Gamma_2} \widehat{a}^{(k)}(w-\zeta,z)(\zeta-z)^k\widehat{b}(\zeta-z,z)d\zeta$$
$$= \frac{1}{2\pi i}\int_{\Gamma_2} \widehat{a}^{(k)}(w-\zeta,z)\widehat{b}_{(k)}(\zeta-z,z)d\zeta$$
$$= M_{t\to w-z}(a^{(k)}(t,z)b_{(k)}(t,z))$$

resulting in $\mathrm{Op}(a \circ b|_N)u$. So, the operator C_N corresponds to the remainder term in (1.18)

$$\widehat{C_N u}(w) = (2\pi i)^{-2} \int_{\Gamma_2} d\zeta \int_{\Gamma_1} dz T_N \widehat{a}(w-\zeta,\zeta,z)\widehat{b}_{(N)}(\zeta - z, z)\widehat{u}(z). \tag{1.20}$$

Observe that the function $\widehat{b}_{(N)}(\zeta - z, z)$ at $N \geq 1$ is holomorphic in $\Re\zeta > -1$, so the integration line Γ_2 may be shifted arbitrary within the strip $|\Re\zeta| < \varepsilon$. It means that the restriction $\Re\Gamma_1 < \Re\Gamma_2 < \Re w$ is needed no more, and the only remaining restriction is $\Re\Gamma_2 < \Re w$. It will be convenient to take

$$\Re\Gamma_2 < \Re w \leq \Re\Gamma_1. \tag{1.21}$$

To prove that (1.20) belongs to the trace class we again represent it as a composition of two Hilbert-Schmidt operators

$$H^{0,(n+1)/2} \overset{C_N}{\to} H^{s,(n+1)/2+\delta} \hookrightarrow H^{0,(n+1)/2}$$

with some $\delta > 0$ and some $s > (n+1)/2$ which may be taken as an even integer number. The second operator in this sequence is an embedding. Note that in virtue of lemma 1.2 the weight gain δ is necessary to have a Hilbert-Schmidt embedding. So, we choose $\Re\Gamma_1 = 0$, $\Re w = -\delta$, $\Re\Gamma_2 < \Re w$ in (1.21).

Lemma 1.3 *Let $s > (n+1)/2$ be an even integer. Then the operator*

$$C_N : H^{0,(n+1)/2} \to H^{s,(n+1)/2+\delta}$$

is a Hilbert-Schmidt operator provided N is sufficiently large.

Proof. The assumption that s is an even integer serves only to simplify the proof. Using representation (1.20) for C_N we are to estimate the Hilbert-Schmidt norm of the operator between L^2-spaces on the lines Γ_0 and $\Gamma_{-\delta}$ with the Schwarz kernel

$$K(w,z) = \int_{\Gamma_2} R^s(w) T_N \widehat{a}(w-\zeta,\zeta,z)\widehat{b}_{(N)}(\zeta - z, z) d\zeta.$$

The Hilbert-Schmidt norm of the integrand may be estimated as

$$\begin{aligned} &||R^s(w)T_N\widehat{a}(w-\zeta,\zeta,z)||_2\, ||\widehat{b}_{(N)}(\zeta - z, z)|| \\ &\leq \frac{C_p}{|\langle\zeta - z\rangle|^p} ||R^s(w)T_N\widehat{a}(w-\zeta,\zeta,z)||_2. \end{aligned}$$

Here we have used (1.8) for $\widehat{b}_{(N)}$ with p arbitrary large. In virtue of (1.19) our next step is to estimate the norm

$$||R^s(w)\widehat{a}^{(N)}(w-\zeta, z+\theta(\zeta - z))||_2. \tag{1.22}$$

Denoting $z+\theta(\zeta-z)$ by η and putting $s=2\sigma,\ \ \sigma\in\mathbb{N}$, we may rewrite $R^s(w)$ using the binomial formula

$$R^s(w)=(1-\Delta-\eta^2+\eta^2-w^2)^\sigma=\sum_{k+l+m=\sigma}C_{klm}R^{2k}(\eta)\eta^{2l}w^{2m}$$

with some constant coefficients C_{klm}. Finally, using (1.10) for the operator $R^{2k}(\eta)$ $a^{(N)}(w-\zeta,\eta)$ of order $\mu-N+2k$, we obtain

$$||R^{2k}(\eta)a^{(N)}(w-\zeta,\eta)||_2\leq\frac{C_q}{|\langle w-\zeta\rangle|^q}|\langle\eta\rangle|^{\mu-N+2k+\frac{n}{2}},$$

provided $\mu-N+2k+\frac{n}{2}<0$, q may be any positive number. It implies the following rough estimate of (1.22)

$$||R^s(w)a^{(N)}(w-\zeta,\eta)||_2\leq\frac{C}{|\langle w-q\rangle|^q}|\langle\eta\rangle|^{\mu_1}|\langle w\rangle|^s$$

where $\mu_1\leq\mu-N+s+n/2$ is supposed to be negative.

Next, writing w as a sum

$$w=(w-\zeta)+(\zeta-z)+z$$

and applying binomial formula, we get

$$|\langle w\rangle|^s\leq C|\langle w-\zeta\rangle|^s|\langle\zeta-z\rangle|^s|\langle z\rangle|^s.$$

These rough estimates result in the following estimate for the kernel $K(w,z)$

$$||K(w,z)||_2\leq C\int_0^1d\theta\int_{\Gamma_2}\frac{|\langle z\rangle|^s|d\zeta|}{|\langle w-\zeta\rangle|^{q_1}|\langle\zeta-z\rangle|^{p_1}|\langle z+\theta(\zeta-z)\rangle|^{\mu_2}}\tag{1.23}$$

where q_1,p_1 are arbitrary positive numbers and μ_2 is negative and may be made larger in magnitude at the expense of N. The needed Hilbert-Schmidt norm is

$$\int_{\Gamma_0}\int_{\Gamma_{-\delta}}||K(w,z)||_2^2|dz||dw|.\tag{1.24}$$

After integration over $w\in\Gamma_{-\delta}$ we get the following integral which estimates (1.24)

$$C\int_0^1d\theta\int_{\Gamma_0}|dz|\int_{\Gamma_2}|d\zeta|\frac{|\langle z\rangle|^{2s}}{|\langle\zeta-z\rangle|^{2p_1}|\langle z+\theta(\zeta-z)\rangle|^{2\mu_2}}.$$

First consider the region $|\zeta-z|<|z|/2$. Then

$$\frac{1}{|\langle\zeta-z\rangle|^{2p_1}}\leq 1$$

$$\frac{1}{|\langle z+\theta(\zeta-z)\rangle|^{2\mu_2}} \le \frac{C}{|\langle z\rangle|^{2\mu_2}}$$

and

$$\int_{\Gamma_0} |dz| \int_{|\zeta-z|<\frac{|z|}{2}} \frac{|d\zeta|}{|\langle z\rangle|^{2\mu_2-2s}} \le \int_{\Gamma_0} \frac{|z|}{|\langle z\rangle|^{2\mu_2-2s}}|dz|$$

which is convergent for μ_2 large enough. Now, for $|\zeta - z| \ge |z|/2$ we estimate

$$\frac{1}{|\langle z+\theta(\zeta-z)\rangle|^{2\mu_2}} \le 1$$

and

$$\int_{\Gamma_0} |dz| \int_{|z|/2}^{\infty} |\langle z\rangle|^{2s} \frac{d|\eta|}{|\eta|^{2p_1}} = \frac{1}{2p_1+1}\int_{\Gamma_0} \frac{|\langle z\rangle|^{2s}}{|z|^{2p_1+1}} \le \infty.$$

So, the Hilbert-Schmidt norm of C_N is finite. □

To prove the theorem, it remains to observe that the functions of the type $C_N u$ have support in a finite interval $t \in [0,T]$ since $a \circ b|_N$ vanish at t large enough independent of z. Thus, lemma 1.2 may be applied implying that C_N belongs to the trace class.

Case 4 ($i = 0, j = 2$).

Here we have a pseudolocallity property similar to case 2, but the proof runs slightly different. Again we have that $(\rho_0 a) \circ (\rho_2 b) = 0$, so we need to prove that $\mathrm{Op}(\rho_0 a)\rho_2\mathrm{Op}(b)$ belongs to the trace class. Since $b(t,z)$ has compact support in t, we may assume that $\rho_2(t)$ is also compactly supported. Then since $\mathrm{Op}(b)$ is bounded in $H^{0,(n+1)1/2}$, we again need to prove that $\mathrm{Op}(\rho_0 a)\rho_2$ belongs to the trace class. But multiplication operator $\rho_2(t)$ may be regarded as Mellin p.d.o. with a holomorphic symbol, so we are in assumptions of case 3. This completes the proof of the theorem. □

2 A regularized trace of a product

For operators $A = \mathrm{Op}(a) \in ML_0^{\mu}(X^\wedge)$ and $B = \mathrm{Op}(b) \in ML_0^{\nu}(X^\wedge)$ with $\mu, \nu \le 0$ define a regularized trace of a product by

$$\mathrm{Tr}_N AB = \mathrm{Tr}(AB - \mathrm{Op}(a \circ b|_N)).$$

By theorem 1.1 this trace exists provided N is sufficiently large.

Theorem 2.1 *The regularized trace of a product does not depend on the order, that is*

$$\mathrm{Tr}_N AB = \mathrm{Tr}_N BA \tag{2.1}$$

Proof. There are several cases corresponding to those listed in the proof of theorem 1.1.

Case 1 ($i, j \neq 0$).

The assertion reduces to the theorem on a regularized trace of a product of Fourier p.d.o. [6].

Case 2 ($i = 2, j = 0$) or case 4 ($i = 0, j = 2$).

For $A = \mathrm{Op}(\rho_i a), B = \mathrm{Op}(\rho_j b)$ we have by theorem 1.1 that AB and BA belong to the trace class. Then their traces are equal by Lidskij's theorem.

Case 3 ($i, j = 0, 1$).

Using (1.20) with

$$\Re\Gamma_2 < \Re w = \Re\Gamma_1 = 0$$

(cf. (1.21)) we get

$$\mathrm{Tr}_N AB = \mathrm{Tr} C_N = (2\pi i)^{-2} \int_{\Gamma_2} d\zeta \int_{\Gamma_1} dz \ \mathrm{tr} T_N \widehat{a}(z-\zeta, \zeta, z) \widehat{b}_{(N)}(\zeta - z, z) \tag{2.2}$$

Recall that Tr denotes the trace of operators on the cone $X^\wedge$ while tr is the trace of operators on the base X. The function

$$T_N \widehat{a}(z-\zeta,\zeta,z) = \frac{1}{z-\zeta} T_N \widehat{a}_{(1)}(z-\zeta,\zeta,z)$$

has a pole of the first order at $\zeta = z$. So, we may shift the lines Γ_1, Γ_2 within the strips $|\Re z| < \varepsilon, |\Re\zeta| < \varepsilon$, provided $|\Re\Gamma_2|$ remains less then $|\Re\Gamma_1|$. Moreover, we may shift Γ_2 crossing Γ_1, but then we must take into account the residue at $\zeta = z$. It is equal to

$$-\frac{1}{2\pi i} \int_{\Gamma_1} \mathrm{tr}\, T_N \widehat{a}_{(1)}(0,z,z) \widehat{b}_{(N)}(0,z) dz.$$

By (1.19) we have

$$T_N \widehat{a}_{(1)}(0,z,z) = \frac{1}{N!} \widehat{a}_{(1)}^{(N)}(0,z),$$

so that the residue is equal to

$$-\frac{1}{2\pi i N!} \int_{\Gamma_1} \mathrm{tr}\, \widehat{a}_{(1)}^{(N)}(0,z) \widehat{b}_{(N)}(0,z) dz.$$

But for $N > 1$

$$\widehat{b}_{(N)}(0,z) = \int_0^\infty -\frac{\partial}{\partial t} D^{N-1} b(t,z) dt = -D^{N-1} b(t,z)|_0^\infty = 0.$$

Thus, for $N > 1$ integral (2.2) *does not depend on a position of the lines* Γ_1, Γ_2 within the strip $|\Re z| < \varepsilon, |\Re\zeta| < \varepsilon$. A similar assertion is true for the integral

$$\mathrm{Tr}_N BA = (2\pi i)^{-2} \int_{\Gamma_2} d\zeta \int_{\Gamma_1} dz \ \mathrm{tr}\, T_N \widehat{b}_{(1)}(z-\zeta,\zeta,z) \widehat{a}_{(N)}(\zeta - z, z). \tag{2.3}$$

To prove that (2.2) and (2.3) are equal, we consider families

$$a_s(t,z) = a(t,z)R^{-s}(z)$$

$$b_s(t,z) = b(t,z)R^{-s}(z)$$

where s is a complex parameter ranging in the half-strip $\Re s \geq 0$, $|\Im s| \leq \varepsilon$.

Let A_s, B_s be the corresponding operators on the cone. The estimates of theorem 1.1 show that $\mathrm{Tr}_N A_s B_s$ and $\mathrm{Tr}_N B_s A_s$ are holomorphic functions for s belonging to the half-strip. Thus, it is sufficient to prove the equality

$$\mathrm{Tr}_N A_s B_s = \mathrm{Tr}_N B_s A_s$$

for $\Re s$ sufficiently large. To put it differently, we need to verify (2.1) for operators A, B of sufficiently large negative orders. To this end we write

$$\frac{T_N \widehat{a}_{(1)}(z-\zeta,\zeta,z)}{(z-\zeta)} = -\frac{\widehat{a}_{(1)}(z-\zeta,\zeta) - \sum_{k=0}^{N-1} \widehat{a}_{(1)}^{(k)}(z-\zeta,z)\frac{(\zeta-z)^k}{k!}}{(\zeta-z)^{N+1}}$$

and then

$$\begin{aligned} \mathrm{Tr}_N\, AB &= -(2\pi i)^{-2}\int_{\Gamma_2} d\zeta \int_{\Gamma_1} dz \frac{\mathrm{tr}\,\widehat{a}_{(1)}(z-\zeta,\zeta)\widehat{b}_{(1)}(\zeta-z,z)}{(\zeta-z)^2} \\ &+ \sum_{k=0}^{N-1}(2\pi i)^{-2}\int_{\Gamma_2} d\zeta \int_{\Gamma_1} dz \frac{\mathrm{tr}\,\widehat{a}_{(1)}^{(k)}(z-\zeta,z)(\zeta-z)^k\widehat{b}_{(1)}(\zeta-z,z)}{(\zeta-z)^2 k!}. \end{aligned} \tag{2.4}$$

Each summand in (2.4) makes sense if a and b have large negative orders and $\Re\Gamma_2 < \Re\Gamma_1$ are fixed. Interchanging z and ζ in the first integral and using that $\mathrm{tr}\,\widehat{a}_{(1)}\widehat{b}_{(1)} = \mathrm{tr}\,\widehat{b}_{(1)}\widehat{a}_{(1)}$ for trace class operators, we obtain

$$-(2\pi i)^{-2}\int_{\Gamma_1} d\zeta \int_{\Gamma_2} dz \frac{\mathrm{tr}\,\widehat{b}_{(1)}(z-\zeta,\zeta)\widehat{a}_{(1)}(\zeta-z,z)}{(\zeta-z)^2} \tag{2.5}$$

Now, the remaining summands may be transformed as follows

$$\begin{aligned} &\int_{\Gamma_{-\delta}} d\eta \int_{\Gamma_1} dz \frac{\mathrm{tr}\,\widehat{a}_{(1)}^{(k)}(-\eta,z)\eta^k\widehat{b}_{(1)}(\eta,z)}{\eta^2} \\ &= (-1)^k\int_{\Gamma_\delta} d\eta \int_{\Gamma_1} dz \frac{\mathrm{tr}\,\widehat{a}_{(1)}^{(k)}(\eta,z)\eta^k\widehat{b}_{(1)}(-\eta,z)}{\eta^2}. \end{aligned}$$

Here we have changed η by $-\eta$. Now, integrating by parts with respect to z and permuting $\widehat{a}_{(1)}$ and $\widehat{b}_{(1)}$ under the trace sign, we obtain

$$\int_{\Gamma_\delta} d\eta \int_{\Gamma_1} dz \frac{\operatorname{tr} \widehat{b}^{(k)}_{(1)}(-\eta, z)\eta^k \widehat{a}_{(1)}(\eta, z)}{\eta^2}$$
$$= \int_{\Gamma_1} d\zeta \int_{\Gamma_2} dz \frac{\operatorname{tr} \widehat{b}^{(k)}_{(1)}(z-\zeta, z)(\zeta - z)^k \widehat{a}_{(1)}(\zeta - z, z)}{(\zeta - z)^2}. \tag{2.6}$$

Here we have shifted the lines of integrations keeping $\Re\Gamma_2 < \Re\Gamma_1$. Now, summing (2.5) and (2.6) for $k = 0, 1, \ldots, N-1$, we get

$$\operatorname{Tr}_N AB = (2\pi i)^{-2} \int_{\Gamma_1} d\zeta \int_{\Gamma_2} dz \frac{\operatorname{tr} T_N \widehat{b}_{(1)}(z-\zeta, \zeta, z)\widehat{a}_{(N)}(\zeta - z, z)}{\zeta - z}.$$

This expression coincides with the corresponding expression (2.2) for $\operatorname{Tr}_N BA$ except that the lines Γ_1 and Γ_2 are interchanged. But, as we have seen, we may interchange Γ_1 and Γ_2 not affecting the value of the integral.

This completes the proof. □

3 An algebraical index

First we introduce an algebra of formal Mellin symbols on $\mathbb{R}_+$, define elliptic symbols and introduce an algebraical index of elliptic elements. Then constructing a parametrix and applying the theorem on a regularized trace of a product, we prove that the analytical and algebraical index coincide.

A formal symbol is a formal power series

$$a(t,z) = \sum_{k=0}^{\infty} h^k a_k(t,z)$$

where coefficients $a_k(t,z) \in C^\infty(\mathbb{R}_+, L^{\mu-k}(X, \Gamma_0))$, $\mu \leq 0$ satisfy the following conditions:

1. $$\frac{\partial^m}{\partial z^m} a_k(t,z) \in C^\infty(\mathbb{R}_+, L^{\mu-k-m}(X, \Gamma_0)),$$
2. for $0 < t < c$ or $t > C$ $a_0(t,z)$ does not depend on t while $a_k(t,z) = 0$ for $k > 0$.

The powers of a formal parameter h serve for ordering the series terms. Define a product $\circ$ of two symbols by

$$a \circ b = \sum_{k,l,m=0}^{\infty} h^{k+l+m} \frac{1}{m!} \frac{\partial^m}{\partial z^m} a_k(t,z) D^m b_l(t,z).$$

It is easy to check that the symbols form an associative algebra with a unit $a(t, z) \equiv 1$ consisting of the leading term only. We denote this algebra by $\mathcal{A}$. Introduce a *trace ideal* $\mathcal{J}$ consisting of symbols with $\mu < -(n+1)$ and with all the functions $a_k(t, z)$ vanishing at $t \in [0, c]$ and $t \geq C$. A *trace* for $a \in \mathcal{J}$ is defined by

$$\operatorname{Tr} a = \sum_{k=0}^{\infty} h^{k-1} \frac{1}{2\pi i} \int_{\Gamma_0} dz \int_0^{\infty} \operatorname{tr} a_k(t, z) \frac{dt}{t}.$$

This is a formal series with constant coefficients and the exponents of h ranging from -1 to $+\infty$. Using integrations by parts, one can check that

$$\operatorname{Tr} a \circ b = \operatorname{Tr} b \circ a$$

if one of the symbols belongs to $\mathcal{J}$.

A symbol $a \in \mathcal{A}$ is called elliptic if there exists a symbol r such that $1 - r \circ a$ and $1 - a \circ r$ belong to $\mathcal{J}$. In particular for leading terms a_0 and r_0 we obtain

$$1 - r_0 a_0 \in \mathcal{J}, \quad 1 - a_0 r_0 \in \mathcal{J}. \tag{3.1}$$

Such symbol r is called a (formal) *parametrix of* a. The following construction is well-known.

Lemma 3.1 *Let there exist a function $r_0(t, z)$ satisfying* (3.1). *Then for N large enough the symbol*

$$r = r_0 \circ \sum_{k=0}^{N} (1 - a \circ r_0)^k = \sum_{k=0}^{N} (1 - r_0 \circ a)^k \circ r_0 \tag{3.2}$$

(*the powers are understood with respect to the product* $\circ$) *is a parametrix of* a.

Proof. By direct calculation we have

$$1 - r \circ a = (1 - r_0 \circ a)^{\circ(N+1)}, \tag{3.3}$$

$$1 - a \circ r = (1 - a \circ r_0)^{\circ(N+1)} \tag{3.4}$$

where exponent $\circ(N+1)$ means the $(N+1)$-th power with respect to $\circ$-product. Clearly, these symols belong to $\mathcal{J}$.

□

We define the algebraical index of an elliptic symbol a by

$$\operatorname{ind} a = \operatorname{Tr}(1 - r \circ a) - \operatorname{Tr}(1 - a \circ r). \tag{3.5}$$

By definition it is a formal series in h with constant coefficients. It turns out, however, that all the coefficients vanish except a constant term, so we can treat it as a number. Moreover, the index does not depend on the choice of a parametrix. All these properties are standard consequences of the *stability* of the index.

Lemma 3.2 *Let $a(\lambda)$ be a family of elliptic symbols, $r(\lambda)$ a family of parametrices. Then*

$$\operatorname{Tr}(1 - r(\lambda) \circ a(\lambda)) - \operatorname{Tr}(1 - a(\lambda) \circ r(\lambda))$$

is independent of λ.

Proof. We have

$$\begin{aligned}(1 - r \circ a)' &= (1 - r \circ a)' \circ (1 - r \circ a) + (1 - r \circ a)' \circ r \circ a = \\ &= \{(1 - r \circ a) \circ r \circ a\}' - (1 - r \circ a) \circ (r \circ a)' - (r \circ a)' \circ (1 - r \circ a)\end{aligned}$$

where prime means derivation by λ. Thus,

$$\operatorname{Tr}(1 - r \circ a)' = \frac{d}{d\lambda} \operatorname{Tr}(1 - r \circ a) \circ r \circ a - 2 \operatorname{Tr}(1 - r \circ a) \circ (r' \circ a + r \circ a').$$

Similarly,

$$\operatorname{Tr}(1 - a \circ r)' = \frac{d}{d\lambda} \operatorname{Tr}(1 - a \circ r) \circ a \circ r - 2 \operatorname{Tr}(1 - a \circ r) \circ (a' \circ r + a \circ r').$$

But

$$\begin{aligned}\operatorname{Tr}(1 - a \circ r) \circ a \circ r &= \operatorname{Tr} r \circ (1 - a \circ r) \circ a = \operatorname{Tr}(1 - r \circ a) \circ r \circ a, \\ \operatorname{Tr}(1 - a \circ r) \circ a' \circ r &= \operatorname{Tr} r \circ (1 - a \circ r) \circ a' = \operatorname{Tr}(1 - r \circ a) \circ r \circ a', \\ \operatorname{Tr}(1 - a \circ r) \circ a \circ r' &= \operatorname{Tr} a \circ (1 - r \circ a) \circ r' = \operatorname{Tr}(1 - r \circ a) \circ r' \circ a.\end{aligned}$$

So, both expressions $\operatorname{Tr}(1 - r \circ a)'$, $\operatorname{Tr}(1 - a \circ r)'$ coincide. □

In particular, given two parametrices r_1 and r_2 of the same elliptic symbol a, we consider a linear homotopy $r(\lambda) = (1-\lambda)r_1 + \lambda r_2$ which gives a family of parametrices. Then, lemma 3.2 implies that index does not depend on the choice of a parametrix.

Now, for a real $\lambda > 0$ define a homomorphism $H_\lambda : \mathcal{A} \to \mathcal{A}$ by

$$H_\lambda a = \sum_{k=0}^{\infty} \lambda^k h^k a_k(t, \lambda z).$$

It is straightforward to check that H_λ is in fact a homomorphism:

$$H_\lambda(a \circ b) = (H_\lambda a) \circ (H_\lambda b).$$

If $a \in \mathcal{J}$, then

$$\operatorname{Tr} H_\lambda a = H_\lambda \operatorname{Tr} a \tag{3.6}$$

where H_λ acts on formal series with constant coefficients replacing h by λh. Equality (3.6) follows by change of variables in the integral

$$\operatorname{Tr} H_\lambda a = \sum_{k=0}^{\infty} \lambda^k h^{k-1} \frac{1}{2\pi i} \int_{\Gamma_0} dz \int_0^\infty \operatorname{tr} a_k(t, \lambda z) \frac{dt}{t} =$$

$$= \sum_{k=0}^{\infty} \lambda^{k-1} h^{k-1} \frac{1}{2\pi i} \int_{\Gamma_0} dz \int_0^\infty \operatorname{tr} a_k(t, z) \frac{dt}{t}.$$

Lemma 3.3 *The formal series* ind *consists of the constant term only.*

Proof. For $\lambda > 0$ consider a family $a(\lambda) = H_\lambda a$ of elliptic symbols. Then $r(\lambda) = H_\lambda r$ is a family of parametrices since

$$1 - H_\lambda r \circ H_\lambda a = H_\lambda(1 - r \circ a) \in \mathcal{J}.$$

So,

$$\operatorname{ind} a(\lambda) = \operatorname{Tr} H_\lambda(1 - r \circ a) - \operatorname{Tr} H_\lambda(1 - a \circ r) = H_\lambda \operatorname{ind} a.$$

On the other hand, $\operatorname{ind} a(\lambda)$ does not depend on λ by lemma 3.2. □

We are going to compare analytical and algebraical indices. Given an elliptic operator

$$A = \operatorname{Op}(a(t,z)),$$

we may treat its symbol $a(t,z)$ as a formal one consisting of the leading term only. The ellipticity conditions listed in the introduction imply that there exists an $r_0(t,z)$ such that $1 - r_0 a$ and $1 - a r_0$ belong to $\mathcal{J}$. Indeed, for $0 < t < c$ and $t > C$ a^{-1} exists by definition. For $t \in [c, C]$ $a(t,z)$ is parameter-dependent elliptic where $z \in \Gamma_0$ is considered as a parameter. In particular it implies that $a(t,z)$ is also invertible at $|z| > M$ for sufficiently large M. As for $t \in [c, C]$, $z \in [-M, M]$ there exists a parametrix $b(t,z)$ since $a(t,z)$ is elliptic at any t, z. Now, using a cut-off function $\varphi(t,z)$ which is equal to 1 in a rectangle $t \in [c, C]$, $|z| \le M$ and vanishes outside a compact in $\mathbb{R}_+ \times \Gamma_0$ define

$$r_0(t,z) = (1 - \varphi) a^{-1}(t,z) + \varphi b(t,z).$$

This function may serve as a leading term of a formal parametrix given by lemma 3.1. Thus, the algebraical index is defined. To compute the analytical index of A, we need an operator parametrix R inverting A up to the trace class operators. Then the analytical index of A is given by the formula

$$\operatorname{ind} A = \operatorname{Tr}(1 - RA) - \operatorname{Tr}(1 - AR). \tag{3.7}$$

Introduce a notation

$$r|_N = \sum_{k=0}^{N-1} r_k$$

for a formal symbol

$$r = \sum_{k=0}^{\infty} h^k r_k.$$

Then the following theorem holds.

Theorem 3.4 *Let r be a formal parametrix (3.2) of the elliptic symbol a. Then for N large enough the operator $R = \mathrm{Op}(r|_N)$ is an operator parametrix of $A = \mathrm{Op}\,(a)$ and*

$$\operatorname{ind} A = \mathrm{Tr}(1 - RA) - \mathrm{Tr}(1 - AR) = \mathrm{Tr}(1 - r \circ a) - \mathrm{Tr}(1 - a \circ r). \tag{3.8}$$

Proof. Denoting $r|_N$ by b and taking M sufficiently large, we get

$$1 - \mathrm{Op}(b)\mathrm{Op}(a) = \mathrm{Op}(\,(1 - b \circ a)|_M) + \{\mathrm{Op}(b)\mathrm{Op}(a) - \mathrm{Op}(b \circ a|_M)\}$$
$$1 - \mathrm{Op}(a)\mathrm{Op}(b) = \mathrm{Op}(\,(1 - a \circ b)|_M) + \{\mathrm{Op}(a)\mathrm{Op}(b) - \mathrm{Op}(a \circ b|_M)\}.$$

By theorems 1.1 and 2.1 the operators in curly brackets belong to the trace class and their traces are equal. Taking $M \geq N$ and writing

$$b = r|_N = \sum_{k=0}^{N-1} r_k,$$

$$b \circ a|_M = \sum_{l=0}^{M-1} \sum_{k=0}^{N-1} \frac{1}{l!} \frac{\partial^l}{\partial z^l} r_k D^l a,$$

$$r \circ a|_N = \sum_{0 \leq k+l < N} \frac{1}{l!} \frac{\partial^l}{\partial z^l} r_k D^l a,$$

we see that $\mathrm{Op}(\,(b \circ a)|_M) - \mathrm{Op}(\,(r \circ a)|_N)$ is a finite sum of terms

$$\mathrm{Op}\left(\frac{\partial^l}{\partial z^l} r_k D^l a\right)$$

with $k + l \geq N, \;\; k < N, \;\; l < M$. If $N > n + 1$, this operator belongs to the trace class since its order is less than $-(n+1)$ and its symbol vanishes at $t \in [0, c]$ and $t \in [C, \infty)$. The same is true for $\mathrm{Op}(a \circ b|_M) - \mathrm{Op}(a \circ r|_N)$, which is the sum of

$$\mathrm{Op}\left(\frac{\partial^l a}{\partial z^l} D^l r_k\right)$$

with $k + l \geq N, \;\; k < N, \;\; l < M$. The traces of such operators are equal because

$$\int_{\Gamma_0} dz \int_0^{\infty} \mathrm{tr}\, \frac{\partial^l}{\partial z^l} r_k D^l a \frac{dt}{t} = \int_{\Gamma_0} dz \int_0^{\infty} \mathrm{tr}\, \frac{\partial^l a}{\partial z^l} D^l r_k \frac{dt}{t},$$

as one can see integrating by parts. Thus,

$$\mathrm{Tr}(1 - RA) - \mathrm{Tr}(1 - AR) = \mathrm{Tr}\;(1 - r \circ a)|_N - \mathrm{Tr}\;(1 - a \circ r)|_N$$

which is precisely the algebraical index. □

4 A topological index

Following [6], we introduce one more algebra which permits to simplify significantly various calculations with noncommutative differential forms. We will use a real variable τ instead of $z = i\tau$. An element a of our new algebra $\widehat{\mathcal{A}}$ is an operator-valued nonhomogeneous differential form of even degrees on the half-plane $\mathbb{R}^2_+$. So,

$$a = a_0(t,\tau) + a_1(t,\tau)d\tau \wedge dt \tag{4.1}$$

where a_0 and a_1 are pseudodifferential operators on X of nonpositive orders. A product $\widehat{\circ}$ of two elements $a, b \in \widehat{\mathcal{A}}$ is defined by

$$a\widehat{\circ}b = a \wedge b + \frac{i}{2} da \wedge db. \tag{4.2}$$

One immediately checks that this product is associative.

Any function $a(t,\tau)$ may be considered as an element of $\widehat{\mathcal{A}}$ consisting of 0-component only. So, for functions a, b we have three products:

- ab is the usual point-wise operator product of functions,
- $$a \circ b = ab + iht\frac{\partial a}{\partial \tau}\frac{\partial b}{\partial t} + \dots \tag{4.3}$$

  is a product in $\mathcal{A}$ as formal symbols,
- $$a\widehat{\circ}b = ab + \frac{i}{2} da \wedge db \tag{4.4}$$

  is a product in $\widehat{\mathcal{A}}$.

We may also consider the powers of a function a with respect to any of these products using notations $a^k, a^{\circ k}, a^{\widehat{\circ} k}$ to distinguish the three possibilities. One can verify a simple rule to pass from $\circ$-product to $\widehat{\circ}$-product of functions: keep the terms linear in h, then alternate derivations $\partial/\partial\tau, \partial/\partial t$ and then write $d\tau \wedge dt$ instead of ht. This rule is valid for any number of functions $a_1 \circ a_2 \circ \dots \circ a_k$ and $a_1 \widehat{\circ} a_2 \widehat{\circ} \dots \widehat{\circ} a_k$.

Similarly to $\mathcal{J}$ we introduce a *trace ideal* $\widehat{\mathcal{J}} \subset \widehat{\mathcal{A}}$. It consists of forms (4.1) where a_0, a_1 are operators of order $\mu < -(n+1)$ with regard for a parameter $\tau \in \mathbb{R}$, vanishing at $t \in [0, c]$ and $t \in [C, \infty)$. For $a \in \widehat{\mathcal{J}}$ define a *trace*

$$\operatorname{Tr} a = \frac{1}{2\pi}\int_{\mathbb{R}^2_+} \operatorname{tr} a = \frac{1}{2\pi}\int_{\mathbb{R}^2_+} \operatorname{tr} a_1(t,\tau)d\tau dt$$

(the orientation of $\mathbb{R}^2_+$ is given by the form $d\tau \wedge dt$). The trace property

$$\operatorname{Tr} a\widehat{\circ}b = \operatorname{Tr} b\widehat{\circ}a$$

is obviously satisfied if a or b belongs to $\mathcal{J}$.

With this definitions we have the following topological index formula.

Theorem 4.1 *For any $N \geq 1$*

$$\operatorname{ind} A = \operatorname{Tr}(1 - r_0\widehat{\circ}a)^{\widehat{\circ}(N+1)} - \operatorname{Tr}(1 - a\widehat{\circ}r_0)^{\widehat{\circ}(N+1)} \tag{4.5}$$

where r_0 is the leading term of the parametrix of a.

Proof. We start with the algebraical index formula (3.8) taking

$$r = r_0 \circ \sum_{k=0}^{N}(1 - a \circ r_0)^k = \sum_{k=0}^{N}(1 - r_0 \circ a)^k \circ r_0.$$

with N large enough. Then

$$1 - r \circ a = (1 - r_0 \circ a)^{\circ(N+1)}$$

$$1 - a \circ r = (1 - a \circ r_0)^{\circ(N+1)},$$

so

$$\operatorname{ind} A = \operatorname{Tr}(1 - r_0 \circ a)^{\circ(N+1)} - \operatorname{Tr}(1 - a \circ r_0)^{\circ(N+1)}. \tag{4.6}$$

According to lemma 3.3 we need to extract a constant term in (4.6). It means that we may calculate $(1 - r_0 \circ a)^{\circ(N+1)}$ keeping the terms linear in h. Thus,

$$1 - r_0 \circ a = 1 - r_0 a - iht\frac{\partial r_0}{\partial \tau}\frac{\partial a}{\partial t} + \dots$$

where dots mean higher-degree terms in h. Using induction one easily obtains a formula

$$\begin{aligned}
&(1 - r_0 \circ a)^{\circ(N+1)} \sim (1 - r_0 a)^{N+1} \\
&-iht\left\{\sum_{k=0}^{N}(1 - r_0 a)^k\frac{\partial r_0}{\partial \tau}\frac{\partial a}{\partial t}(1 - r_0 a)^{N-k} - \sum_{k+p+q=N-1}(1 - r_0 a)^k \times\right. \\
&\left.\times\ \frac{\partial(1 - r_0 a)}{\partial \tau}(1 - r_0 a)^p\frac{\partial(1 - r_0 a)}{\partial t}(1 - r_0 a)^q\right\}
\end{aligned} \tag{4.7}$$

where $\sim$ means that the linear terms coincide. The second sum may be written as

$$\sum_{k=0}^{N} \frac{\partial(1-r_0a)^k}{\partial\tau}\frac{\partial(1-r_0a)}{\partial t}(1-r_0a)^{N-k} \tag{4.8}$$

or

$$\sum_{k=0}^{N}(1-r_0a)^k\frac{\partial(1-r_0a)}{\partial\tau}\frac{\partial(1-r_0a)^{N-k}}{\partial t}. \tag{4.9}$$

Using "integration by parts" transform (4.9) to the form

$$\begin{aligned}&\frac{\partial}{\partial t}\sum_{k=0}^{N}(1-r_0a)^k\frac{\partial(1-r_0a)}{\partial\tau}(1-r_0a)^{N-k}\\&-\sum_{k=0}^{N}\frac{\partial(1-r_0a)^k}{\partial t}\frac{\partial(1-r_0a)}{\partial\tau}(1-r_0a)^{N-k}\\&-\sum_{k=0}^{N}(1-r_0a)^k\frac{\partial^2(1-r_0a)}{\partial\tau\partial t}(1-r_0a)^{N-k}.\end{aligned} \tag{4.10}$$

If $N \geq 1$ all the written terms belong to the trace ideal $\mathcal{J}$ since they contain a factor $1-r_0a \in \mathcal{J}$ or its derivatives.

Let us now write down a constant term of the trace of (4.7). We represent the second sum in (4.7) as a half-sum of expression (4.8) and (4.10). We may drop the first sum in (4.10) since complete derivatives vanish under integration and permute cyclically the factors under trace sign. Finally we obtain

$$\begin{aligned}&\frac{1}{2\pi i}\int_{\mathbb{R}^2_+}(N+1)\mathrm{tr}(1-r_0a)^N\left(\frac{\partial r_0}{\partial\tau}\frac{\partial a}{\partial t}-\frac{1}{2}\frac{\partial^2(r_0a)}{\partial\tau\partial t}\right)\\&-\frac{1}{2}\sum_{k=0}^{N}\left\{\frac{\partial(1-r_0a)^k}{\partial\tau}\frac{\partial(1-r_0a)}{\partial t}-\frac{\partial(1-r_0a)}{\partial t}^k\frac{\partial(1-r_0a)}{\partial\tau}\right\}\times\\&\times(1-r_0a)^{N-k}d\tau dt.\end{aligned}$$

Next,

$$\frac{\partial r_0}{\partial\tau}\frac{\partial a}{\partial t}-\frac{1}{2}\frac{\partial^2(r_0a)}{\partial\tau\partial t}=\frac{1}{2}\frac{\partial r_0}{\partial\tau}\frac{\partial a}{\partial t}-\frac{1}{2}\frac{\partial r_0}{\partial t}\frac{\partial a}{\partial\tau}-\frac{1}{2}\frac{\partial^2 r}{\partial t\partial\tau}a-\frac{1}{2}r\frac{\partial^2 a}{\partial t\partial\tau}$$

and

$$\begin{aligned}&\left(\frac{\partial r_0}{\partial\tau}\frac{\partial a}{\partial t}-\frac{\partial r_0}{\partial t}\frac{\partial a}{\partial\tau}\right)d\tau\wedge dt=dr_0\wedge da,\\&\left\{\frac{\partial(1-r_0a)^k}{\partial\tau}\frac{\partial(1-r_0a)}{\partial t}-\frac{\partial(1-r_0a)^k}{\partial t}\frac{\partial(1-r_0a)}{\partial\tau}\right\}d\tau\wedge dt\\&=d(1-r_0a)^k\wedge d(1-r_0a).\end{aligned}$$

Thus, for the constant term of $\mathrm{Tr}(1 - r_0 \circ a)^{\circ(N+1)}$ we get an expression

$$
\begin{aligned}
&\frac{1}{2\pi i}\int_{\mathbb{R}^2_+}\left\{\frac{(N+1)}{2}\mathrm{tr}(1-r_0a)^N dr_0 \wedge da\right.\\
&\left.\frac{1}{2}\sum_{k=0}^{N}\mathrm{tr}\, d(1-r_0a)^k \wedge d(1-r_0a)(1-r_0a)^{N-k}\right\}\\
&-\frac{1}{2\pi i}\frac{N+1}{2}\int_{\mathbb{R}^2_+}\mathrm{tr}(1-r_0a)^N\left(\frac{\partial^2 r_0}{\partial\tau\partial t}a + r_0\frac{\partial^2 a_0}{\partial\tau\partial t}\right)d\tau dt. \qquad (4.11)
\end{aligned}
$$

A similar expression may be written for the constant term of $\mathrm{Tr}(1 - a \circ r_0)^{\circ(N+1)}$ by interchanging a and r_0. Note that

$$
\begin{aligned}
&\mathrm{tr}(1-ar_0)^N\frac{\partial^2 a}{\partial\tau\partial t}r_0 = \mathrm{tr}\, r_0(1-ar_0)^N\frac{\partial^2 a}{\partial\tau\partial t}\\
&= \mathrm{tr}(1-r_0a)^N r_0\frac{\partial^2 a}{\partial\tau\partial t}.
\end{aligned}
$$

It implies that the last integral in (4.11) does not change under permutation of a and r_0. Thus, taking the difference of (4.11) and the corresponding expression obtained by interchanging a and r_0, we find

$$
\begin{aligned}
\mathrm{ind}\, A = &\frac{1}{2\pi i}\int_{\mathbb{R}^2_+}\left\{\frac{(N+1)}{2}\mathrm{tr}(1-r_0a)^N dr_0\wedge da\right.\\
&-\frac{1}{2}\sum_{k=0}^{N}\mathrm{tr}\, d(1-r_0a)^k\wedge d(1-r_0a)(1-r_0a)^{N-k}\\
&-\frac{N+1}{2}\mathrm{tr}(1-ar_0)^N da\wedge dr_0\\
&\left.+\frac{1}{2}\sum_{k=0}^{N}\mathrm{tr}\, d(1-ar_0)^k\wedge d(1-ar_0)(1-ar_0)^{N-k}\right\}. \qquad (4.12)
\end{aligned}
$$

Taking into account the rule for passing from $\circ$-product to $\widehat{\circ}$-product one easily recognizes formula (4.5) in (4.12). In particular, we obtain that the right hand side of (4.12) or (4.5) is independent of N provided N is large enough. Using $\widehat{\circ}$-product, we prove now that (4.5) is valid for $N \geq 1$. Indeed,

$$
\begin{aligned}
&\mathrm{Tr}(1-r_0\widehat{\circ}a)^{\widehat{\circ}(N+1)} - \mathrm{Tr}(1-a\widehat{\circ}r_0)^{\widehat{\circ}(N+1)} =\\
&= \mathrm{Tr}(1-r_0\widehat{\circ}a)^{\widehat{\circ}N} - \mathrm{Tr}(1-a\widehat{\circ}r_0)^{\widehat{\circ}N} -\\
&-\mathrm{Tr}(1-r_0\widehat{\circ}a)^{\widehat{\circ}N}\widehat{\circ}r_0\widehat{\circ}a + \mathrm{Tr}(1-a\widehat{\circ}r_0)^{\widehat{\circ}N}\widehat{\circ}a\widehat{\circ}r_0.
\end{aligned}
$$

But by the associativity of the $\widehat{\circ}$-product

$$
\mathrm{Tr}(1-r_0\widehat{\circ}a)^{\widehat{\circ}N}\widehat{\circ}r_0\widehat{\circ}a = \mathrm{Tr}\, r_0\widehat{\circ}(1-a\widehat{\circ}r_0)^{\widehat{\circ}N}\widehat{\circ}a = \mathrm{Tr}(1-a\widehat{\circ}r_0)^{\widehat{\circ}N}\widehat{\circ}a\widehat{\circ}r_0.
$$

Here we have used that $(1-a\hat{\circ}r_0)^{\hat{\circ}N} \in \widehat{\mathcal{J}}$ for $N \geq 1$, so a cyclic permutation of factors under a trace sign is possible. □

For $N = 1$ (4.12) becomes

$$\operatorname{ind} A = \frac{1}{2\pi i}\int_{\mathbb{R}^2_+} \operatorname{tr}(1-r_0 a)dr_0 \wedge da - \operatorname{tr}(1-ar_0)da \wedge dr_0$$

since

$$\operatorname{tr} d(1-r_0 a) \wedge d(1-r_0 a) = \operatorname{tr} d(1-ar_0) \wedge d(1-ar_0) = 0.$$

Integrating by parts in the first summand, we get

$$-\int_{\mathbb{R}^2_+} \operatorname{tr}\,(d(1-r_0 a))r_0 da = \int_{\mathbb{R}^2_+} \operatorname{tr}(dr_0 a r_0 \wedge da + r_0 d a r_0 \wedge da).$$

The second summand may be transformed as follows

$$-\operatorname{tr}(1-ar_0)da \wedge dr = \operatorname{tr} dr(1-ar_0) \wedge da.$$

So, (4.12) at $N = 1$ gives formula (0.5).

5 Example

Consider the simplest example of a singular integral operator on a half–line (a cone over a point)

$$(Au)(x) = \alpha(x)u(x) + \frac{\beta(x)}{\pi i}\int_0^\infty \frac{u(y)}{y-x}dy \tag{5.1}$$

where $\beta(x)$ and $1-\alpha(x)$ belong to $C_0^\infty(\overline{\mathbb{R}}_+)$. For $x > 0$ this operator has a principal (Fourier) symbol

$$a_0(x,\xi) = \alpha(x) + \beta(x)\operatorname{sgn}\xi$$

and we assume as usual that $a_0(x, \pm 1) \neq 0$.

Let us pass to the Mellin representation. Taking

$$u(y) = \frac{1}{2\pi i}\int_\Gamma y^{-z}\hat{u}(z)dz$$

with $0 < \Re\Gamma < 1$, substituting it into (5.1) and calculating the singular integral

$$\frac{1}{\pi i}\int_0^\infty \frac{y^{-z}}{y-x}dy = x^{-z}\frac{1+e^{2\pi i z}}{1-e^{2\pi i z}}$$

we represent (5.1) in the form

$$(Au)(x) = \frac{1}{2\pi i}\int_\Gamma x^{-z}a(x,z)\hat{u}(z)dz \tag{5.2}$$

where the Mellin symbol is

$$a(x,z) = \alpha(x) + \beta(x)\frac{1+e^{2\pi iz}}{1-e^{2\pi iz}}. \tag{5.3}$$

The ellipticity conditions for the operator (5.2) besides $\alpha(x) \pm \beta(x) \neq 0$ mean that the line $\Gamma = \Gamma_\sigma$ does not contain the poles of $a(t,z)$ as well as the zeros of $a(0,z)$, so that $a^{-1}(0,z)$ is holomorphic in a strip containing Γ_σ. Now, since $rda \wedge rda = 0$ for scalar functions we obtain by the Stokes formula

$$\text{ind}A = \frac{1}{2\pi i}\int_{\Gamma\times\mathbb{R}_+} dr \wedge da = \frac{1}{2\pi i}\int_{\partial(\Gamma\times\mathbb{R}_+)} a^{-1}da$$

which is equal to the variation of $\arg a$ along the "boundary" $\partial(\Gamma \times \mathbb{R}_+)$. The latter consists of the line $z \in \Gamma$, $x = 0$ and two half–lines $z = \sigma \pm i\infty$, $x \in \mathbb{R}_+$. This gives a final formula

$$\text{ind}A = \frac{1}{2\pi i}\int_{\sigma-i\infty}^{\sigma+i\infty} a^{-1}(0,z)a'(0,z)dz + \frac{1}{2\pi}\Delta\arg\frac{\alpha(x)+\beta(x)}{\alpha(x)-\beta(x)}$$

where Δ means variation along the positive half–axis $x > 0$. It is precisely formula (0.3) from [2].

References

[1] B.–W.Schulze. *Pseudo–Differential Operators on Manifolds with Singularities*, North–Holland, Amsterdam 1991.

[2] B.A. Plamenevskij and V.G. Rosenblum. *On the index of pseudodifferential operators with isolated singularities in the symbol,* Leningrad Math. J. **2** (1991), 1085–1110.

[3] M.F. Atiyah and I.M. Singer. *The index of elliptic operators III,* Ann. of Math. **87** (1968), 546–604.

[4] M.F. Atiyah and I.M. Singer. *The index of elliptic operators IV,* Ann. of Math. **93** (1971), 119–138.

[5] B.V. Fedosov. *Periodicity theorem in the algebra of symbols*, Mat. Sbornik **105** (1978), 622–637, (Russian).

[6] B.V. Fedosov. *Analytic Index Formulae*, Trudy Mosk. Mat. Obshch. **30** (1974), 159 – 240 (in Russian).

[7] M.A. Shubin. *Pseudodifferential Operators and Spectral Theory*, Springer-Verlag, Berlin–Heidelberg–New York–London–Paris–Tokyo 1987.

On Composition Series in Algebras of Pseudodifferential Operators and in Algebras of Wiener–Hopf Operators

B. A. Plamenevskij and V. N. Senichkin[1]

1 Introduction

1.1 An operator $A : L_2(\Lambda) \to L_2(\Lambda)$ of the form

$$(Au)(x) = \int_{\Lambda} \psi(x-y)u(y)\,dy, x \in \Lambda,$$

where Λ is a cone in $\mathbb{R}^n$ and $\psi \in L_1(\mathbb{R}^n)$, is called a Wiener–Hopf operator. We assume the cone to be 'piecewise' and discuss the C^*–algebras generated by Wiener–Hopf operators. We also consider C^*–algebras of pseudodifferential operators (PDO) on smooth compact manifolds. The symbols of operators may have discontinuities of rather general type.

All the equivalence classes of irreducible representations of the above algebras were found in [Dyn 2], [MR], [S], [PS 2]. It was established that these algebras are of type 1. Dynin [1,2] introduced a more narrow class of 'solvable' algebras possessing specific 'solving' series. He outlined some applications of the properties of solvable algebras (definitions are given below).

[1]This work was supported by ISF (R40000), INTAS (93–185), and RFFI.

The purpose of this paper is to prove the solvability of the algebras of Wiener–Hopf operators on cones and of the algebras of PDO's with discontinuous symbols. In other words, we present the solving composition series of the algebras and describe some their properties.

The solving composition series for the algebras of PDO's have been constructed for the first time in such setting (a specific case that the symbols may have discontinuities along a smooth submanifold was studied in [PS 1]). The composition series for the algebras of Wiener–Hopf operators were found in [S] (and in [Dyn 2], [MR] in more special situation). However in the general case the solvability of the algebras of Wiener–Hopf operators in the sense of the 'narrow' definition has been proved for the first time (see explanation in Section 1.2. In brief the narrow definition requires that the subquotients of the composition series should be the algebras determined by trivial fields of elementary algebras, and not only by locally trivial ones as in the broad definition).

The paper consists of five sections. The statement of the problem is given in Subsection 1.2. In Section 2 we recall some general facts about continuous fields of spaces and algebras and deduce sufficient conditions of the triviality of a field of elementary algebras which will be used in the sequel. Section 3 is devoted to the algebras of Wiener–Hopf operators, and Section 4 to the algebras of PDO's.

1.2 Let $\mathcal{A}$ be a solvable C^*–algebras, and let

$$\{0\} = I_{-1} \subset I_0 \subset \cdots \subset I_l = \mathcal{A}$$

be a solving composition series of its ideals. By definition, it means that

$$I_j/I_{j-1} \simeq C_0(T_j) \otimes \mathcal{K}H, j = 1, \ldots, l, (1.1)$$

where T_j is a locally compact topological space and $\mathcal{K}H$ is the ideal of compact operators on a Hilbert space $H = H_j$. It is known that every irreducible representation of the algebra $C_0(T_j) \otimes \mathcal{K}H \simeq C_0(T_j, \mathcal{K}H)$ is equivalent to a representation of the form $\pi_t A = A(t)$, $t \in T_j$.Therefore the spectrum of the above algebra (i.e., the set of all equivalence classes of irreducible representations) is identified with the space T_j. In view of the isomorphism (1.1), $(I_j/I_{j-1})^\wedge = T_j$. Thus, $\hat{I}_j = G_j$ where $G_j = \cup_{i=0}^{j} T_i$. Being the spectrum of an ideal, every set G_j is an open subset of the spectrum of the algebra $\mathcal{A}$.

Let us have a look at the situation from another standpoint. Let $\mathcal{A}$ be an arbitrary C^*–algebra. To each element $a \in \mathcal{A}$ there corresponds the function $\hat{a}$ given on the spectrum $\hat{\mathcal{A}}$ by $\hat{a}(\pi) = \pi(a) \in \mathcal{B}H_\pi$. The functions $\hat{a}$ form an algebra isomorphic to $\mathcal{A}$ since

$$\|\hat{a}\| = \sup\{\|\hat{a}(\pi); \mathcal{B}H_\pi\|; \pi \in \hat{\mathcal{A}}\} = \|a\|.$$

Generally, the field of the algebras $(\{\pi(\mathcal{A})\}, \Theta), \pi \in \hat{\mathcal{A}}$, is not continuous for any choice of the set $\Theta \subset \{\prod \mathcal{B}H_\pi, \pi \in \hat{\mathcal{A}}\}$ of vector fields on $\hat{\mathcal{A}}$ containing $\{\hat{a}, a \in \mathcal{A}\}$. (In Section 2 we recall definitions related to continuous fields of spaces and algebras.

They originate from [D].) The reason is that as a rule the space $\hat{A}$ endowed with the Jacobson topology is not Hausdorff, so the functions $\pi \mapsto \|a(\pi); \mathcal{B}H_\pi\|$ are not continuous (one can ensure the continuity only from below [D], 3.3.2). Let

$$\emptyset = G_{-1} \subset G_0 \subset \cdots \subset G_l = \hat{A}$$

be open subsets of $\hat{A}$ such that $T_j = G_j \setminus G_{j-1}$ is a Hausdorff space for every $j \geq 0$ (the existence of G_j is supposed). To each G_j we associate the ideal

$$I_j = \{a \in \mathcal{A} : \hat{a}(\pi) = 0 \quad \text{for} \quad \pi \in \hat{A} \setminus G_j\}.$$

The spectrum of the ideal can be identified with G_j and the spectrum of I_j/I_{j-1} with T_j. Every element a in I_j determines a vector field $\hat{a}$ on T_j depending only on the coset representative $a + I_{j-1}$. Let Θ_j be the local closure of the set of such fields. Then $(\{\pi(I_j)\}, \Theta_j), \pi \in T_j$, is a continuous field of C^*–algebras (the continuity follows from the fact that T_j is Hausdorff [D], 3.3.9).

The solvability of an algebra $\mathcal{A}$ means that the sets subject to the aforementioned conditions can be chosen for all j to satisfy the requirements:

(i) $\pi(I_j) = \mathcal{K}H_\pi$ for $\pi \in T_j$.

(ii) The field of the algebras $\mathcal{L}_j = (\{\mathcal{K}H_\pi\}_{\pi \in T_j}, \Theta_j)$ is trivial.

In this case the algebra $\mathcal{A}_j$ determined by the field $\mathcal{L}_j$ is isomorphic to $C_0(T_j) \otimes \mathcal{K}H$. On the other hand, $\mathcal{A} \simeq I_j/I_{j-1}$. Therefore the isomorphism (1.1) takes place for $j \in \{0, \ldots, l\}$.

We shall call the above definition of the solvability temporarily the definition in the narrow sense. We obtain the definition in the broad sense by replacing (ii) with the following condition

(ii)′ The fields $\mathcal{L}_j$ are locally trivial. (Then the $\mathcal{A}_j$ are algebras with continuous trace.)

Clearly every algebra solvable in the narrow sense turns out to be solvable in the broad one. Generally, the converse is wrong. If $H^3(T_j, \mathbb{Z}) = 0$ for all j then the broad and narrow definitions coincide. If $H^3(T_j, \mathbb{Z}) \neq 0$ for some j and if it has already been known that $\mathcal{A}$ is solvable in the broad sense then we need further arguing related to specific properties of $\mathcal{A}$ to prove the solvability in the narrow sense. In the cases of interest (algebras of PDO's, algebras of Wiener–Hopf operators) these properties lie in the fact that the union $\{\sqcup H_\pi, \pi \in T_j\}$ of the spaces of representations together with a family Γ_j of 'vector fields' $T_j \ni \pi \mapsto x(\pi) \in H_\pi$ form the locally trivial field $\mathcal{E}_j = (\{H_\pi\}, \Gamma_j)$ of Hilbert spaces. Local trivializations (ϕ, Ω) of the field $\mathcal{E}_j$ generate the local trivializations (Φ, Ω) of the field $\mathcal{L}_j$ (here $\Omega \subset T$ is an open set, ϕ is an isomorphism $\Omega \times H \to \mathcal{E}_j|\Omega$, and the mapping $\Phi : \Omega \times \mathcal{K}H \to \mathcal{L}_j|\Omega$ is determined by the equality $\Phi(\hat{a})(\pi) = \phi(\pi)^{-1}\hat{a}(\pi)\phi(\pi)$. It implies the triviality of the field $\mathcal{L}_j$ (see Section 2.2).

2 Sufficient conditions of the triviality of fields of elementary C^*–algebras

We shall use the following notation: $\mathcal{B}(E, E')$ is the space of linear bounded operators acting from a Banach space E to a Banach space E' (in the case $E = E'$ we write $\mathcal{B}(E)$); $U(H)$ is the group of unitary operators on a Hilbert space H; $\mathcal{K}H$ is the ideal of compact operators in the algebra $\mathcal{B}H$; $C(T, L)$ is the space of continuous functions given on T with values in L where T and L are topological spaces.

2.1. Continuous fields of Banach spaces. Let T be a topological space to every point of which there corresponds a Banach space E_t. A mapping $x : T \to \{\prod E_t, t \in T\}$ satisfying the condition $x(t) \in E_t$ for all $t \in T$ is called a vector field on T. It is convenient sometimes to identify the vector fields x with their images $\{x(t)\} \subset \prod E_t$.

A continuous field $\mathcal{E}$ of Banach spaces on T is a family $\{E_t\}_{t\in T}$ of Banach spaces together with a set Γ of vector fields subject to the following conditions:

(i) Γ is a linear space;

(ii) the set $\{x(t) : x \in \Gamma\}$ is dense in E_t for each $t \in T$;

(iii) the function $t \mapsto \|x(t)\|$ is continuous for every $x \in \Gamma$;

(iv) let $y \in \{\prod E_t, t \in T\}$ be a vector field.

If for any $t_0 \in T$ and $\epsilon > 0$ there exists $x \in \Gamma$ such that $\|y(t) - x(t)\| < \epsilon$ in a neighborhood of t_0 then $y \in \Gamma$.

Elements of Γ are called continuous vector fields in $\mathcal{E}$.

An isomorphism of continuous vector fields $\mathcal{E} = (\{E_t\}, \Gamma)$ and $\mathcal{E}' = (\{E_t'\}, \Gamma')$ is a family $\phi = \{\phi_t\}_{t\in T}$ of linear operators $\phi_t : E_t \to E_t'$ such that

(1) the operator ϕ_t is an isometric isomorphism for every t;

(2) $\phi(\Gamma) \subset \Gamma'$.

It can be shown that (2) is equivalent to the requirement $\phi(\Gamma) = \Gamma'$ [D], 10.2.4.

Example. Let us consider the product $T \times E$ where E is a Banach space. We set $E_t = \{t\} \times E$ and denote the projection $T \times E \to E$ by q with $q_t = q|E_t$. The mapping q enables us to provide every 'fibre' E_t with a structure of Banach space, while the operator $q_t : E_t \to E$ becomes an isometric isomorphism. We take as a family Γ the set of all vector fields of the form $t \mapsto (t, f(t))$, where $f \in C(T, E)$. Then $(\{E_t\}, \Gamma)$ is a continuous field of Banach spaces which is called the constant field. We denote this field somewhat incorrectly by $T \times E$ and write $\Gamma(T \times E)$ instead of Γ. A field isomorphic to the constant one is called trivial.

Let $\mathcal{E} = (\{E_t\}_{t\in T}, \Gamma)$ be a continuous field of Banach spaces. A set $\Lambda \subset \Gamma$ is said total if for every $t \in T$ the set $\{x(t) : x \in \Lambda\}$ is total in E_t. The field $\mathcal{E}$ is called separable if Γ contains a countable total subset.

The next assertion proved in [D],10.8.7, plays an important role in the sequel.

Proposition 2.1. Let T be a finite-dimensional paracompact set and let $\mathcal{E} = (\{E_t\}_{t\in T}, \Gamma)$ be a separable continuous field of Hilbert spaces where the dimension of each E_t is equal to $\aleph_0$. Then the field $\mathcal{E}$ is trivial.

We consider the isomorphisms of trivial fields in more detail. Assume that $\phi : T \times E \to T \times E'$ is such an isomorphism. Then $\phi_t(t,e) = (t, A_t e)$ for every $t \in T$ where A_t is a linear isometric isomorphism $E \to E'$. The operator A_t is connected with ϕ_t by the formula $A_t = q'_t \phi_t q_t^{-1}$.

Lemma 2.2. Let $\phi : T \times E \to T \times E'$ be an isomorphism of trivial fields. Then the function $t \mapsto A_t \in \mathcal{B}(E, E')$ with $A_t = q'_t \phi_t q_t^{-1}$ is continuous relative to the strong operator topology.

Proof. We fix $e \in E$. Since the vector field $t \mapsto (t, e)$ belongs to $\Gamma(T \times E)$, its image under ϕ belongs to $\Gamma(T \times E')$. Consequently, the function $t \mapsto A_t e$ is continuous.

We turn now to the definition of locally trivial fields. Let $\mathcal{E} = (\{E_t\}_{t\in T}, \Gamma)$ be a continuous field of Banach spaces on T and let $\Omega \subset T$ be an open set. We denote by Γ_Ω the set of vector fields on Ω that are the limits of elements in the set $\Gamma|\Omega$ with respect to the local uniform convergence. Then $(\{E_t\}_{t\in\Omega}, \Gamma_\Omega)$ is a continuous field of Banach spaces on Ω which is denoted by $\mathcal{E}|\Omega$ and called the field induced by $\mathcal{E}$ on Ω.

A field $\mathcal{E}$ is called locally trivial if for every point $t \in T$ there exists a neighborhood Ω such that the field $\mathcal{E}|\Omega$ is trivial.

Let $\mathcal{E}$ be a locally trivial field of Banach spaces on T, $\{\Omega_i\}_{i\in I}$ an open covering of T such that all the fields $\mathcal{E}|\Omega_i$ are trivial, and let $\phi_i : T \times E_i \to \mathcal{E}|\Omega_i$ be isomorphisms of the trivial fields. The family $\{(\phi_i, \Omega_i)\}_{i\in I}$ of the charts is called a trivializing atlas or simply an atlas for $\mathcal{E}$.

To close the section we consider continuous fields of C^*–algebras.

A continuous field $(\{\mathcal{A}_t\}, \Theta)$ of Banach spaces on T is called a continuous field of C^*–algebras if every $\mathcal{A}_t$ is a C^*–algebra while Θ is invariant under multiplication and involution. An isomorphism of fields $(\{\mathcal{A}_t\}, \Theta)$ and $(\{\mathcal{A}'_t\}, \Theta')$ is determined by a family $\{\phi_t\}_{t\in T}$ where $\phi_t : \mathcal{A}_t \to \mathcal{A}'_t$ is an isomorphism for every t and $\phi(\Theta) \subset \Theta'$. In accordance with this definition we introduce the notions of the triviality and local triviality of fields of C^*–algebras.

Suppose that $(\{\mathcal{A}_t\}, \Theta)$ is a continuous field of C^*-algebras such that every $\mathcal{A}_t$ is an elementary C^*–algebra (i.e., $\mathcal{A}_t$ is isomorphic to the ideal of compact operators on a Hilbert space). Then $(\{\mathcal{A}_t\}, \Theta)$ is called a continuous field of elementary C^*–algebras.

Let T be a locally compact space, $\mathfrak{A} = (\{\mathcal{A}_t\}, \Theta)$ a continuous field of C^*–algebras on T, and let $\mathcal{A}$ be the set of $x \in \Theta$ such that $\|x(t)\|$ tends to zero as $t \to \infty$. Then $\mathcal{A}$ is a subalgebra in Θ invariant under involution. For $x \in \mathcal{A}$ we set

$$\|x\| = \sup\{\|x(t)\|; t \in T\}.$$

The algebra $\mathcal{A}$ equipped with this norm turns out to be a C^*–algebra which is called the C^*–algebra determined by the field $\mathfrak{A}$. For example, if $\mathfrak{A}$ is a trivial field then

$\mathcal{A} \simeq C_0(T, \mathcal{B}) \simeq C_0(T) \otimes \mathcal{B}$ where $\mathcal{B}$ is a C^*–algebra. In the case that $\mathfrak{A}$ is a trivial field of elementary algebras we have $\mathcal{A} \simeq C_0(T) \otimes \mathcal{K}H$.

2.2. Sufficient condition of the triviality of fields of elementary C^*–algebras. Let T be a finite-dimensional paracompact space and let H be a Hilbert space with $\dim H = \aleph_0$. When considering the trivial fields $\Omega \times H$ and $\Omega \times \mathcal{K}H$ where Ω is an open subset of T, we denote the 'fibres' $\{t\} \times H$ and $\{t\} \times \mathcal{K}H$ by H_t and $(\mathcal{K}H)_t$.

We assume that $\mathcal{E} = (\{E_t\}_{t\in T}, \Gamma)$ is a separable continuous field of Hilbert spaces, $\dim E_t = \aleph_0$, and $\mathcal{L} = (\{\mathcal{K}E_t\}_{t\in T}, \Theta)$ a continuous field of elementary C^*–algebras (to unify notation we shall write $\Gamma(\mathcal{L})$ instead of Θ). According to Proposition 2.1, there exists an isomorphism $\phi : T \times H \to \mathcal{E}$. It is easy to see that if for any $K(\cdot) \in \Gamma(\mathcal{L})$ the vector field $t \mapsto \phi(t)^{-1}K(t)\phi(t)$ belongs to $\Gamma(T \times \mathcal{K}H)$ then the field $\mathcal{L}$ is trivial (we used the notation ϕ_t earlier instead of $\phi(t)$). This simple condition of the triviality of $\mathcal{L}$ is little used because one can not construct the isomorphism ϕ explicitly. In applications, as a rule, local trivializations of the field $\mathcal{E}$ arise. In the next proposition a sufficient condition of the triviality of $\mathcal{L}$ is given in terms of such trivializations.

Proposition 2.3. Let T, H, $\mathcal{E}$, and $\mathcal{L}$ be the same as before and let $\{(\phi_i, \Omega_i)\}_{i\in I}$ be an atlas for $\mathcal{E}$. We suppose that for all $i \in I$ and $K \in \Gamma(\mathcal{L})$ the vector field $t \mapsto \phi_i(t)^{-1}K(t)\phi_i(t)$ on Ω_i belongs to $\Gamma(\Omega_i \times \mathcal{K}H)$. Then the field $\mathcal{L}$ is trivial and, consequently, the C^*–algebra determined by this field is isomorphic to $C_0(T) \times \mathcal{K}H$.

We first prove the following lemma.

Lemma 2.4. Let $t \mapsto R(t) \in \mathcal{K}H$ and $t \mapsto u(t) \in U(H)$ be operator-valued functions on an open set $\Omega \subset$ while R is continuous with respect to the uniform operator topology and u is continuous relative to the strong operator topology. Then the function $t \mapsto u(t)R(t)u(t)^{-1}$ is continuous relative to the uniform topology.

Proof. The continuity of uR with respect to the uniform topology is evident. Clearly the function uRu^{-1} is continuous, too.

Proof of the proposition. Let $\phi = \{\phi(t)\}_{t\in T}$ be an isomorphism $T \times H \to \mathcal{E}$. To every vector field $K(\cdot) \in \Gamma(\mathcal{L})$ there corresponds the vector field

$$\Phi(K) : t \mapsto \phi(t)^{-1}K(t)\phi(t) \in (\mathcal{K}H)_t.$$

It suffices to establish that for any $K(\cdot) \in \Gamma(\mathcal{L})$ the function $Q : t \mapsto q_t\Phi(K)q_t^{-1}$ belongs to $C(T, \mathcal{K}H)$ (as before, we denote by q_t the restriction of the projection $q : T \times H \to H$ to the fibre H_t). Indeed, the inclusion $Q \in C(t, \mathcal{K}H)$ means that $\Phi(K) \in \Gamma(t \times \mathcal{K}H)$, i.e., the family $\{\Phi_t\}$ of the isometric isomorphisms

$$\Phi_t : \mathcal{K}E_t \to \mathcal{K}H_t : \mathfrak{K} \mapsto \phi(t)^{-1}\mathfrak{K}\phi(t)$$

sends $\Gamma(\mathcal{L})$ to $\Gamma(T \times \mathcal{K}H)$.

To check the inclusion $Q \in C(T, \mathcal{K}H)$ it suffices to show that $Q|\Omega_i \in C(\Omega_i, \mathcal{K}H$ for all $i \in I$. By Lemma 2.4, the continuity of $Q|\Omega_i$ follows from the continuity of the function

$$R_i : t \mapsto q_t\psi_i(t)^{-1}\Phi(K)\psi_i(t)q_t^{-1},\ t \in \Omega_i,$$

where $\psi_i = \phi^{-1} \circ \phi_i$. (Indeed, since ψ_i is an isomorphism of the fields $\Omega_i \times H$ and $(T \times H)|\Omega_i$, the function $u_i : t \mapsto q_t\psi_i(t)q_t^{-1}$ is continuous relative to the strong topology by Lemma 2.2). It remains to note that $R_i(t) = q_t\phi_i(t)^{-1}K(t)\phi_i(t)q_t^{-1}$ so the continuity of R_i follows from the hypothesis of the proposition.

The next modification of Proposition 2.3 will be of use in applications.

Proposition 2.5. Suppose that for every $i \in I$ there exists $K_i \in \Gamma(\mathcal{L})$ such that $\phi_i^{-1}K_i\phi_i \in \Gamma(\Omega_i \times \mathcal{K}H)$ and $K_i(t) \neq 0$ for all $t \in \Omega_i$. We also assume that for any $Q \in \Gamma(\mathcal{L})$ the function

$$t \mapsto q_t\phi_i(t)^{-1}Q(t)\phi_i(t)q_t^{-1} \in \mathcal{K}H$$

is continuous with respect to the strong topology. Then the field $\mathcal{L}$ is trivial.

Proof. Since the space T is paracompact, the atlas $\{(\phi_i, \Omega_i)\}$ can be viewed as locally finite. Let $\{\chi_i\}$ be a partition of unity subject to the covering $\{\Omega_i\}$ of T where $\chi_i \in C(T)$ and $\chi_i \geq 0$. Set

$$R(t) = \sum_{j \in J} \chi_i(t)K_j(t)^*K_j(t),\ t \in T.$$

Then $R(t) > 0$ for all $t \in T$ and $R \in \Gamma(\mathcal{L})$ [D], 10.1.9. We show that for every $i \in I$ the field $\phi_i^{-1}R\phi_i$ belongs to $\Gamma(\Omega_i \times \mathcal{K}H)$. Indeed,

$$q_t\phi_i(t)^{-1}R(t)\phi_i(t)q_t^{-1} = \sum \chi_j(t)u_j(t)R_j(t)u_j(t)^{-1}, \tag{2.1}$$

where $R_j(t) = q_t\phi_i(t)^{-1}K_j(t)^*K_j(t)\phi_i(t)q_t^{-1}$ and $u_i(t) = q_t\phi_i(t)^{-1}\phi_j(t)q_t^{-1}$ We have supposed that $R_j \in C(\Omega_j, \mathcal{K}H)$. Besides, since the family of unitary operators $\phi_i(t)^{-1}\phi_j(t)$ with $t \in \Omega_i \cap \Omega_j$ determines an isomorphism $(\Omega_i \cap \Omega_j) \times H \to (\Omega_i \cap \Omega_j) \times H$, the function $\Omega_i \cap \Omega_j \ni t \mapsto u_j \in U(H)$ is continuous relative to the strong topology (Lemma 2.2). By Lemma 2.4, $u_jR_ju_j^{-1} \in C(\Omega_i \cap \Omega_j, \mathcal{K}H)$. In view of (2.1) it follows that $\phi_i^{-1}R\phi_i \in \Gamma(\Omega_i \times \mathcal{K}H)$.

Let $\Gamma_0(\mathcal{L})$ denote the local closure of the linear hull of vector fields of the form QKQ' where $Q, Q' \in \Gamma(\mathcal{L})$. Let $t \in T$ be an arbitrary point. The set $J_t = \{S(t) : S \in \Gamma_0(\mathcal{L})\}$ is a nonzero two-sided ideal in the algebra $\mathcal{K}E_t$. There is no nontrivial closed two-sided ideals in $\mathcal{K}E_t$ so $\overline{J_t} = \mathcal{K}E_t$. Consequently, $(\{\mathcal{K}E_t\}, \Gamma_0(\mathcal{L}))$ is a continuous field of C^*–algebras. Furthermore, for every $i \in I$ all the fields $\phi_i^{-1}QKQ'\phi_i$ belong to $\Gamma(\Omega_i \times \mathcal{K}H)$ (cf. Proof of Lemma 2.4). Therefore $\phi_i^{-1}S\phi_i \in \Gamma(\Omega \times \mathcal{K}H)$ for all $S \in \Gamma_0(\mathcal{L})$. Then according to Proposition 2.3, the field $(\{\mathcal{K}E_t\}, \Gamma_0(\mathcal{L}))$ is trivial.

It remains to check the equality $\Gamma_0(\mathcal{L}) = \Gamma(\mathcal{L})$. Let $Q \in \Gamma(\mathcal{L})$, $t \in T$ and let $S \in \Gamma_0(\mathcal{L})$ be chosen so that $\|Q(t) - S(t)\| < \epsilon$ We have $\|Q(\tau) - S(\tau)\| < \epsilon$ in a neighborhood of t because the function $\tau \mapsto \|Q(\tau) - S(\tau)\|$ is continuous. Therefore $Q \in \Gamma_0(\mathcal{L})$. Thus we obtained that $\Gamma(\mathcal{L}) \subset \Gamma_0(\mathcal{L})$. The converse inclusion follows from the definition of the set $\Gamma_0(\mathcal{L})$.

3 Algebras of Wiener–Hopf operators

3.1. The algebra $\mathcal{W}(\Lambda)$. Let $\Lambda \subset \mathbb{R}^n$ be a closed convex cone with nonempty interior ($\operatorname{Int}\Lambda \neq \emptyset$). An integral operator of the form

$$W_\psi u(x) = \int_\Lambda \psi(x-y)u(y)\,dy, \psi \in \mathrm{L}_1(\mathbb{R}^n) \tag{3.1}$$

is called a Wiener–Hopf operator. We consider such operators on $L_2(\Lambda)$. All the operators W_ψ are bounded:

$$\|W_\psi\| = \|F\psi; C_0(\mathbb{R}^n)\| \leq \|\psi; L_1(\mathbb{R}^n)\|.$$

Furthermore, $W_\psi^* = W_{\psi^*}$ where $\psi^*(x) = \overline{\psi(-x)}$. Let $\mathcal{W}(\Lambda)$ stand for the C^*–algebra generated by the operators W_ψ. The algebra $\mathcal{W}(\Lambda)$ is commutative in the case $\Lambda = \mathbb{R}^n$ only.

Let $P : L_2(\mathbb{R}^n) \to L_2(\Lambda)$ be the orthogonal projector, and let $\psi*$ with $\psi \in L_1(\mathbb{R}^n)$ be the convolution operator $v \mapsto \psi * v$ on $L_2(\Lambda)$. Then (3.1) can be rewritten in the form $W_\psi = P(\psi*)|L_2(\Lambda)$ (we shall omit the sign of the restriction sometimes).

The linear part of a cone Λ, denoted by $l(\Lambda)$, is defined to be the largest linear subspace of $\mathbb{R}^n$ that is contained in Λ. A cone Λ is called acute if $l(\Lambda)$. The following two assertion were proved in [Dyn 2].

Proposition 3.1. The algebra $\mathcal{W}(\Lambda)$ is irreducible if and only if Λ is an acute cone.

Proposition 3.2. 1) If Λ is an acute cone then the algebra $\mathcal{W}(\Lambda)$ contains all the compact operators on $L_2(\Lambda)$. 2) If a cone Λ is not acute then there is no nonzero compact operators in the algebra $\mathcal{W}(\Lambda)$.

3.2. Irreducible representations of the algebra $\mathcal{W}(\Lambda)$. Let $\omega \in \Lambda$. By definition, the support cone Λ_ω of the cone Λ is the least closed convex cone K satisfying the conditions $\Lambda \subset K$ and $\omega \in l(K)$. The plane $T_\omega = l(\Lambda_\omega)$ is called the tangent plane at the point ω. For example, if $\omega \in l(\Lambda)$ then $\Lambda_\omega = \Lambda$ and $T_\omega = l(\Lambda)$; if $\omega \in \operatorname{Int}\Lambda$ then $\Lambda_\omega = T_\omega = \mathbb{R}^n$.

Support cones and tangent planes introduced just now will be called primary. A cone that is a support one for a certain primary cone is called a secondary support cone while its linear part is, by definition, a secondary tangent plane. Subsequent generations of support cones and tangent planes are defined in a similar way. The collection of all the tangent planes of the k–th generation is denoted by $\mathcal{T}^{(k)}(\Lambda)$. It is clear that $\mathcal{T}^{(1)}(\Lambda) \subset \mathcal{T}^{(2)}(\Lambda) \subset \dots$ and $\mathcal{T}^{(n-1)}(\Lambda) = \mathcal{T}^{(n)}(\Lambda) = \dots$. We set $\mathcal{T}(\Lambda) = \cup_{k\geq 1}\mathcal{T}^{(k)}(\Lambda)$. If $T \in \mathcal{T}(\Lambda)$ then Λ_T denotes the support cone with $l(\Lambda_T) = T$. It is easy to see that the correspondence $\Lambda_T \leftrightarrow T$ is a bijection. Indeed, Λ_T can be described as the least cone of the form $K \otimes T$ with $K \subset T^{\perp}$ containing Λ.

Example 3.3. Let Λ be a polyhedral cone, i.e., the intersection of a finite family $\{G_i\}_{i\in I}$ of closed half-spaces, and let $\omega \in \Lambda$ be an arbitrary point. If $\omega \in \operatorname{Int}\Lambda$ then as has already been noted $\Lambda_\omega = \mathbb{R}^n$. If $\omega \in \partial\Lambda$ then $\Lambda_\omega = \cap_{i\in I(\omega)} G_i$ and $T_\omega = \cap_{i\in I(\omega)}\partial G_i$ where $I(\omega) = \{i \in I : \omega \in \partial G_i\}$. Every secondary cone coincides with a primary one. Thus there exists only one generation of support cones (or equivalently only one generation of tangent planes).

Example 3.4. Let $\Lambda \subset \mathbb{R}^n$ be a smooth cone, $n \geq 2$. It means that the intersection $\Lambda \cap S^{n-1}$ has smooth boundary on the sphere. If the cone Λ is acute then the set $\mathcal{T}(\Lambda)$ consists of $\{0\}$, $\mathbb{R}^n$ and the ordinary tangent planes H_ω with $\omega \in \partial\Lambda \setminus \{0\}$ to the smooth surface $\partial\Lambda \setminus \{0\}$. The correspondent support cones are Λ, $\mathbb{R}^n$, and G_ω, where G_ω is the closed half-space with boundary H_ω containing Λ. Since any support cone for a half-space coincides with either $\mathbb{R}^n$ or the half-space itself, there is again only one generation of support cones.

Example 3.5. This example generalizes both the preceding ones. Let Λ be a piecewise smooth cone,i.e., the intersection of a finite family $\{K_i\}_{i\in I}$ of smooth cones, and let $\omega \in \partial\Lambda \setminus \{0\}$ be an arbitrary point. We set $I(\omega) = \{i \in I : \omega \in \partial K_i\}$ and denote by $H_{i,\omega}$ with $i \in I_\omega$ the hyperplane tangent to ∂K_i at ω while $G_{i,\omega}$ is the half-space with boundary $H_{i,\omega}$ containing Λ. Then $\Lambda_\omega = \cap_i H_{i,\omega}$ and $T_\omega = H_{i,\omega}$ where $i \in I(\omega)$. Thus all the primary support cones are polyhedral. By virtue of what was said in Example 3.3 it follows that a piecewise smooth cone gives at most two generations of support cones. In certain situations there exist two generations (see [PS 3]).

The conormal cone to Λ corresponding to a plane $T \in \mathcal{T}(\Lambda)$ is defined to be the acute cone $C_T = \Lambda_T \cap T^{\perp}$. We denote by P_T the orthogonal projector $L_2(T^{\perp}) \to L_2(C_T)$. Introduce the set $\mathcal{W}^{(0)}(\Lambda)$ of operators on $L_2(\Lambda)$ of the form

$$A = \sum_j \prod_k P(\psi_{jk}*)|L_2(\Lambda), \psi_{jk} \in S(\mathbb{R}^n) \tag{3.2}$$

where j and k run over finite sets, and $S(\mathbb{R}^n)$ stands for the Schwartz class of smooth functions rapidly decreasing at infinity. For each point $x \in T$ and every operator (3.2) we set

$$\pi(T,x)A = \sum_j \prod_k P_T((F_T\psi_{jk})_x*)|L_2(C_T) \in \mathcal{W}^{(0)}(C_T);$$

here F_T is the 'partial' Fourier transform on the plane T and the functions $(F_T\psi_{jk})_x$ are defined by $(F_T\psi_{jk})_x(y) = F_T\psi_{jk}(x+y)$, $y \in T^{\perp}$.

Proposition 3.6. ([Dyn 2], [S]). (1) For any plane $T \in \mathcal{T}(\Lambda)$ and point $x \in T$ the mapping $\pi(T, x)$ extends to the representation of the algebra $\mathcal{W}(\Lambda)$ on the space $\mathrm{L}_2(C_T)$.

(2) All the representations $\pi(T, x)$ are irreducible.

(3) If $(T_1, x_1) \neq (T_2, x_2)$ then the representations are nonequivalent.

Remark. The mapping $\pi(T, x) : \mathcal{W}^{(0)}(\Lambda) \to \mathcal{W}^{(0)}(C_T)$ is surjective and the set $\mathcal{W}^{(0)}(C_T)$ is dense in $\mathcal{W}(C_T)$, therefore $\pi(T, x)\mathcal{W}(\Lambda) = \mathcal{W}(C_T)$.

3.3. Completeness of the list of irreducible representations. Let $\mathfrak{S}(\Lambda) = \{(T, x) : T \in \mathcal{T}(\Lambda), x \in T\}$. According to Proposition 3.6, 3) the mapping $\iota : (T, x) \to [\pi(T, x)]$ of $\mathfrak{S}(\Lambda)$ into $\mathcal{W}(\Lambda)^\wedge$ is injective; as usual, $[\pi]$ stands for the equivalence class of the representation π. However, this mapping is not surjective in general. To provide any irreducible representation of $\mathcal{W}(\Lambda)$ with an equivalent representation of the form $\pi(T, x)$ one has to require further conditions on the cone Λ.

We say that a cone Λ satisfies the condition $(\mathcal{T}_\omega)$ with $\omega \in \Lambda$ if there exists conical neighborhoods U and V of the point ω and a continuous 'straightening' mapping $q : U \to V$ such that

(i) $q(\omega) = \omega$;

(ii) $q(U \cap \Lambda) = V \cap \Lambda_\omega$;

(iii) $q(x) - q(y) - (x - y) = o(|x - y|) \quad \text{as} \quad x, y \to \omega$;

(iv) q is homogeneous of degree one.

The condition $(\mathcal{T}_\omega)$ is fulfilled, for instance, if $\omega \in l(\Lambda)$ or $\omega \in \operatorname{Int}\Lambda$; in these cases q is the identity mapping. It is clear that if $\omega_2 = t\omega_1$ for $t > 0$ then the conditions $(\mathcal{T}_{\omega_1})$ and $(\mathcal{T}_{\omega_2})$ can be fulfilled only simultaneously.

A cone Λ will be said to satisfy the condition $(\mathcal{T})$ if it is subject to $(\mathcal{T}_{\omega_1})$ for all $\omega \in \Lambda$ or,equivalently, for all $\Omega \in S^{n-1} \cap \partial\Lambda$. Finally, if all the conormal cones satisfy $\mathcal{T}$ we say that Λ subject to the condition $(\mathfrak{CT})$ (i.e., Λ is a 'completely tangible cone' according to [Dyn 2]). For example, polyhedral and smooth cones satisfy the condition $(\mathfrak{CT})$ as well as any closed convex cone in $\mathbb{R}^3$. Piecewise cones that are the intersections of smooth cones K_i satisfy $(\mathfrak{CT})$ under the additional condition of the regularity of the intersections of the surfaces ∂K_i (see [S], [PS 3]) for details).

The next assertion is one of the main results in [Dyn 2].

Theorem 3.7. If a cone Λ satisfies the condition $(\mathfrak{CT})$ then every irreducible representation of the algebra $\mathcal{W}(\Lambda)$ is equivalent to one of the representations $\pi(T, x)$.

Corollary 3.8. $\mathcal{W}(\Lambda)$ is an algebra of type 1.

Proof. Let π be an irreducible representation of $\mathcal{W}(\Lambda)$ on a Hilbert space H. Due to Theorem 3.7, $\pi \sim \pi(T,x)$ for a point $(T,x) \in \mathfrak{S}(\Lambda)$. Since $\pi(T,x)\mathcal{W}(\Lambda) = \mathcal{W}(C_T)$ and the cone C_T is acute, we obtain that $\pi(T,x)\mathcal{W}(\Lambda) \supset \mathcal{K}L_2(C_T)$ (see Proposition 3.2, 1)). Therefore $\pi\mathcal{W}(\Lambda) \supset \mathcal{K}H$.

Sections 3.2 and 3.3 may be summed up as follows. Let Λ be a closed convex cone in $\mathbb{R}^n$ with nonempty interior, subject to $(\mathfrak{CT})$. Then the set $\mathfrak{S}(\Lambda)$ of all the pairs (T,x) where $T \in \mathcal{T}(\Lambda)$ and $x \in T$ and the spectrum $\mathcal{W}(\Lambda)^\wedge$ are connected by a bijection $\iota : (T,x) \mapsto \pi(T,x)$. In the case $\dim T < n$ the representation $\pi(T,x)$ is infinite-dimensional; the algebra $\pi(T,x)\mathcal{W}(\Lambda)$ coincides with the algebra $\mathcal{W}(C_T)$ of Wiener–Hopf operators on the conormal cone C_T while $\mathcal{W}(C_T) \supset \mathcal{K}L_2(C_T)$. If the cone Λ is acute then the set $\mathfrak{K}(\Lambda)$ contains the pair $(\{0\},0)$ and $\pi(\{0\},0)$ is the identity representation of $\mathcal{W}(\Lambda)$ on $L_2(\Lambda)$. In the case $T = \mathbb{R}^n$ there arise the one-dimensional representations $\pi(\mathbb{R}^n,x) : W_\psi \mapsto (F\psi)(x), x \in \mathbb{R}^n$.

3.4. The Jacobson topology on $\mathcal{W}(\Lambda)^\wedge$. On the set of all compact subsets of $\mathbb{R}^n$ we define a semimetric ρ_0 by $\rho_0(E_1,E_2) = \sup\{\operatorname{dist}(x,E_2); x \in E_1\}$. The semimetric ρ_0 is neither symmetric (generally $\rho_0(E_1,E_2) \neq \rho_0(E_2,E_1)$) nor nondegenerate (it is easy to see that the equality $\rho_0(E_1,E_2) = 0$ is equivalent to the inclusion $E_1 \subset E_2$). Nevertheless, the triangle inequality $\rho_0(E_1,E_3) \le \rho_0(E_1,E_2) + \rho_0(E_2,E_3)$ is fulfilled. We recall that the standard distance between sets E_1 and E_2 is $\tilde{\rho}(E_1,E_2) = \max\{\rho_0(E_1,E_2), \rho_0(E_1,E_2)\}$.

Let us fix a closed convex cone $\Lambda \subset \mathbb{R}^n$ with $\operatorname{Int}(\Lambda) \neq \emptyset$. On $\mathfrak{S}(\Lambda) \times \mathfrak{S}(\Lambda)$ we define the function

$$r((S,y),(T,x)) = \rho(\Lambda_S,\Lambda_T) + |y - p_S x|;$$

here $\rho(\Lambda_S,\Lambda_T) = \rho_0(\Lambda_S \cap B^n, \Lambda_T \cap B^n)$, $B^n = \{x \in \mathbb{R}^n : |x| \le 1\}$, and $p_S x$ is the projection of x onto the plane S. For every $(T,x) \in \mathfrak{S}(\Lambda)$ and $\epsilon > 0$ we set $B(T,x;\epsilon) = \{(S,y) \in \mathfrak{S}(\Lambda) : r((S,y),(T,x)) < \epsilon\}$. It is easy to check that if $(S,y) \in B(T,x;\epsilon)$ then $B(S,y;\delta) \subset B(T,x;\epsilon)$ for all small $\delta > 0$. Therefore the 'balls' $B(T,x;\epsilon)$ with $(T,x) \in \mathfrak{S}(\Lambda)$ and $\epsilon > 0$ form a base of a topology on $\mathfrak{S}(\Lambda)$ which is called the r–topology. We denote by J the topology on $\mathfrak{S}(\Lambda)$ induced by the Jacobson topology by means of the mapping $\iota^{-1} : \mathcal{W}(\Lambda)^\wedge \to \mathfrak{S}(\Lambda)$. If $\mathcal{Z}$ is an everywhere dense subset of $\mathcal{W}(\Lambda)$ then sets of the form $\{(T,x) \in \mathfrak{S}(\Lambda) : \|\pi(T,x)A; \mathcal{B}L_2(C_T)\| > 1\}$ with $A \in Z$ comprise a base of the J-topology [D], 3.3.3.

Theorem 3.9. ([S]). Let $\Lambda \subset \mathbb{R}^n$ be a closed convex cone with nonempty interior satisfying the condition $(\mathfrak{CT})$. Then J– and r–topology on $\mathfrak{S}(\Lambda)$ coincide.

3.5. The solving composition series for the algebra $\mathcal{W}(\Lambda)$. As before, Λ denotes a closed convex cone with nonempty interior. Let $\{m_0,\dots,m_l\}$ be the set of all values of the function $\mathcal{T}(\Lambda) \ni T \mapsto \dim T$ ordered increasingly, and let $\mathfrak{S}_j = \{(T,x) \in \mathfrak{S}(\Lambda) : \dim T = m_j\}$. Using the bijection $\iota : (T,x) \mapsto \pi(T,x)$ we identify $\mathcal{W}(\Lambda)^\wedge$ with the set $\mathfrak{S}(\Lambda)$ endowed with the J–topology. It can be shown [S] that on every set $\mathfrak{S}_j$ the topology induced by the J–topology is generated by the metric

$$\tilde{r}((S,y),(T,x)) = \tilde{\rho}(\Lambda_S,\Lambda_T) + |y-x|,$$

where $\tilde{\rho}(\Lambda_S,\Lambda_T) = \tilde{\rho}_0(\Lambda_S \cap B^n, \Lambda_T \cap B^n)$ (see the preceding section). All the metric spaces $(\mathfrak{S}_j, \tilde{r})$ are locally compact, finite-dimensional, and second countable. Consequently, they are paracompact.

For each $j = -1, 0, \dots, l$ we define an ideal J_j in the algebra $\mathcal{W}(\Lambda)$ by $J_j = \cap \ker \pi(T,x)$, $(T,x) \in \cup_{k>j} \mathfrak{S}_k$. Then

$$\{0\} = J_{-1} \subset J_0 \subset \cdots \subset J_l = \mathcal{W}(\Lambda) \tag{3.3}$$

so the ideals J_j form a composition series. It is easily seen that $(J_j/J_{j-1})^\wedge = \mathfrak{S}_j, j = 0, \dots, l$.

We fix $j \in \{0,\dots,l\}$. To every point $(T,x) \in \mathfrak{S}_j$ we associate the algebra $\mathcal{K}_{T,x} = \mathcal{K}L_2(C_T)$. If $A \in J_j$ then $\pi(T,x)A \in \mathcal{K}_{T,x}$ for all $(T,x) \in \mathfrak{S}_j$. Indeed, from the definition of J_j it follows that $\pi(T,x)A$ is a compact operator on $L_2(C_T)$ because it belongs to the intersection of the kernels of all irreducible representations of $\mathcal{W}(C_T)$ with the possible exception of the identity one. Thus to every element $A \in J_j$ there corresponds the vector field $Q_A : (T,x) \mapsto \pi(T,x)A \in \mathcal{K}_{T,x}$ on $\mathfrak{S}_j$. The collection of all such fields is denoted by $\mathcal{A}_j$. The field Q_A depends only on the coset representative $[A]$ of A in the quotient algebra J_j/J_{j-1}. We introduce a norm in $\mathcal{A}_j$ by the equality $\|Q;\mathcal{A}_j\| = \sup\{\|Q(T,x);\mathcal{K}_{T,x}\|;(T,x) \in \mathfrak{S}_j\}$. Since $(J_j/J_{j-1})^\wedge = \mathfrak{S}_j$, we obtain $\|[A]\| = \sup\{\|\pi(T,x)A\|;(T,x) \in \mathfrak{S}_j\} = \|Q_A\|$. Therefore the mapping $J_j/J_{j-1} \ni [A] \mapsto Q_A \in \mathcal{A}_j$ where $A \in J_j$, is an isomorphism of C^*–algebras.

Let Θ_j be the set of vector fields on $\mathfrak{S}_j$ that are the limits of elements of $\mathcal{A}_j$ with respect to the local uniform convergence. Then $\mathcal{L}_j = (\{\mathcal{K}_{T,x}\}, \Theta_j)$ with $(T,x) \in \mathfrak{S}_j$ is a continuous field of elementary C^*–algebras. The algebra determined by this field coincides with $\mathcal{A}_j$ [D], 10.5.4. The algebra $\mathcal{W}(\Lambda)$ is separable so all the fields $\mathcal{L}_j$ are separable too. In what follows we shall write $\Gamma(\mathcal{L}_j)$ instead of Θ_j.

In [S] it was shown that the field $\mathcal{L}_j$ is locally trivial for any $j \in \{0,\dots,l\}$ (the proof was based on Dynin's idea [Dyn 3]). Actually the fields $\mathcal{L}_j$ are trivial, so the series (3.3) is a solving composition series in $\mathcal{W}(\Lambda)$. The triviality of $\mathcal{L}_l$ is evident because $\mathcal{L}_l \simeq C(\mathbb{R}^n)$. The case $j < l$ we consider in Section 3.6 and 3.7.

3.6. The triviality of the field $\mathcal{L}_{l-1}$. We recall that elements of the space $\mathfrak{S}_j$, $j = 0,\dots,l$, are pairs (T,x) with $T \in \mathcal{T}(\Lambda)$, $x \in T$, and $\dim T = m_j$. For $\sigma = (T,x) \in \mathfrak{S}_j$ we set $C_\sigma = x + C_T$. Somewhat in contrast to the preceding section we shall consider $\pi(\sigma) \equiv \pi(T,x)$ as a representation of $\mathcal{W}(\Lambda)$ on $L_2(C_\sigma)$:

$$\pi(\sigma)W_\psi = P_\sigma((F_T\psi)_x *)|L_2(C_\sigma), \tag{3.4}$$

where $P_\sigma : L_2(x+T^\perp) \to L_2(C_\sigma)$ is the orthogonal projection, $(F_T\psi)_x$ the restriction of $F_T\psi$ to the plane $x + T^\perp$. We identify the algebra $\mathcal{K}_\sigma \equiv \mathcal{K}_{T,x}$ with $\mathcal{K}L_2(C_\sigma)$ and write E_σ instead of $L_2(C_\sigma)$.

Let us turn to the space $\mathfrak{S}_{l-1}$. It consists of points $\sigma = (T, x)$ with $\dim T = n-1$. All the cones C_σ are rays. From the definition of the metric $\tilde{r}$ it follows that there exists an open covering $\{\Omega_i\}_{i\in I}$ of $\mathfrak{S}_{l-1}$ such that if σ and σ_1 lie in the same Ω_i then the angle between C_σ and C_{σ_1} is acute. By means of the mapping $\mathbb{R} \ni t \mapsto te_1 \in \mathbb{R}^n$ where $e_1 = (1, 0, \ldots, 0)$ we identify $\mathbb{R}_+$ with the half-line $\{te_1 : t > 0\}$ and $\mathbb{R}$ with the corresponding line. Denote by $V_i(\sigma)$ with $\sigma \in \Omega_i$ an affine isometric transformation of $\mathbb{R}^n$ that sends C_σ onto $\mathbb{R}_+$. The $V_i(\sigma)$ can be chosen in such a way that the function $\Omega_i \ni \sigma \mapsto V_i(\sigma)y$ proves to be continuous for any $y \in \mathbb{R}^n$.

We consider the trivial field $\Omega_i \times L_2(\mathbb{R}_+)$ of Hilbert spaces. Let $H_\sigma = \{\sigma\} \times L_2(\mathbb{R}_+)$ for $\sigma \in \Omega_i$ and let $Q_i(\sigma)$ be the restriction to H_σ of the projection $\Omega_i \times L_2(\mathbb{R}_+) \to L_2(\mathbb{R}_+)$. To every transform $V_i(\sigma)$ there corresponds the unitary operator

$$\phi_i(\sigma) : H_\sigma \to E_\sigma : u \mapsto (q_i(\sigma)u) \circ (V_i(\sigma)|C_\sigma).$$

The family $\{\phi_i(\sigma)\}_{\sigma\in\Omega_i}$ determines a bijection

$$\phi_i : \Omega_i \times L_2(\mathbb{R}_+) \to \bigsqcup_{\sigma\in\Omega_i} E_\sigma.$$

Let $h : \sigma \mapsto (\sigma, f(\sigma))$ with $f \in C(\Omega_i \times L_2(\mathbb{R}_+))$ be an arbitrary vector field in $\Gamma(\Omega_i \times L_2(\mathbb{R}_+))$. Its image under ϕ_i is the vector field

$$\Omega_i \ni \sigma \mapsto \phi_i(\sigma)h(\sigma) \in E_\sigma.$$

The collection of all such fields is denoted by Γ_i. Then $(\{E_\sigma\}_{\sigma\in\Omega_i}, \Gamma_i)$ is a continuous Hilbert space field on Ω_i. Let Γ be the set of all the vector fields satifying $\gamma|\Omega_i \in \Gamma_i$ for every $i \in I$. It is easy to check (see the proof of Proposition 10.1.13 in [D]) that Γ is subject to (i) - (iv) in 2.1. Therefore $\mathcal{E} = (\{E_\sigma\}_{\sigma\in\mathfrak{S}_{l-1}}, \Gamma)$ is a continuous vector field of Hilbert spaces on $\mathfrak{S}_{l-1}$, while the family $\{(\phi_i, \Omega_i)\}$ is a trivializing atlas of the field. Since $\mathfrak{S}_{l-1}$ is a separable metric space, the field $\mathcal{E}$ is separable too [D], 10.2.7.

Let T' denote the hyperplane $\{(0, x') : x' \in \mathbb{R}^{n-1}\}$ in $\mathbb{R}^n$, F' the Fourier transform on T', and let P' be the orthogonal projection $L_2(\mathbb{R}) \to L_2(\mathbb{R}_+)$. By (3.4), we have

$$\phi_i(\sigma)^{-1}(\pi(\sigma)W_\psi)\phi_i(\sigma) = (\sigma, P'(\psi'_{i,\sigma} *)) \tag{3.5}$$

for an arbitary generator W_ψ in $\mathcal{W}(\Lambda)$ and any $i \in I$ and $\sigma \in \Omega_i$ where $\psi'_{i,\sigma} = (F'\psi_{i,\sigma})|\mathbb{R}^+$,
$\psi_{i,\sigma} = \psi \circ V_i(\sigma)^{-1}$. It is easily seen that the function

$$\Omega_i \ni \sigma \mapsto P'(\psi'_{i,\sigma} *) \in \mathcal{W}(\mathbb{R}_+) \tag{3.6}$$

is continuous. Indeed, the continuity of the mapping $\sigma \mapsto V_i(\sigma)$ implies $\|\psi_{i,\sigma} - \psi_{i,\sigma_1}; L_1(\mathbb{R}^n)\| \to 0$ as $\sigma \to \sigma_1$. We consider the function $F'\psi_{i,\sigma}$ as an element of the space $C_0(T', L_1(\mathbb{R}))$ and conclude that $F'\psi_{i,\sigma} \to F'\psi_{i,\sigma_1}$ in $C_0(T', L_1(\mathbb{R}))$. Then $\psi'_{i,\sigma}(\cdot) = (F'\psi_{i,\sigma})(\cdot, 0') \to (F'\psi_{i,\sigma_1})(\cdot, 0') = \psi'_{i,\sigma_1}(\cdot)$ in $L_1(\mathbb{R})$, consequently, $P'(\psi'_{i,\sigma} *) \to P'(\psi'_{i,\sigma_1} *)$ in $\mathcal{W}(\mathbb{R}_+)$. Thus the function (3.6) is continuous.

From (3.5) it follows now that the vector field $\Omega_i \ni \sigma \mapsto \phi_i(\sigma)(\pi(\sigma) \times W_\psi)\phi_i(\sigma) \in \mathcal{B}H_\sigma$ belongs to $\Gamma(\Omega_i \times \mathcal{W}(\mathbb{R}_+))$. In view of the ψ is an arbitrary function in $L_1(\mathbb{R}^n)$ the field

$$\Omega_i \ni \sigma \mapsto \phi_i(\sigma)^{-1}(\pi(\sigma)A)\phi_i(\sigma) \in \mathcal{B}H_\sigma, \tag{3.7}$$

too, belongs to $\Gamma(\Omega_i \times \mathcal{W}(\mathbb{R}_+))$ for any $A \in \mathcal{W}(\Lambda)$. In particular, the field (3.7) is an element of $\Gamma(\Omega_i \times \mathcal{K}L_2(\mathbb{R}_+))$ for $A \in J_{l-1}$. The last assertion is still valid if $\pi(\cdot)A$ in (3.7) is replaced by an arbitrary element in $\Gamma(\mathcal{L}_{l-1})$. Applying Proposition 2.3 with $\mathfrak{S}_{l-1}$, $\mathcal{L}_{l-1}$, $L_2(\mathbb{R}_+)$ instead of T, $\mathcal{L}$, H, we conclude that the field $\mathcal{L}_{l-1}$ is trivial.

3.7. The triviality of fields $\mathcal{L}_j$ for $j < l-1$. In this section we describe the main stages by which the proof of the triviality of $\mathcal{L}_j$ proceeds in the case $j < l-1$ omitting some technical details. Let us fix $j \in \{0, \dots, l-2\}$. Recall that $\mathcal{L}_j$ is defined on the metric space $\mathfrak{S}_j$ consisting of points $\sigma = (T, x)$ with $\dim T = m_j$.

In the subspace $\mathbb{R}^{n-m_j} = \{x = (x_1, \dots, x_{n-m_j}, 0, \dots, 0\} \subset \mathbb{R}^n$ we choose an acute closed convex cone C' with nonempty interior containing the ray $\mathbb{R}_+ = \{te_1 : t > 0\}$. Let $\sigma^{(0)} = (T^{(0)}, x^{(0)}) \in \mathfrak{S}_j$. There exists a neighborhood $\Omega = \Omega(\sigma^{(0)}) \subset \mathfrak{S}_j$ of $\sigma^{(0)}$ such that if $(T, x) \in \Omega$ then the orthogonal projection $p_T : (T^{(0)})^\perp \to T^\perp$ is a bijection. Let us fix a ray $l_{\sigma^{(0)}}$ in $\operatorname{Int} C_\sigma^{(0)}$ with origin at x_0. (The definition of the cones C_σ was given at the beginning of Section 3.6). If the neighborhood Ω is sufficiently small then the ray $l_\sigma = x + p_T(-x^{(0)} + l_{\sigma^{(0)}})$ is situated in $\operatorname{Int} C_\sigma$ for all $\sigma = (T, x) \in \Omega$.

We denote by $V(\sigma)$ with $\sigma = (T, x) \in \Omega$ an affine isometric transform of $\mathbb{R}^n$ that maps $x + T^\perp$ on $\mathbb{R}^{n-m_j}$ and l_σ on $\mathbb{R}^n$. The $V(\sigma)$ can be chosen continuously depending on σ. Set $C'_\sigma = V(\sigma)^{-1}(C')$. We assume that $y \in l_\sigma$, $|y - x| = 1$, and $S \subset x + T^\perp$ is a plane of dimension $n - m_j - 1$ that contains y and is orthogonal to l_σ. Let G_σ and G'_σ stand for the intersections $S \cap C_\sigma$ and $S \cap C'_\sigma$. Finally, denote by $p_\sigma^{(0)}$ the projection along rays with origin at y lying in S. Then $p_\sigma^{(0)}$ sends G_σ onto G'_σ. We extend $p_\sigma^{(0)}$ to a bijection p_σ by setting $p_\sigma(z) = k^{-1}p_\sigma^{(0)}(x + k(z - x))$ where $z \in C_\sigma$ and k is a positive number such that $x + k(z - x) \in S$. One can check that the mappings p_σ and p_σ^{-1} are Lipschitz, so they are differentiable almost everywhere. Then the same is true for $U(\sigma) = V(\sigma) \circ p_\sigma : C_\sigma \to C'$ and for the inverse $U(\sigma)^{-1}$. Denote by $\Delta_\sigma^{\pm 1}$ the Jacobian of $U(\sigma)^{\pm 1}$. The function Δ_σ (Δ_σ^{-1}) is defined on the cone C_σ (C'), measurable, bounded, separated from zero, while $\Delta_{\sigma^{(k)}}^{-1} \to \Delta_\sigma^{-1}$ almost everywhere on C' as $\sigma^{(k)} \to \sigma \in \Omega$.

We consider the trivial field $\Omega \times L_2(C')$. Let $H_\sigma = \{\sigma\} \times L_2(C')$ with $\sigma \in \Omega$ and let $q(\sigma)$ be the restriction of the projection $\Omega \times L_2(C') \to L_2(C')$ to H_σ. To every transform $U(\sigma)$ there corresponds the unitary operator

$$\phi(\sigma) : H_\sigma \to E_\sigma : u \mapsto \sqrt{\Delta_\sigma}(q(\sigma)u) \circ (U(\sigma)|C_\sigma). \tag{3.8}$$

The family $\{\phi_\sigma\}_{\sigma\in\Omega}$ determines a bijection

$$\phi : \Omega \times L_2(C') \to \bigsqcup_{\sigma\in\Omega} E_\sigma .$$

Let $h : \sigma \mapsto (\sigma, f(\sigma))$ with $f \in C(\Omega, L_2(C'))$ be an arbitrary vector field in $\Gamma \in C(\Omega, L_2(C'))$. The collection of images of all such fields under the mapping ϕ is denoted by Γ_Ω. Then $\mathcal{E}_\Omega = (\{E_\sigma\}_{\sigma\in\Omega}, \Gamma_\Omega)$ is a continuous field of Hilbert spaces on Ω.

For every generator W_ψ of the algebra $\mathcal{W}(\Lambda)$

$$\phi(\sigma)^{-1}(\pi(\sigma)W_\psi)\phi(\sigma) = q(\sigma)^{-1}A(\sigma)q(\sigma) = (\sigma, A(\sigma)) \tag{3.9}$$

where $A(\sigma)$ with $\sigma \in \Omega$ is a bounded operator on $L_2(C')$. In view of (3.4) and (3.8)

$$A(\sigma)u(z) = \int_{C'} \sqrt{\Delta_\sigma^{-1}\Delta_\sigma^{-1}(w)}(F_T\psi)_x(U(\sigma)^{-1}(z) - U(\sigma)^{-1}(w))u(w)\,dw. \tag{3.10}$$

From the properties of $U(\sigma)$ and $\Delta_\sigma^{\pm1}$ indicated above it follows that $\|A(\sigma^{(k)})u - A(\sigma)u;$
$L_2(C')\| \to 0$ as $\sigma^{(k)} \to \sigma$. Taking into account (3.9) we conclude that the function

$$\Omega \ni \sigma \mapsto q(\sigma)\phi(\sigma)^{-1}(\pi(\sigma)W_\psi)\phi(\sigma)q(\sigma)^{-1} \in \mathcal{B}L_2(C')$$

is continuous with respect to the strong operator topology. Since ψ is an arbitrary element in $L_1(\mathbb{R}^n)$, the same is true after changing from W_ψ to any $A \in \mathcal{W}(\Lambda)$. Therefore

(a) if $Q \in \Gamma(\mathcal{L}_j)$ then the function

$$\Omega \ni \sigma \mapsto q(\sigma)\phi(\sigma)^{-1}Q(\sigma)\phi(\sigma)q(\sigma)^{-1} \in \mathcal{K}L_2(C')$$

is continuous with respect to the strong operator topology.

Let $k = n - m_j$ and let $C \subset \mathbb{R}^k$ be an acute closed convex cone with $\operatorname{Int} C \neq 0$. The polar cone C^0 is defined by the equality $C^0 = \{\xi \in \mathbb{R}^k : x\xi \leq 0\ \forall x \in C\}$. Note that $\operatorname{Int} C^0 \neq 0$ because of $l(C) = \{0\}$. We denote the characteristic function of the cone C by χ and choose functions ψ_1 and ψ_2 in $S(\mathbb{R}^k)$ so that their Fourier transform belong to $C_c^\infty(\mathbb{R}^k)$ while

$$\{\xi - \eta : (\xi, \eta) \in \operatorname{supp}\hat{\psi}^{(1)} \times \operatorname{supp}\hat{\psi}^{(2)}\} \subset \operatorname{Int} C^0. \tag{3.11}$$

Simple calculation shows that the distribution $\hat{\chi}$ on the domain $\operatorname{Int} C^0$ coincides with the function

$$\xi \mapsto c_k \int_{C\cap S^{k-1}} (\xi\omega)^{-k}\,d\omega.$$

We set $R = \chi(\psi^{(1)}*)\chi(\psi^{(2)}*)|L_2(C')$. The operator $F(\psi^{(1)}*)\chi(\psi^{(2)}*)F^{-1}$ is integral, and its kernel $\mathcal{H}(\xi,\eta) = \hat{\psi}^{(1)}(\xi)\hat{\chi}(\xi-\eta)\hat{\psi}^{(2)}(\eta)$ is smooth and has compact support. Therefore R is a Hilbert-Schmidt operator with norm

$$\|R\|_2 = (\iint_{C\times C} |\mathcal{R}(x,y)|^2\,dx\,dy)^{1/2},$$

where

$$\mathcal{R}(x,y) = (2\pi)^{-k}\iint e^{i(x\xi-y\eta)}\mathcal{H}(\xi,\eta)\,d\xi\,d\eta.$$

In what follows we set $\psi_\sigma^{(i)} = (F_T\Psi^{(i)})_x$ for $\Psi^{(i)} \in S(\mathbb{R}^n)$, where $i = 1,2$, and $\sigma = (T,x) \in \Omega$. Let $R(\sigma) = \pi(\sigma)(W_{\psi^{(1)}}W_{\psi^{(2)}})$ with $\sigma \in \mathfrak{S}_j$. The functions $\Psi(1)$ and $\Psi(2)$ can be chosen so that $\psi_\sigma^{(1)}$ and $\psi_\sigma^{(2)}$ satisfy (3.11) with $C_\sigma^0 = x + C_T^0$ instead of C^0 for all $\sigma \in \Omega$ (conceivably to this end one has to diminish the neighborhood Ω) while $R(\sigma^{(0)}) \neq 0$. We have $R(\sigma) = \chi_\sigma(\psi_\sigma^{(1)}*)\chi_\sigma(\psi_\sigma^{(2)}*)|L_2(C_\sigma)$ where χ_σ is the characteristic function of the cone C_σ. According to what has been proved, the operator R_σ is Hilbert-Schmidt on $L_2(C_\sigma)$ for any $\sigma \in \Omega$. Let $\mathcal{R}_\sigma$ stands for its kernel. Setting

$$B(\sigma)u(z) = \int_{C'} \sqrt{\Delta_\sigma^{-1}(z)\Delta_\sigma^{-1}(w)}\mathcal{R}_\sigma(U(\sigma)^{-1}(z),U(\sigma)^{-1}(w))u(w)\,dw$$

for $u \in L_2(C')$ we obtain

$$\phi(\sigma)^{-1}R(\sigma)\phi(\sigma) = q(\sigma)^{-1}B(\sigma)q(\sigma) = (\sigma, B(\sigma))$$

(cf. (3.10) and (3.9)). It can be shown that the function $\Omega \ni \sigma \mapsto B(\sigma)$ is continuous with respect to the Hilbert-Schmidt norm and, consequently, relative to the operator norm, too.

We choose a vector field $Q \in \Gamma(\mathcal{L}_j)$ such that $Q(\sigma^{(0)})R(\sigma^{(0)}) \neq 0$ and $Q(\sigma) = 0$ for $\sigma \in \mathfrak{S}_j \setminus \Omega$. Further, we set $K = QR$. By virtue of the assertion (a) the function

$$\begin{aligned}\mathfrak{S}_j \ni \sigma &\mapsto q(\sigma)\phi(\sigma)^{-1}K(\sigma)\phi(\sigma)q(\sigma)^{-1}\\ &= q(\sigma)\phi(\sigma)^{-1}Q(\sigma)\phi(\sigma)q(\sigma)^{-1}B(\sigma) \in \mathcal{K}L_2(C')\end{aligned}$$

is continuous with respect to the norm. Thus we establish the existence of $K \in \Gamma(\mathcal{L}_j)$ such that

(b) the vector field $\Omega \ni \sigma \mapsto \phi(\sigma)^{-1}K(\sigma)\phi(\sigma)$ belongs to $\Gamma(\Omega \times \mathcal{K}L_2(C'))$.

To every point $\sigma^{(0)} = (T^{(0)}, x^{(0)}) \in \mathfrak{S}_j$ we have associated its neighborhood $\Omega = \Omega(\sigma^{(0)})$, an isomorphism $\phi : \Omega \times L_2(C') \to \mathcal{E}_\Omega$ and a vector field $K \in \Gamma(\mathcal{L}_j)$ satisfying (a) and (b). The neighborhoods $\Omega(\sigma)$ form a covering of the space $\mathfrak{S}_j$. To complete the proof it remains to apply Proposition 2.5.

4 Algebras of pseudodifferential operators

In this section we construct the solving composition series for algebras generated by zero order PDO's on a smooth compact m-dimensional manifold $\mathcal{M}$ without boundary. The symbols of the operators may have discontinuities supported, roughly speaking, by submanifolds intersecting at nonzero angles. More precisely, the structure of admissible discontinuities are described by means of a stratification of the manifold $\mathcal{M}$. The 'strata' of positive codimension play the role of the supports of discontinuities.

The construction of the solving composition series is divided into two stages. We first produce an auxiliary composition series. If the algebra $\mathcal{A}/\mathcal{K}L_2(\mathcal{M})$ is noncommutative then the series is not solving. However, its subquotients are structured in a similar fashion. They are comparatively simple which enable us to obtain the solving series.

4.1. Stratification of the manifold $\mathcal{M}$. Let $\mathcal{T} = \{s^j_\alpha\}$ be a finite partition of $\mathcal{M}$ into 'strata' subject to the following conditions. The stratum s^j_α is a connected j-dimensional submanifold of $\mathcal{M}$. We do not exlude the partition consisting of the only element - the manifold itself. The boundary $\overline{s^j_\alpha} \setminus s^j_\alpha$ of each stratum is a union of strata of smaller dimension (or the empty set). The collection of all strata with boundary containing s^j_α is called the star $\mathrm{st}\,(s^j_\alpha)$ of s^j_α.

It is required that for every point $x \in s^j_\alpha$ there exist a neighborhood U in $\mathcal{M}$ and a diffeomorphism $\kappa : U \to \mathbb{R}^m$ that locally rectifies s^j_α and all $s^k_\beta \in \mathrm{st}\,(s^j_\alpha)$. It means that $\kappa(s^j_\alpha \cap U)$ is a neighborhood of the origin in $\mathbb{R}^j = \{y \in \mathbb{R}^m : y_1 = \cdots = y_{m-j} = 0\}$ and $\kappa(s^k_\beta \cap U)$ coincides near $\kappa(s^j_\alpha \cap U)$ with the product $\mathbb{R}^j \times \mathbb{K}^{k-j}$ where $\mathbb{K}^{k-j}$ is a $(k-j)$-dimensional cone in the subspace $\mathbb{R}^{m-j}$ orthogonal to $\mathbb{R}^j$.

Let S^{m-j-1} be the unit sphere in this subspace $\mathbb{R}^{m-j}$. It is assumed that the projections $p(s^k_\beta)$ of the sets $\kappa(s^k_\beta \cap U)$ onto S^{m-j-1} where $s^k_\beta \in \mathrm{st}\,(s^j_\alpha)$ form an admissible partition of S^{m-j-1} in the sense of the above requirements. We give to the partition $\mathcal{T}(s^j_\alpha)$ and the stratum $p(s^k_\beta)$ with $k \le m-1$ the roles of $\mathcal{T}$ and s^j_α. This determines a partition of S^{m-j-1}; denote it by $\mathcal{T}(s^j_\alpha, s^k_\beta)$. In general, let $s_1, \ldots, s_k$ be a chain of strata such that $\dim s_j = d_j, d_k \le m-1, s_i \in \mathrm{st}\,(s_j)$ for $i > j$ (simplifying the notation, we write strata with a single index). Continuing the procedure, we obtain the partitions $\mathcal{T}(s_1)$, $\mathcal{T}(s_1, s_2)$, ... ,$\mathcal{T}(s_1, \ldots, s_k)$. If $\mathcal{T}(s_1, \ldots, s_k)$ consists only of a single element — the sphere S^{m-d_k-1} (for $d_k < m-1$) — or of two points — the sphere S^0 (for $d_k = m-1$) — then the chain $s_1, \ldots, s_k$ is said to be complete.

4.2. The algebra of PDO's. The algebra under consideration is generated on $L_2(\mathcal{M})$ by operators of two types. One of them consists of PDO's whose symbols are homogeneous functions of zero degree defined on nonzero cotangent vectors. It is supposed that the symbols belong to $C^\infty(S^*(\mathcal{M}))$ where $S^*(\mathcal{M})$ is the bundle of unit cotangent vectors. The other type of generators includes the multiplication operators by functions in the class $\mathfrak{M}(\mathcal{T})$ defined below.

We first define the corresponding classes of functions on spheres endowed with admissible partitions. Let $\mathfrak{M}_0$ stand for the set of all functions on the sphere S^0 of dimension zero. Assume that the classes $\mathfrak{M}_k$ on S^k have been defined for $k \le n-2$ while $\mathfrak{M}_k = C(S^k)$ in the case that the partition of S^k contains the only element (the sphere itself). On S^{n-1} with a partition $\mathcal{T}$ we introduce the class $\mathfrak{M}_{n-1} = \mathfrak{M}_{n-1}(\mathcal{T}) \subset L_\infty(S^{n-1})$. Let $|\mathcal{T}|$ be the union of all strata (as subsets of the sphere) whose dimension does not exceed $n-2$. A continuous function f given on $S^{n-1} \setminus |\mathcal{T}|$ belongs to $\mathfrak{M}_{n-1}(\mathcal{T})$ if the representations

$$(f \circ \kappa_{-1})(y,z) = \Psi(y,z) + \Psi_1(y,z) \tag{4.1}$$

holds in a neighborhood U of any point of every stratum $s_j, \dim s_j = d_j \le n-2$; here $(y,z) \in \kappa(U) \subset \mathbb{R}^{n-1}$, $y = (y_1, \dots, y_{n-j-1})$, $z = (z_1, \dots, z_j)$, $\Psi(y,z) = \Psi(y/|y|, z)$ (for $y \ne 0$), and $\Psi(\cdot, z) \in \mathfrak{M}_{n-j-2}(\mathcal{T}(s_j))$ for all z. The terms in (4.1) depend continuously on z with respect to the norm in L_∞ and $\Psi_1(y,z) = o(1)$ as $y \to 0$. We call the function $\Psi(\cdot, z)$ the limiting value at $z \in s_j$ (we write simply z instead of $\kappa^{-1}(0,z)$).

Finally, let us turn to $\mathfrak{M}(\mathcal{T})$ where $\mathcal{T}$ is an admissible partition of the manifold $\mathcal{M}$. We denote by $|\mathcal{T}|$ the union of all strata of dimension $j \le m-1$, $m = \dim \mathcal{M}$. A continuous function Ψ given on $\mathcal{M} \setminus |\mathcal{T}|$ is included in $\mathfrak{M}(\mathcal{T})$ if there exists the limiting value $\psi(\cdot, z)$ at any point of every j-dimensional stratum s and $\psi(\cdot, z)$ belongs to $\mathfrak{M}_{m-j-1}(\mathcal{T}(s))$ on the corresponding sphere S^{m-j-1}. Thus the algebra $\mathcal{A}$ is generated by the PDO's that were described at the beginning of Section 4.2 and by the operators aI of multiplication by functions $a \in \mathfrak{M}(\mathcal{T})$.

4.3. Operators implementing irreducible representations. First we introduce the operation of 'freezing' coefficients along a chain of strata. Let $f \in \mathfrak{M}(\mathcal{T})$ where $\mathcal{T}$ is an admissible partition of $\mathcal{M}$ and let $s_1, \dots, s_k$ be a chain of strata $s_j \in \mathcal{T}$. We assume that $f(z^1; \cdot) \in \mathfrak{M}_{m-d_1-1}(\mathcal{T}(s_1))$ is the limiting value of f at $z^1 \in s_1$. To the stratum s_2 there corresponds the $(d_1 - d_2 - 1)$-dimensional stratum $p(s_2)$ of the partition $\mathcal{T}(s_1)$. We denote the limiting value of $f(z^1; \cdot)$ at a point $z^2 \in p(s_2)$ by $f(z^1, z^2; \cdot) \in \mathfrak{M}_{m-d_2-1}(\mathcal{T}(s_1, s_2))$. Continuing this process, we arrive at the collection of functions $f(z^1; \cdot)$, $\dots$, $f(z^1, \dots, z^k); \cdot)$ where $f(z^1, \dots, z^j); \cdot) \in \mathfrak{M}_{m-d_j-1}(\mathcal{T}(s_1, \dots, s_j))$. We say that this collection is obtained by freezing f along the chain $s_1, \cdot, s_k$. If the chain is complete, then $f(z^1, \dots, z^k; \cdot) \in C(S^{m-d_k-1})$.

In what follows we denote by F the Fourier transform on $\mathbb{R}^k$,

$$(Fu)(\xi) = (2\pi)^{-k/2} \int_{\mathbb{R}^k} e^{-ix\xi} u(x)\, dx \tag{4.2}$$

and by $E(\lambda)$ an operator on the sphere S^{k-1} of the form

$$(E(\lambda)v)(\omega) = (2\pi)^{-k/2} e^{i(k/2+i\lambda)\pi/2} \Gamma(k/2 + i\lambda) \int_{S^{k-1}} (-\omega\psi + i0)^{-k/2-i\lambda} v(\psi)\, d\psi \tag{4.3}$$

where $\psi, \omega \in \mathbb{R}^k$, $|\psi| = |\omega| = 1$ and $\psi\omega = \psi_1\omega_1 + \cdots + \psi_k\omega_k$. The operator $E(\lambda)$ is unitary on $L_2(S^{n-1})$ for all real λ, while

$$(E(\lambda)^{-1}v)(\psi) = (2\pi)^{-k/2}e^{i(k/2-i\lambda)\pi/2}\Gamma(k/2-i\lambda)\int_{S^{k-1}}(-\psi\omega+i0)^{i\lambda-k/2}v(\omega)\,d\omega \tag{4.4}$$

(The operator $E(\lambda)$ was studied in [P].) We shall use the operators (4.2) – (4.4) for various k keeping the same notation for them; the dimension will be clear from the context.

Now we introduce operators to implement irreducible representations of the algebra $\mathcal{A}$. Starting with the operator

$$A = F^{-1}_{\xi\to x}\Phi(\xi)F_{y\to\xi} : L_2(\mathbb{R}^m) \to L_2(\mathbb{R}^m)$$

and an admissible partition $\mathcal{T}$ of S^{n-1} we construct series of operators of two different types. We assume that Φ is a smooth function of zero degree on $\mathbb{R}^m \setminus 0$, $1 \le n \le m$, and

$$S^{n-1} = \{x = (x^{(1)}, 0) : |x^{(1)}| = (\sum_{j=1}^{n} x_j^2)^{1/2} = 1\}.$$

Let

$$A(\theta) = F^{-1}_{\eta\to x}\Phi(\eta,\theta)F_{y\to\eta} : L_2(\mathbb{R}^n) \to L_2(\mathbb{R}^n) \tag{4.5}$$

where $\theta \in \mathbb{R}^{m-n}$, $|\theta| = 1$, $x, y, \theta \in \mathbb{R}^n$;

$$\mathfrak{A}(\lambda) = E_{\omega\to\phi}(\lambda)^{-1}\Phi(\omega, 0)E_{\psi\to\omega}(\lambda) : L_2(S^{n-1}) \to L_2(S^{n-1}) \tag{4.6}$$

where $\lambda \in \mathbb{R}$, $\phi, \omega, \psi \in \mathbb{R}^n$, $|\phi| = |\psi| = |\omega| = 1$. For $m = n$ the operators (4.5) are not defined.

Let $s_1, \ldots, s_k$ be a chain of strata of a partition $\mathcal{T}$ of the sphere S^{n-1}. We introduce in $\mathbb{R}^m$ new orthogonal coordinates $\tilde{x}_1, \ldots, \tilde{x}_m$ with the same origin. The axes $\tilde{x}_{n+1}, \ldots, \tilde{x}_m$ coincide with the old $x_{n+1}, \ldots, x_m$. The axis $\tilde{x}_n$ is directed toward the point $z^1 \in s_1$ and the axes $\tilde{x}_{n-d_1}, \ldots, \tilde{x}_{n-1}$ are parallel to the d_1–dimensional space tangent to s_1 at z^1. The axes $\tilde{x}_1, \ldots, \tilde{x}_{n-d_1-1}$ turn out to be parallel to the space tangent to S^{m-1} and orthogonal to s_1 at z^1. The coordinate transformation with block-diagonal matrix will be written in the form $x = \operatorname{diag}(J(z^1), I_{m-n})\tilde{x}$.

We introduce the operators

$$A(z^1; \theta, \theta^1) = F^{-1}_{\eta^1\to x^1}\Phi(J(z^1)(\eta^1, \theta^1), \theta)F_{y^1\to\eta^1} \tag{4.7}$$

on $L_2(\mathbb{R}^{n-d_1-1})$ where $\theta^1 \in \mathbb{R}^{d_1+1}$, $\theta \in \mathbb{R}^{m-n}$, $|\theta|^2 + |\theta^1|^2 = 1$, $x^1, y^1, \theta^1 \in \mathbb{R}^{n-d_1-1}$,

$$\mathfrak{A}(z^1; \lambda_1) = E_{\omega^1\to\phi^1}(\lambda)^{-1}\Phi(J(z^1)(\omega^1, 0), 0)E_{\psi^1\to\omega^1}(\lambda) \tag{4.8}$$

on $L_2(S^{n-d_1-2})$ where $\phi^1, \omega^1, \psi^1 \in \mathbb{R}^{n-d_1-1}$, $|\phi^1| = |\psi^1| = |\omega^1| = 1$, $\lambda_1 \in \mathbb{R}$.

Giving the role of $A(\theta)$ to the operator $A(z^1; \theta, \theta^1)$, we define a new pair of operators $A(z^1, z^2; \theta, \theta^1, \theta^2)$ and $\mathfrak{A}(z^1, z^2; \lambda_2)$. Continuing this process, we arrive at the collection of operators

$$A(z^1 \dots, z^j; \theta, \theta^1, \dots, \theta^j),\ \mathfrak{A}(z^1, \dots, z^j; \lambda_j), 1 \le j \le k,$$

where k is the length of the chain of strata, $\lambda_j \in \mathbb{R}$, $\theta \in \mathbb{R}^{m-n}$, $\theta^1 \in \mathbb{R}^{d_1+1}$, $\theta^i \in \mathbb{R}^{d_i - d_{i-1}}$ for $i \ge 2$, and $|\theta|^2 + |\theta^1|^2 + \dots + |\theta^j|^2 = 1$.

4.4. Auxiliary composition series. In this section we list all the equivalence classes of irreducible representations of the algebra $\mathcal{A}$. It is convenient to do with the help of a certain composition series which is not solving in general. Such a series turns out to be useful for construction of solving composition series.

Let $\mathcal{T}$ be an admissible partition of the manifold $\mathcal{M}$ and let $\mathcal{T}_i$ be the j–dimensional skeleton of the partition, i.e., the union of all strata of dimension not exceeding j. The sequence $\emptyset \equiv \mathcal{T}_{-1}, \mathcal{T}_0, \dots, \mathcal{T}_{m-1}$ can contain coinciding elements. We form the subsequence $\mathcal{T}_{-1}, \mathcal{T}_{m_1}, \dots, \mathcal{T}_{m_k}$ with distinct terms. For this, we move from left to right and include in the subsequence only the elements not encountered earlier.

For ease of notation we shall write simply $\mathcal{T}_j$ instead of $\mathcal{T}_{m_j}$ assuming that $\dim \mathcal{T}_j = m_j$. Introduce all possible collections $(\mathcal{T}_{j_1}, \dots, \mathcal{T}_{j_q})$ of skeletons of the partition $\mathcal{T}$ such that $j_1 < \dots < j_q$. We simplify notation again and write $(j_1, \dots, j_q)$ in place of $(\mathcal{T}_{j_1}, \dots, \mathcal{T}_{j_q})$. Put in order these collections according to the rule: $(j_1, \dots, j_q) \prec (l_1, \dots, l_p)$ if and only if either $j_1 = l_1, \dots, j_h = l_h$ and $j_{h+1} > l_{h+1}$ or $p > q$ and $j_1 = l_1, \dots, j_q = l_q$. Clearly one of any two different collections is subject to the other. We say that collections T and T' are neighboring if $T \prec T'$ and there is no collection T'' such that $T \prec T'' \prec T'$.

To every collection $T = (j_1, \dots, j_q)$ we associate the ideal $\mathcal{A}_0(T)$ in the algebra $\mathcal{A}$ generated by the operators of multiplication by functions a in $\mathfrak{M}(\mathcal{T})$ satisfying the following requirement. Let $s_1, \dots, s_q$ be an arbitrary chain of strata such that $\dim s_k = m_{j_k}$, $k = 1, \dots, q$, $s_i \in \operatorname{st}(s_j)$ for $i > j$. A function $a \in \mathfrak{M}(\mathcal{T})$ belongs to the ideal $\mathcal{A}_0(T)$ if it vanishes on $\mathcal{T}_j$ for $i < j_1$ $(\dim \mathcal{T}_j = m_j)$ and

$$a(z^1, \dots, z^q; \cdot) \equiv 0 \tag{4.9}$$

for all chains of strata with the above properties; as usual, $a(z^1, \dots, z^q; \cdot)$ is obtained by freezing a along the chain $s_1, \dots, s_q$.

Proposition 4.1. If $T \prec T'$ then $\mathcal{A}_0(T) \subset \mathcal{A}_0(T')$.

Proof. For arbitrary T and T' there exists a chain of collections $T_0, T_1, \dots, T_\mu$ satisfying the conditions $T = T_0 \prec T_1 \prec \dots \prec T_\mu = T'$ while T_j, T_{j+1} are neighboring, $j = 0, 1, \dots \mu - 1$. Therefore, it suffices to check the inclusion $\mathcal{A}_0(T) \subset \mathcal{A}_0(T')$ for neighboring collections only.

The passage from a collection $T = (j_1, \dots, j_q)$ to the neighboring collection $T' \succ T$ can be done by one of the following operations:

(1) If $j_q < k$ then the 'coordinate' $j_{q+1} = k$ has been added, i.e., $T' = (j_1, \dots, j_q, k)$.

(2) If $j_q = k > 1$ and either $j_{q-1} < k-1$ or $q = 1$ then $T' = (j_1, \dots, j_{q-1}, k-1)$ (the number of coordinates has not changed).

(3) If $j_q = k$, $j_{q-1} = k-1$, ... ,$j_{q-p} = k-p > 1$ and either $j_{q-p-1} = k-p-1$ or $q - p = 1$ then $T' = (j_1, \dots, j_{q-p-1}, j_{q-p} - 1)$ (the number of coordinates has decreased).

In the case that none of the above three operations is applicable the collection T coincides with the maximal collection $(1, \dots, k)$.

Now we turn to the ideals $\mathcal{A}_0(T)$. Suppose that the passage from $T = (j_1, \dots, j_q)$ to the neighboring collection $T' = (j_1, \dots, j_q, k)$ can be implemented by the first operation. We also assume that $s_1, \dots, s_q, s_{q+1}$ is one of the chains participating in definition of $\mathcal{A}_0(T')$ where $\dim s_p = m_{j_p}$ for $p = 1, \dots, q$ and $\dim s_{q+1} = m_k$. Then the chain $s_1, \dots, s_q$ participates in definition of $\mathcal{A}_0(T)$. If $a \in \mathcal{A}_0(T)$ then the condition (4.9) is fulfilled. In such a case $a(z^1, \dots, z^q, z^{q+1}; \dot{}) \equiv 0$ (freezing along the chain $s_1, \dots, s_q, s_{q+1}$), i.e., $\mathcal{A}_0(T) \subset \mathcal{A}_0(T')$. This inclusion is equally easy to check under passage from T to T' by means of the second or third operation.

Introduce the composition series

$$0 \subset \mathcal{K}L_2(\mathcal{M}) \subset \mathcal{A}_0(k) \subset \dots \mathcal{A}_0(T) \subset \mathcal{A}_0(T') \subset \dots \mathcal{A}_0(1, \dots, k) \subset \mathcal{A} \tag{4.10}$$

consisting of the ideal $\mathcal{K}K_2(\mathcal{M})$ of compact operators, all ideals of the form $\mathcal{A}_0(T)$ and the algebra $\mathcal{A}$. We show that the subquotients of the series (4.10) are of like construction (except the first two of them). Namely, they have sets of 'analogous' irreducible representations. This fact will be of use in constructing a solving composition series.

Let A be a PDO with symbol $\sigma \in C^\infty(S^*(\mathcal{M}))$, and let $a \in \mathfrak{M}(\mathcal{T})$ while $a|\mathcal{T}_k = 0$, i.e., $a \in \mathcal{A}_0(k)$. We introduce the mappings

$$\pi(\tau_x) : A \mapsto \sigma(\tau_x), \ \pi(\tau_x) : a \mapsto a(x) \tag{4.11}$$

where $x \in \mathcal{M} \setminus |\mathcal{T}_k|$ and $\tau_x \in S^*(\mathcal{M})_x$. Recall that $\dim \mathcal{T}_k = m_k$; in other words, $\mathcal{T}_k$ is the skeleton of maximal dimension, not containing m–dimensional strata, $m = \dim \mathcal{M}$.

Proposition 4.2. The following assertions are valid:

(1) The spectrum of $\mathcal{K}K_2(\mathcal{M})$ consists of the only element which is the identity representation.

(2) The mappings (4.11) extend to irreducible nonequivalent representations of $\mathcal{A}_0(k)/\mathcal{K}K_2(\mathcal{M})$. Any irreducible representation of this quotient algebra is equivalent to one of those in (4.11).

Proof. (1) is the well-known fact (see [D]). Let us verify (2). Clearly that the functions in $\mathfrak{M}(\mathcal{T})$ vanishing on $\mathcal{T}_k$ are continuous. Therefore $\mathcal{A}_0(k)$ can be viewed as ideal in the algebra generated on $L_2(\mathcal{M})$ by PDO's with smooth symbols. It is common knowledge that this algebra is irreducible and contains the ideal of compact operators. Every its representation is equivalent either to $\pi(\tau_x)$ or to the identity. All the representations $\pi(\tau_x)$ annulate $\mathcal{K}L_2(\mathcal{M})$. It remains to note that each irreducible representation of an ideal is the restriction of the irreducible representation of the algebra.

The ideals $\mathcal{A}_0(k-1)$ and $\mathcal{A}_0(k)$ are neighboring in (4.10). The representations of $\mathcal{A}_0(k-1)/\mathcal{A}_0(k)$ are described in Proposition 4.3 below. By definition, the ideal $\mathcal{A}_0(k-1)$ is generated in $\mathcal{A}$ by the multiplication operators by functions in $\mathfrak{M}(\mathcal{T})$ that vanish at the skeleton $\mathcal{T}_{k-1}$.

Let z be a point of m_k- dimensional stratum s. On the tangent plane $T_z(\mathcal{M})$ we introduce orthogonal coordinates $x=(x^1,x^2)$, x^2 being all axes directed along s. Assume that A is a PDO with symbol $\sigma \in C^\infty(S^*(\mathcal{M}))$ and $a \in \mathfrak{M}(\mathcal{T}) \cap \mathcal{A}_0(k-1)$. We consider the mappings

$$\pi(z,\theta) \quad : \quad A \mapsto A(z,\theta),\ \pi(z,\theta) : a \mapsto a(z;\cdot)\mathrm{Id}\,(L_2(\mathbb{R}^{m-m_k})), \tag{4.12}$$

$$\pi(z,\lambda) \quad : \quad A \mapsto \mathfrak{A}(z,\lambda),\ \pi(z,\lambda) : a \mapsto a(z;\cdot)\mathrm{Id}\,(L_2(S^{m-m_k-1})) \tag{4.13}$$

Here $A(z,\theta)$ and $\mathfrak{A}(z,\lambda)$ are operators of the form (4.5) and (4.6) with $\Phi(\eta,\theta)$ replaced by $\sigma(z;\eta,\theta)$ where $\theta \in \mathbb{R}^{m_k}$; $a(z;\cdot)$ is obtained by freezing a at $z \in s$. So in (4.12) we have the operator of multiplication by the function $\mathbb{R}^{m-m_k} \setminus 0 \ni x \mapsto a(z;x/|x|)$ on $L_2(\mathbb{R}^{m-m_k})$, and in (4.13) the multiplication by the function $S^{m-m_k-1} \ni \psi \mapsto a(z;\psi)$ on $L_2(S^{m-m_k-1})$.

Proposition 4.3. For every $z \in |\mathcal{T}_k| \setminus |\mathcal{T}_{k-1}|$ the mappings (4.12), (4.13) extend to irreducible representations of $\mathcal{A}_0(k-1)/\mathcal{A}_0(k)$. All these representations are pairwise nonequivalent. Each irreducible representation of $\mathcal{A}_0(k-1)/\mathcal{A}_0(k)$ is equivalent either to one of the representations in (4.12), (4.13) or to a one-dimensional representation of the form

$$\pi(z,\phi;\xi) : A \mapsto \sigma(z;\xi),\ \pi(z;\phi,\xi) : a \mapsto a(z;\phi) \tag{4.14}$$

where ϕ is the direction of approach to s orthogonal to s, $|\phi|=1$ and $\xi \in S^*(\mathcal{M})_z$.

The proof of this proposition is divided into two steps. First we reduce the matter to description of the representations of 'local' algebras (cf., for example, [PS 3] where the proof of an appropriate localization principle was given. Close principles were used earlier in [Dyn 2], [Si], [Do]). For $\mathcal{A}_0(k-1)/\mathcal{A}_0(k)$ it means that every irreducible representation of this quotient algebra is equivalent to an irreducible representation of a local algebra $\mathcal{A}_z$ generated on $L_2(\mathbb{R}^m)$ by operators of the form

$$F^{-1}_{\xi\to x}\sigma(z;\xi)F_{y\to\xi},\ z \in |\mathcal{T}_k| \setminus |\mathcal{T}_{k-1}|$$

and the operators of multiplication by functions $\mathbb{R}^m \ni x \mapsto a(z; x^1/|x^1|)$ with $a \in \mathfrak{M}(\mathcal{T})$.

The second step is to find all the irreducible representations of $\mathcal{A}_z$. The list of such representations was given in [PS 2].

Now let T be an arbitrary collection of skeletons participating in (4.10) and T' the neighboring collection, $T \prec T'$. We shall study the spectrum of $\mathcal{A}_0(T')/\mathcal{A}_0(T)$. We assume that the first operation leads from T to T', i.e., $T = (j_1, \dots, j_q) \mapsto (j_1, \dots, j_q, j_{q+1}) = T'$ with $j_{q+1} = k$. Further, we assume that T does not coincide with the collection that contains only the zero-dimensional skeleton (this requirement will be explained in Remark 4.5). We also suppose that $s_1, \dots, s_q$ is an arbitrary chain of strata 'belonging' to T. It means that $\dim s_l \equiv d_l = m_{j_l}$, $l = 1, \dots, q+1$, $s_j \in \operatorname{st}(s_i)$ for $j > i$. Finally, let $z^1, \dots, z^q$ be a point chosen in the process of freezing along the chain $s_1, \dots, s_q$. For every PDO A with symbol $\sigma \in C^\infty(S^*(\mathcal{M}))$ and each function $a \in \mathfrak{M}(\mathcal{T}) \cap \mathcal{A}_0(T')$ we introduce the mappings

$$\pi(z^1, \dots, z^q; \theta^1, \dots \theta^q) : A \mapsto A(z^1, \dots, z^q; \theta^1, \dots, \theta^q), \tag{4.15}$$

$$\pi(z^1, \dots, z^q; \theta^1, \dots \theta^q) : a \mapsto a(z^1, \dots, z^q; \cdot)I \tag{4.16}$$

The operator $A(z^1, \dots, z^q; \theta^1, \dots, \theta^q)$ acting on $L_2(\mathbb{R}^{m-m_k})$ is defined according to the rule in Section 4.3 (an operator of the form (4.12) with $z = z^1(s_1)$ plays the role of the original operator $A(\theta)$). The operator $a(z^1, \dots, z^q; \cdot)I$ in (4.16) is the multiplication operator by a function $\mathbb{R}^{m-d_q} \ni x \mapsto a(z^1, \dots, z^q; x/|x|)$ on the space $L_2(\mathbb{R}^{m-d_q})$. We also introduce the mappings

$$\pi(z^1, \dots, z^q; \lambda_q) : A \mapsto \mathfrak{A}(z^1, \dots, z^q; \lambda_q), \tag{4.17}$$

$$\pi(z^1, \dots, z^q; \lambda_q) : A \mapsto a(z^1, \dots, z^q; \cdot) \tag{4.18}$$

whose origination were explained in Section 4.3. The operators in the right-nand sides of (4.17), (4.18) act on $L_2(S^{m-d_q-1})$.

Lastly, we define the mappings

$$\pi(z^1, \dots, z^q; \phi^q, \xi) : A \mapsto \sigma(z^1, \xi), \tag{4.19}$$

$$\pi(z^1, \dots, z^q; \phi^q, \xi) : a \mapsto a(z^1, \dots, z^q; \phi^q) \tag{4.20}$$

where $\xi \in S^*(\mathcal{M})_{z_1}$, $\phi^q \in S^{m-d_q-1} \setminus |\mathcal{T}(s_1, \dots, s_q)|$ (see notation in Section 4.3), and $|\mathcal{T}(s_1, \dots, s_q)|$ is the union of all strata of the partition $\mathcal{T}(s_1, \dots, s_q)$ of S^{m-d_q-1} whose dimension is less than $m - d_q - 1$.

Proposition 4.4. The mappings (4.15) and (4.16); (4.17) and (4.18); (4.19) and (4.20) extend to irreducible representations of the quotient algebra $\mathcal{A}_0(T')/\mathcal{A}_0(T)$. These representations are pairwise noneqivalent. Every irreducible representation of $\mathcal{A}_0(T')/\mathcal{A}_0(T)$ is equivalent to one of them.

Proof. According to the localization principle it suffices to enumerate all the irreducible representations of the local algebras. For $\mathcal{A}_0(T')/\mathcal{A}_0(T)$ the local algebras are enumerated by points $z^1, \dots, z^q; \cdot$ (in the case that the passage from T to T' is executed by a first type operation). Every local algebra $\mathcal{A}(z^1, \dots, z^q)$ is generated on $L_2(\mathbb{R}^m)$ by operators of the form

$$F^{-1}_{\xi\to x}\Psi(z^1, \dots, z^q; \xi)F_{y\to\xi}$$

and the multiplication operators by functions $\mathbb{R}^m \ni x \mapsto a(z^1, \dots, z^{q+1}; x^1/|x^1|)$. The function $\Psi(z^1, \dots, z^q)$ can be obtained from the restriction $\sigma|S^*(\mathcal{M})_z$ of the symbol by a chain of changes of variables (rotations), which are connected with the point $(z^1, \dots, z^q)$; $x = (x^1, x^2)$ where $x^1 = (x_1, \dots, x_{m-d_q})$ and the directions of the axes x^1 are determined in accordance with the rule in 4.3. The inclusion $a \in \mathcal{A}_0(T')$ means that the function $S^{m-d_q-1} \ni \psi \mapsto a(z^1, \dots, z^q; \psi)$ vanishes on the set $|\mathcal{T}(s_1, \dots, s_q)|$, i.e., on all the strata of the partition of S^{m-d_q-1} whose dimension less than the dimension of the sphere. In other words the function $a(z^1, \dots, z^q; \cdot)$ is continuous on S^{m-d_q-1}. Therefore, to describe representations of $\mathcal{A}(z^1, \dots, z^q)$ one can use the results in [PS 1] (cf. Proposition 4.3).

Remark 4.5. If the collection T of skeletons contains only the zero-dimensional skeleton then either $T' = (T, k)$ or T is the maximal collection (in the latter case coefficients a in the algebra $\mathcal{A}$ may have discontinuities at finitely many isolated points only). The quotient $\mathcal{A}_0(T')/\mathcal{A}_0(T)$ (or $\mathcal{A}/\mathcal{A}_0(T)$ if T is the maximal collection) has no representations of the form (4.15), (4.16). Proposition 4.4 holds with such a change.

We have discussed the structure of the spectrum of $\mathcal{A}_0(T')/\mathcal{A}_0(T)$ under the condition that T and T' are connected by a first type passage. The modifications needed for a second or third type passage are evident and left to the reader. All along, the spectrum of $\mathcal{A}_0(T')/\mathcal{A}_0(T)$ consists of the three series of representations: 'θ–series' (representations of the form $\pi(z^1, \dots, z^q; \theta^1, \dots, \theta^q)$), '$\lambda$–series' (representations of the form $\pi(z^1, \dots, z^q; \lambda_q)$), and the series of one-dimensional representations of the form $\pi(z^1, \dots, z^q; \phi^q, \xi)$. The spectrum with such a structure is called regular.

Proposition 4.6. Each subquotient $\mathcal{A}_0(T')/\mathcal{A}_0(T)$ of the composition series (4.10) has regular spectrum (except for the case indicated in Remark 4.5).

It remains to analyse the last subquotient $\mathcal{A}/\mathcal{A}_0(1, \dots, k)$. Clearly any chain of strata $s_1, \dots, s_k$ that belongs to the collection $(1, \dots, k)$ is complete. In particular, it means that the functions $a(z^1, \dots, z^k; \cdot)$ obtained by freezing along the chains are continuous on the corresponding sphere S^{m-m_k-1}. Thus we arrive at the following assertion.

Proposition 4.7. The spectrum of the quotient algebra $\mathcal{A}/\mathcal{A}_0(1, \dots, k)$ is regular (except for the case in Remark 4.5). It consists of $\pi(z^1, \dots, z^q; \theta^1, \dots, \theta^q)$), $\pi(z^1, \dots, z^q; \lambda_q)$, and $\pi(z^1, \dots, z^k; \phi^k, \xi)$ obtained by freezing along $s_1, \dots, s_k$. The parameter ϕ^k in the series of one-dimensional representations runs over S^{m-m_k-1}.

4.5. The solving composition series. In general, the images of subquotients of the series (4.10) under irreducible representations contain noncompact operators. Here we construct a new composition series, of which every subquotient Q is a CCR-algebra (i.e., the image $\pi(Q)$ under any irreducible representation π consists of compact operators only). Actually this series turns out to be solving, which will be proved in the next section.

Subquotients of the series (4.10) has regular spectrum (see Proposition 4.6 and 4.7). First we consider the simplest algebra with regular spectrum and construct its composition series. Such a series was constructed in [PS 1]; we gives the corresponding arguing for the completeness of presentation.

Let $\mathcal{L}$ be an algebra generated by operators of the form $A = F^{-1}_{\xi\to x}\Phi(\phi,\xi)F_{y\to\xi}$ on $L_2(\mathbb{R}^m)$ where $x = (x^1, x^2)$, $x^1 = (x_1, \ldots, x_n)$, $x^2 = (x_{n+1}, \ldots, x_m)$, and $\phi = x^1/|x^1|$, while $\Phi(\phi, t\xi) = \Phi(\phi,\xi)$ for $t > 0$ and $\Phi \in C^\infty(S^{n-1}\times S^{m-1})$. The spectrum of $\mathcal{L}$ consists of the representations $\pi(\theta) : A \mapsto A(\theta)$ on $L_2(\mathbb{R}^n)$ with $\theta \in S^{m-n-1}$; representations $\pi(\theta) : a \mapsto \mathfrak{A}(\lambda)$ on $L_2(S^{n-1})$ with $\lambda \in \mathbb{R}$, and one-dimensional representations $\pi(\phi,\xi) : A \mapsto \Phi(\phi,\xi)$ where $(\phi,\xi) \in S^{n-1}\times S^{m-1}$; here $A(\theta)$ and $\mathfrak{A}(\lambda)$ are operators of the form (4.5) and (4.6), respectively, with $\Phi(\phi,\eta,\theta)$ and $\Phi(\phi,\omega,0)$ instead of $\Phi(\eta,\theta)$ and $\Phi(\omega,0)$ (cf. Proposition 4.3).

Proposition 4.8 (see [PS 1]). Let I be the ideal in the algebra $\mathcal{L}$ equal to the intersection of the kernels of all representations of the form $\pi(\lambda)$ and $\pi(\phi,\xi)$ and let $\operatorname{com}\mathcal{L}$ be the ideal generated by the commutators of elements in $\mathcal{L}$. Then the composition series

$$\{0\} \subset I \subset \operatorname{com}\mathcal{L} \subset \mathcal{L} \tag{4.21}$$

is solving, and

$$\begin{aligned} I &\simeq C(S^{m-n-1}) \otimes \mathcal{K}L_2(\mathbb{R}^n), \\ \operatorname{com}\mathcal{L}/I &\simeq C_0(\mathbb{R}) \otimes \mathcal{K}L_2(S_{n-1}), \\ \mathcal{L}/\operatorname{com}\mathcal{L} &\simeq C(S^{n-1}\times S^{m-1}). \end{aligned} \tag{4.22}$$

The length of $\mathcal{L}$ is equal to 2.

We outline the scheme of the proof and refer the reader to [PS 1] for more details. With the help of the Fourier transform $F_{x^2\to\xi^2}$ and changes of variables it is easy to show that the algebra $\mathcal{L}$ is isomorphic to the algebra generated by operator-valued functions

$$S^{m-n-1} \ni \theta \mapsto A(\theta) = F^{-1}_{\xi^1\to x^1}\Phi(\phi,\xi^1,\theta)F_{y^1\to\xi^1} : L_2(\mathbb{R}^n) \to L_2(\mathbb{R}^n) \tag{4.23}$$

with norm

$$\|A(\cdot)\| = \sup\{\|A(\theta); \mathcal{B}L_2(\mathbb{R}^n)\|, \theta \in S^{m-n-1}\}, x^1, y^1, \xi^1 \in \mathbb{R}^n.$$

For a fixed $\theta \in S^{m-n-1}$ we denote by $\mathcal{L}(\theta)$ the algebra spanned by operators $A(\theta)$ of the form (4.23). Introduce the algebra $\mathfrak{S}(\mathbb{R})$ of operator-valued functions

$$\mathbb{R} \ni \lambda \mapsto \mathfrak{A}(\lambda) = E_{\omega\to\phi}(\lambda)^{-1}\Psi(\phi,\omega)E_{\psi\to\omega} : L_2(S^{n-1}) \to L_2(S^{n-1}))$$

with norm $\|\mathfrak{A}(\cdot)\| = \sup\{\|\mathfrak{A}(\lambda); \mathcal{B}L_2(S^{n-1})\|, \lambda \in \mathbb{R}\}$. The mapping p that sends an operator $A(\theta)$ defined by (4.23) to the operator-valued function (4.24) with $\Psi(\phi,\omega) \equiv \Phi(\phi,\omega,0)$ extends to an epimorphism $p : \mathcal{L}(\theta) \to \mathfrak{S}(\mathbb{R})$. We set $J(\theta) = \ker p$. It turns out that $J(\theta)$ contains the ideal $\mathcal{K}L_2(\mathbb{R}^n)$ and the quotient algebra $J(\theta)/\mathcal{K}L_2(\mathbb{R}^n)$ is commutative. We describe the maximal ideal space of this algebra.

Assume that $(\phi,\xi) \mapsto \Phi(\phi,\xi)$ is a function in $C^\infty(S^{n-1} \times (\mathbb{R}^m \setminus 0))$ that is homogeneous of degree zero with respect to ξ and satisfies the condition $\Phi(\phi,\omega,0) \equiv 0$. On the set $Y = S^{n-1} \times \{(\omega,t) : \omega \in S^{n-1}, -\infty \le t < \infty\}$ we consider the function $(\phi,\omega,t) \mapsto \Phi(\phi,\omega,e^{-t}\theta)$ for fixed $\theta \in S^{m-n-1}$. If the set $\{(\omega,-\infty) : \omega \in S^{n-1}\}$ is contracted to a point, then Y becomes a set $\dot{Y}$ homeomorphic to the product $S^{n-1} \times S^n_+$ where S^n_+ is the open hemisphere with pole corresponding to the contracted set. Let $\gamma : \dot{Y} \to S^{n-1} \times S^n_+$ be a homeomorphism determined by the equality

$$\gamma(\phi,\omega,t) = (\phi, \omega/\sqrt{1+e^{-2t}}, e^{-t}\theta\sqrt{1+e^{-2t}}).$$

In view of the condition $\Phi(\phi,\omega,0) \equiv 0$ the function $\Phi \circ \gamma$ can be regarded as element of the space $C_0^\infty(S^{n-1} \times S^n_+)$ of continuous functions vanishing at $S^{n-1} \times \partial S^n_+$.

Suppose that $A(\theta) \equiv A_\Phi(\theta)$ is an operator of the form (4.23) while $\Phi(\phi,\omega,0) \equiv 0$. It can be shown that the mapping $A_\Phi(\theta) \mapsto \Phi \circ \gamma \in C_0(S^{n-1} \times S^n_+)$ implements an isomorphism $J(\theta)/\mathcal{K}L_2(\mathbb{R}^n) \simeq C_0(S^{n-1} \times S^n_+)$.

We turn to verifying the properties of the series (4.21). In [PS 1] it was proved that the algebra generated by functions (4.23) contains all the continuous functions $S^{m-n-1} \ni \theta \mapsto A(\theta) \in J(\theta)$. In particular, $C(S^{m-n-1}) \otimes \mathcal{K}L_2(\mathbb{R}^n)$ is a subalgebra of this algebra. The spectrum $\mathfrak{S}(\mathbb{R})^\wedge$ consists of representations of the form $\pi(\lambda) : \mathfrak{A} \mapsto \mathfrak{A}(\lambda)$ and the one-dimensional representations $\pi(\phi,\omega) : \mathfrak{A} \mapsto \Psi(\phi,\omega)$ (see (4.24)). It follows that if a function $\theta \mapsto G(\theta) \in \mathcal{L}(\theta)$ is annihilated by all the representations $\pi(\lambda)$ and $\pi(\phi,\omega)(\equiv \pi(\phi,\omega,0))$ then $G(\theta) \in \mathcal{L}(\theta)$ for all $\theta \in S^{m-n-1}$ because $\mathcal{L}(\theta)/\mathcal{T}(\theta) \simeq \mathfrak{S}(\mathbb{R})$. Finally, taking into account the isomorphism $J(\theta)/\mathcal{K}L_2(S^{n-1}) \simeq C_0(S^{n-1} \times S^n_+)$ we obtain that if all the rest one-dimensional representations vanish at G then $G(\theta) \in \mathcal{K}L_2(\mathbb{R}^n)$ for all θ. Hence, $I \simeq C(S^{m-n-1}) \times \mathcal{K}L_2(\mathbb{R}^n)$, and the first of the relations (4.22) is thus proven.

Let J be the ideal in $\mathcal{L}$ generated by functions $\theta \mapsto A(\theta) \in J(\theta)$. We introduce a mapping $r : \mathcal{L}/J \to \mathfrak{S}(\mathbb{R})$ as the composition of the morphisms

$$\mathcal{L} \to \mathcal{L}(\theta) \to \mathcal{L}(\theta)/J(\theta) \to \mathfrak{S}(\mathbb{R}).$$

The first arrow means evaluation at a point θ, the second one is the projection, and the third arrow stands for the morphism mentioned above. Then r does not depend on θ, annihilates the ideal and is an isomorphism $\mathcal{L}/J \equiv \mathfrak{S}(\mathbb{R})$. Clearly $r : \operatorname{com}\mathcal{L} \to \operatorname{com}\mathfrak{S}$. We define a morphism $q : \operatorname{com}\mathcal{L}/I \to \operatorname{com}\mathfrak{S}$ by the equality $q([A]) = r(A)$ where $[A]$ is the coset representative of A in $\operatorname{com}\mathcal{L}/I$. This definition is correct due to the inclusion $I \subset J$. It is known [PS 4], [P], that $\operatorname{com}\mathfrak{S}(\mathbb{R}) \simeq C_0(\mathbb{R}) \otimes \mathcal{K}L_2(S^{n-1})$. The mapping q is an isomorphism so we obtain the second relation (4.22). The third formula (4.22) follows from the description of one-dimensional representations of the algebra $\mathcal{L}$.

Thus the series (4.21) is a solving one. We show now that the length of $\mathcal{L}$ is equal to 2, i.e., there is no other solving series that would be shorter than (4.21). Suppose, contrary to our claim, that there exists a solving series $0 \subset Q \subset \mathcal{L}$. Then the spectrum of the ideal q contains either all infinite-dimensional representations or all one-dimensional ones (as a matter of fact, the spectrum $\hat{Q}$ must consist of representations of a single dimension). The first case is impossible, since $\hat{Q}$ is then not a Hausdorff space, and the composition series is not solving (the description of the spectral topology of $\mathcal{L}$ was given in [PS 1], see also [PS 3]). The second case cannot be, because the set of all one-dimensional representations is not open, while the spectrum of an ideal is always an open part of the spectrum of an algebra.

Let us cross out the 'θ–series' in the list $\{\pi(\theta), \pi(\lambda), \pi(\phi, \xi)\}$ of all irreducible representations of $\mathcal{L}$. Then the ideal I in (4.21) turns out to be the intersection of the kernels of all the remaining representations. The ideal $\operatorname{com}\mathcal{L}$ is the intersection of the kernels of all one-dimensional representations $\pi(\phi, \xi)$, i.e., to obtain $\operatorname{com}\mathcal{L}$ we have to cross out both the θ– and λ–series of the representations. We turn now to the algebra $\mathcal{A}$ and generalize the above procedure.

As in Section 4.4, let T and T' be neighboring collections of skeletons, while $T \prec T'$. We denote by $\pi(\theta, T)$ and $\pi(\lambda, T)$ the θ–series and λ–series of irreducible representations of the quotient algebra $\mathcal{A}_0(T')/\mathcal{A}_0(T)$. These representations can be thought of as representations of $\mathcal{A}$ which annihilate the ideal $\mathcal{A}_0(T)$ and do not vanish on $\mathcal{A}_0(T')$. All the one-dimensional representations form the series $\pi_1(\mathcal{A})$. We define $\ker \pi(\cdot, T)$ ($\ker \pi_1(\mathcal{A})$) to be the intersection of the kernels of all representations in the series $\pi(\cdot, T)$ ($\pi_1(\mathcal{A})$).

It should be recorded that if T contains the zero-dimensional skeleton only, then the quotient algebra $\mathcal{A}_0(T')/\mathcal{A}_0(T)$ does not have the θ–series of representations (see Remark 4.5). For the sake of simplicity of descriptions assume (up to Theorem 4.12) that the zero- dimensional skeleton $\mathcal{T}_0$ of the partition $\mathcal{T}$ is empty.

Let

$$T_{min} = (k) = T^1 \prec T^2 \prec \cdots \prec T_l = T_{max} = (1, 2, \ldots, k)$$

be the sequence of all the admissible collections of skeletons of a partition $\mathcal{T}$ of a manifold $\mathcal{M}$ ordered as in Section 4.4. We define ideals $I(T^q)$ and $J(T^q)(q = 1, \ldots, l)$ in $\mathcal{A}$ by the equalities

$$\begin{aligned} I(T^q) &= \{\cap \ker \pi(\cdot, T^p), p > q\} \cap \ker \pi(\lambda, T^q) \cap \ker \pi_1(\mathcal{A}), \\ I(T^q) &= \{\cap \ker \pi(\cdot, T^p), p > q\} \cap \ker \pi_1(\mathcal{A}). \end{aligned} \tag{4.25}$$

It is obvious that

$$0 \subset \mathcal{K}L_2(\mathcal{M}) \subset I(T^1) \subset J(T^1) \subset \ldots \subset I(T^l) \subset J(T^l) = \ker \pi_1(\mathcal{A}) \subset \mathcal{A} \tag{4.26}$$

The aim is to prove that the composition series (4.26) is solving.

4.6. Properties of the series (4.26). We show that the subquotients of the series (4.26) are algebras determined by continuous fields of elementary algebras on locally compact Hausdorff spaces, and we describe these spaces. It is a simple matter to check that the fields are locally trivial. Then Proposition 2.3 provides their global triviality. The latter means that the series (4.26) is solving and the algebra $\mathcal{A}$ is solvable.

We give only a sketch of the proof because of cumbersome descriptions.

Note first that $J(T^j) \subset \mathcal{A}_0(T^{j+1})$ for $j = 1, \dots, l-1$. Indeed, on the one hand for $x \in \mathcal{B}$ in any C^*–algebra $\mathcal{B}$

$$\|x; \mathcal{B}\| = \sup\{\|\pi(x)\|, \pi \in \hat{\mathcal{B}}\}. \tag{4.27}$$

On the other hand all the irreducible representations of $\mathcal{A}$ that vanish on $\mathcal{A}_0(T^{j+1})$ annihilate the ideal $J(T^j)$ too. Therefore the coset representative $[x] \in \mathcal{A}/\mathcal{A}_0(T^{j+1})$ of any $x \in J(T^j)$ proves to be zero.

Thus the ideal $J(T^j)$ in the algebra $\mathcal{A}_0(T^{j+1})$ is the intersection of the kernels of all one-dimensional representations of $\mathcal{A}_0(T^{j+1})$, i.e.,

$$J(T^j) \simeq \operatorname{com} \mathcal{A}_0(T^{j+1}). \tag{4.28}$$

Finally, taking into account that $\mathcal{K}L_2(\mathcal{M}) = \operatorname{com} \mathcal{A}_0(T)$ we can rewrite (4.26) in more expressive form

$$\begin{aligned} 0 \subset \operatorname{com} \mathcal{A}_0(T^1) &\subset I(T^1) \subset \operatorname{com} \mathcal{A}_0(T^2) \subset I(T^2) \subset \\ &\subset \operatorname{com} \mathcal{A}_0(T^3) \subset \dots \subset I(T^l) \subset \operatorname{com} \mathcal{A} \subset \mathcal{A}. \end{aligned} \tag{4.29}$$

Proposition 4.9. The subquotient $I(T^1)/\mathcal{K}L_2(\mathcal{M})$ is the algebra determined by a continuous field of elementary algebras on the cospherical bundle $S^*(|\mathcal{T}_k| \setminus |\mathcal{T}_{k-1}|)$ (see notation at the beginning of Section 4.4. In particular, $\mathcal{T}_j$ is the m_j–dimensional skeleton of a partition $\mathcal{T}$).

Proof. We denote by $\Gamma_0(T^1)$ the ideal in $\mathcal{A}_0(T^2)/\mathcal{A}_0(T^1)$ equal to the intersection of the kernels of all one-dimensional representations and of the representations $\pi(\lambda, T^1)$. Introduce the continuous field of algebras $\{B_0(\cdot), \Gamma_0(T^1)\}$ on $|\mathcal{T}_k| \setminus |\mathcal{T}_{k-1}|$ (at every point z of a stratum $s \in \mathcal{T}_k$ the algebra $B_0(z)$ is formed by the values at the point of elements in $\Gamma_0(T^1)$). Let us check that $I(T^1)/\mathcal{K}L_2(\mathcal{M})$ is the algebra determined by the above field.

Assume that $I(T^1) \to \mathcal{A}_0(T^2)/\mathcal{A}_0(T^1)$ is the composition of the inclusions $I(T^1) \subset J(T^1) \subset \mathcal{A}_0(T^2)$ and the projection $\mathcal{A}_0(T^2) \to \mathcal{A}_0(T^2)/\mathcal{A}_0(T^1)$. Since $\mathcal{K}L_2(\mathcal{M}) \subset \mathcal{A}_0(T^1)$, we obtain the mapping

$$p : I(T^1)/\mathcal{K}L_2(\mathcal{M}) \to \mathcal{A}_0(T^2)/\mathcal{A}_0(T^1).$$

We show that p is a monomorphism and $\operatorname{Im} p$ coincides with the ideal $\Gamma_0(T^1)$. Note that the representations $\pi(\theta, T^1)$, $\pi(\lambda, T^1)$ and $\pi(\phi, \xi; T^1)$ forming the spectrum of

$\mathcal{A}_0(T^2)/\mathcal{A}_0(T^1)$ are defined on $\mathcal{A}_0(T^2)/\mathcal{A}_0(T^1)$ and $I(T^1)/\mathcal{K}L_2(\mathcal{M})$ by the same formulas, while $\pi(\lambda, T^1)$ and $\pi(\phi, \xi; T^1)$ vanish on $I_0(T^1)$. Therefore the equality $p(x) = 0$ implies the inclusion $x \in \ker \pi(\theta, T^1)$ for all θ. The spectrum of $I(T^1)/\mathcal{K}L_2(\mathcal{M})$ consists of the representations $\pi(\theta, T^1)$ which together with (4.27) leads to the equality $x = 0$. Thus, p is a monomorphism. Moreover, it is clear that if $x \in I(T^1)/\mathcal{K}L_2(\mathcal{M})$ then $\pi(\lambda, T^1)p(x) = 0$ and $\pi(\phi, \xi; T^1)p(x) = 0$,i.e., $p(x) \in \Gamma_0(T^1)$. On the other hand, it is easy to see that any generator in $\Gamma_0(T^1)$ has preimage in $I(T^1)/\mathcal{K}L_2(\mathcal{M})$.

From Proposition 4.8 it follows that

$$B_0(z) \simeq C(S^*(|\mathcal{T}_k| \setminus |\mathcal{T}_{k-1}|)_z) \otimes \mathcal{K}L_2(\mathbb{R}^{m-m_k}) \tag{4.30}$$

for every $z \in |\mathcal{T}_k| \setminus |\mathcal{T}_{k-1}|$ where $\mathbb{R}^{m-m_k}$ is the subspace of the tangent space $T_z(\mathcal{M})$ orthogonal to the stratum $s \ni z$ at a point z. So one can pass from the field $\{B_0(\cdot), \Gamma_0(T^1)\}$ to the field of elementary algebras on $S^*(|\mathcal{T}_k| \setminus |\mathcal{T}_{k-1}|)$.

Besides the extreme subquotients $\mathcal{A}/\ker \pi_1(\mathcal{A}) = \mathcal{A}/\mathrm{com}\,\mathcal{A}$ and $I(T^1)/\mathcal{K}L_2(\mathcal{M})$, the series (4.26) contains subquotients of the form $J(T^j)/I(T^j)$ and $I(T^j)/J(T^{j-1})$ for $j = 2, \ldots, l$. We consider $J(T^1)/I(T^1)$ and $I(T^2)/J(T^1)$. The structure of the remaining subquotients can be found in the same way. All one need to do is to describe the corresponding locally compact spaces where the fields of elementary algebras are defined.

Proposition 4.10. The subquotient $J(T^1)/I(T^1)$ is the algebra determined by a continuous field of elementary algebras on the space $(|\mathcal{T}_k| \setminus |\mathcal{T}_{k-1}|) \times \mathbb{R}$.

Proof. Define $\Gamma_1(T^1)$ to be the ideal in the algebra $\mathcal{A}_0(T^2)/\mathcal{A}_0(T^1)$ equal to the intersection of the kernels of all one-dimensional representations. We introduce the continuous field of algebras $\{B_1(\cdot), \Gamma_1(T^1)\}$ on $|\mathcal{T}_k| \setminus |\mathcal{T}_{k-1}|$ where the algebra $B_1(z)$ is formed by the values at z of elements in $\Gamma_1(T^1)$. Then $J(T^1)/I(T^1)$ proves to bc the algebra determined by this field. By Proposition 4.8,

$$B_1(z) \simeq C_0(\mathbb{R}) \otimes \mathcal{K}L_2(S^{m-m_k-1}(\mathcal{M})_z) \tag{4.31}$$

where $S^{m-m_k-1}(\mathcal{M})_z$ is the unit sphere in the subspace $\mathbb{R}^{m-m_k}$ of the tangent space $T_z(\mathcal{M})$ orthogonal to the stratum s at a point z (see the second formula (4.22)). Using (4.21) we pass from the field $\{B_1(\cdot), \Gamma_1(T^1)\}$ to the field of elementary algebras on $(|\mathcal{T}_k| \setminus |\mathcal{T}_{k-1}|) \otimes \mathbb{R}$.

Let us turn to $I(T^2)/J(T^1)$. We have $J(T^1) \subset I(T^2) \subset \mathcal{A}_0(T^3)$ so $I(T^2)/J(T^1)$ is the ideal in $\mathcal{A}_0(T^3)/J(T^1)$. The spectrum of $\mathcal{A}_0(T^3)/\mathcal{A}_0(T^2)$ consists of the representations $\pi(\theta, T^2)$, $\pi(\lambda, T^2)$ and the one-dimensional representations $\pi(\phi, \xi; T^2)$ (see Proposition 4.4 and Remark 4.5). By virtue of (4.28), $J(T^1) = \mathrm{com}\,\mathcal{A}_0(T^2)$. Hence, the spectrum of consists of the one-dimensional representations $\pi(\phi, \xi; T^1)$. It follows that the spectrum of $\mathcal{A}_0(T^3)/J(T^1)$ is formed by the series $\pi(\theta, T^2)$, $\pi(\lambda, T^2)$,

$\pi(\phi,\xi;T^2)$ and $\pi(\phi,\xi;T^1)$ while only the representations $\pi(\theta,T^2)$ vanish on the ideal $I(T^2)/J(T^1)$.

On $|\mathcal{T}_{k-1}|\setminus|\mathcal{T}_{k-2}|$ we introduce the continuous field of algebras $\{B_0(\cdot),\Gamma_0(T^2)\}$ where $\Gamma_0(T^2)=I(T^2)/J(T^1)$ and consists of the values at a point $z\in|\mathcal{T}_{k-1}|\setminus|\mathcal{T}_{k-2}|$ of functions in $\Gamma_0(T^2)$. By Proposition 4.8,

$$B_0(z)\simeq C(S^*(|\mathcal{T}_{k-1}|\setminus|\mathcal{T}_{k-2}|)_z)\otimes\mathcal{K}L_2(\mathbb{R}^{m-m_{k-1}}),\tag{4.32}$$

which enable us to pass from the field $\{B_0(\cdot),\Gamma_0(T^2)\}$ to the field of elementary algebras on the space $S^*(|\mathcal{T}_{k-1}|\setminus|\mathcal{T}_{k-2}|)$. Thus we arrive at the following assertion.

Proposition 4.11. The subquotient $I(T^2)/J(T^1)$ is the algebra determined by a continuous field of elementary algebras on the space $S^*(|\mathcal{T}_{k-1}|\setminus|\mathcal{T}_{k-2}|)$.

We shall take a quick look at $J(T^{j+1})/I(T^{j+1})$ and $I(T^{j+1})/J(T^j)$ for $j=3,\dots,l-1$. We start with $I(T^{j+1})/J(T^j)$. Instead of the series $\pi(\cdot,;T^2)$ (see the proof of Proposition 4.11) there appears the series $\pi(z^1,\dots,z^q);\cdot)$ obtained by freezing along chains of strata $s_1,\dots,s_q$ that belong to T^{j+1}. Clearly the point $z=(z^1,\dots,z^q)$ runs over the locally compact space $\tilde{Z}_{j+1}(S_1,\dots,s_q)$ where $z^1\in s_1$, $z^2\in p(s_2)\in\mathcal{T}(s_1)$, etc; see notation in Section 4.3. Denote by Z_{j+1} the union of all $\tilde{Z}_{j+1}(S_1,\dots,s_q)$ for chains of strata that belong to T^{j+1}. To every point $z=(z^1,\dots,z^q)$ we associate the subspace $T^*(\mathcal{M},z)$ of the cotangent space $T^*_{z_1}(\mathcal{M})$ consisting of the directions cotangent to the stratum s_q at the point z, so $\delta_{j+1}=\dim T^*(\mathcal{M},z)=\dim s_q$. Let $T^*(Z_{j+1})$ be the bundle with base Z_{j+1} and fiber $T^*(Z_{j+1})_z=T^*(\mathcal{M},z)$ over $z\in Z_{j+1}$, and let $S^*(Z_{j+1})$ be the corresponding cospherical bundle, i.e., $S^*(Z_{j+1})_z$ is the unit sphere in the space $T^*(Z_{j+1})_z$. Finally, we denote by $T^\perp(\mathcal{M},z)$ the subspace in the tangent space $T_{z^1}(\mathcal{M})$ orthogonal to $T^*(\mathcal{M},z)$.

In the same way as in the proof of Proposition 4.11 one can show that the subquotient $I(T^{j+1})/J(T^j)$ of the series (4.26) is the algebra determined by a continuous field of algebras $\{B_0(\cdot),\Gamma_0(T^{j+1})\}$ on the locally compact space Z_{j+1}. For $z\in Z_{j+1}$ we have

$$B_0(z)\simeq C(S^*(Z_{j+1})_z)\otimes\mathcal{K}L_2(\mathbb{R}_z^{m-\delta_{j+1}}),\tag{4.33}$$

which enable us to consider $I(T^{j+1})/J(T^j)$ as algebra determined by a continuous field of elementary algebras on the space $S^*(Z_{j+1})$.

The subquotient $J(T^{j+1})/I(T^{j+1})$ turns out to be the algebra determined by a continuous field of algebras $\{B_1(\cdot),\Gamma_1(T^{j+1})\}$ on the space Z_{j+1}. Since

$$B_1(z)\simeq C_0((Z_{j+1}\times\mathbb{R})_z)\otimes\mathcal{K}L_2(S^{m-\delta_{j+1}-1}),\tag{4.34}$$

for $z\in Z_{j+1}$, the quotient $J(T^{j+1})/I(T^{j+1})$ is the algebra determined by the field of elementary algebras on $Z_{j+1}\times\mathbb{R}$. Thus we arrive at the following result.

Theorem 4.12. Let the zero-dimensional skeleton $\mathcal{T}_0$ be empty. Then the following assertions are valid.

(1) The composition series (4.26) (or alternatively (4.29)) is solving, so the algebra $\mathcal{A}$ is solvable.

(2) The following relations hold:

$$\begin{aligned} I(T^1)/\mathcal{K}L_2(\mathcal{M}) &\simeq C_0(S^*(|\mathcal{T}_k| \setminus |\mathcal{T}_{k-1}|) \otimes \mathcal{K}L_2(\mathbb{R}^{m-m_k}), \quad (4.35) \\ J(T^1)/I(T^1) &\simeq C_0((|\mathcal{T}_k| \setminus |\mathcal{T}_{k-1}|) \times \mathbb{R}) \otimes \mathcal{K}L_2(S^{m-m_k-1}), \quad (4.36) \end{aligned}$$

and, in general,

$$\begin{aligned} I(T^{j+1})/J(T^j) &\simeq C_0(S^*(Z_{j+1}) \otimes \mathcal{K}L_2(\mathbb{R}^{m-\delta_{j+1}}), \quad (4.37) \\ J(T^{j+1})/I(T^{j+1}) &\simeq C_0(Z_{j+1} \times \mathbb{R}) \otimes \mathcal{K}L_2(S^{m-\delta_{j+1}-1}), \quad (4.38) \end{aligned}$$

$j = 1, \ldots, l-1$ (Z_{j+1} and δ_{j+1} were introduced before the formula (4.33)), and $\mathcal{T}_k$ were defined at the beginning of Section 4.4).

(3) The set of all one-dimensional representations of $\mathcal{A}$, i.e., the spectrum of the quotient algebra $\mathcal{A}/\operatorname{com}\mathcal{A}$ which is the last subquotient of the series (4.26), form a compact topological manifold.

(4) The series (4.26) is the shortest, i.e., the length of the algebra $\mathcal{A}$ is equal to $2l+1$.

If $\mathcal{T}_0 \neq 0$ then one of the subquotients of the series (4.26) has no θ–series of the representations but only λ–series. The length of the algebra $\mathcal{A}$ is equal to $2l$ (see Remark 4.5). With these changes all the rest is still valid.

A few further remarks on the proof of the theorem. In Section 4.6 we established that all the subquotient of (4.26) were the algebras determined by fields of elementary algebras. the local trivialityof the fields follows from the fact that they are invariably realized as fields of algebras of compact operators acting on the fibers of locally trivial fields of Hilbert spaces. Local trivializations of the field of spaces generate the local trivializations of the fields of algebras in the sense of Section 1.2. Therefore, Proposition 2.3 provides the global triviality of the fields of algebras.

The relations (4.35)–(4.38) follows from the formulas (4.30)–(4.33).

The verification of the assertions 3 and 4 makes use of the spectral topology of $\mathcal{A}$. We refer the reader to [PS 2] (see also [PS 3]) for the description of this topology and recommend to generalize the corresponding argument in the proof of Proposition 4.8.

References

[D] Dixmier, J., *Les C^*–algèbras et Leurs Représentations*, Gauthier-Villars Éditeur, Paris 1969.

[Do] Douglas, R.G., *Banach algebras techniques in Operator Theory*, Academic Press, New York 1972.

[Dyn] Dynin, A.S., *Inverse Problem for Singular Integral Operators: C^*-approach*, Proc. Nat. Acad. Sci. USA **75** (1978), 4668–4670.

[Dyn 2] Dynin, A.S., *Multivariable Wiener–Hopf Operators,* 1.*Representations*, Integr. Eq. Oper. Theory **9** (1986), 537–556.

[Dyn 3] Dynin, A.S. *Multivariable Wiener–Hopf Operators,* 2. *Spectral Topology and Solvability*, Integr. Eq. Oper. Theory **10** (1987), 554–576.

[MR] Muhly, P.S. and Renault, J.N., *C^*-algebras of multivariable Wiener-Hopf operators*, Trans. Amer. Math. Soc. **274** (1982), 1–44.

[P] Plamenevskij, B.A, *Algebras of Pseudodifferential Operators*, Nauka, Moscow 1986 (in Russian) (English translation: Kluwer Academic Publishers, Dordrecht 1989).

[PS 1] Plamenevskij, B.A. and Senichkin, V.N., *On representations of an algebra of pseudodifferential operators with multidimensional discontinuities in the symbols*, Izv. Akad. Nauk SSSR Ser. Math., **51** (1987), 833–859 (in Russian) (English translation: Math. USSR Izv. **31** (1988)).

[PS 2] Plamenevskij, B.A. and Senichkin, V.N., *The spectrum of an algebra of pseudodifferential operators with piecewise smooth symbols*, Izv. Akad. Nauk SSSR Ser. Math., **53** (1989), 147–177 (in Russian) (English translation: Math. USSR Izv., **34** (1990), 147–177).

[PS 3] Plamenevskij, B.A. and Senichkin, V.N., *Solvable Operator Algebras*, Algebra i Analiz, (1994), 1–87 (in Russian).

[PS 4] Plamenevskij, B.A. and Senichkin, V.N., *On the spectrum of C^*-algebras generated by pseudodifferential operators with discontinuous symbols*, Izv. Akad. Nauk SSSR Ser. Math., **47** (1983), 1263–1284 (in Russian) (English translation: Math. USSR Izv., **23** (1984), 524–544).

[PS 5] Plamenevskij, B.A. and Senichkin, V.N., *On composition series in algebras of pseudodifferential operators*, Doklady Russian Acad. Sci.

[S] Senichkin, V.N., *On the Jacobson topology on the spectrum of an algebra of Wiener-Hopf operators in a cone*, Algebra i Analiz, **5** (1993), 194–216 (in Russian).

[Si] Simonenko, I.B., *Operators of convolution type in cones*, Mat.Sb. **74** (1967), 298–313 (in Russian) (English translation: Math. USSR-Sb. **3** (1967) 279–294).

List of Authors

B.V. Fedosov
Department of Mathematics
Moscow Institute of Physics
and Technology
Dolgoprudny
Moscow region 141700
Russia

B. Helffer
UA 762 du CNRS
DMI
Ecole Normale Supérieure
45, rue d'Ulm
75230 Paris Cédex
France

M. Holschneider
Centre Physique Théorique
CNRS
Luminy
Case 907
13288 Marseille Cédex 9
France

V. Liskevich
Institute of Mathematics
The Hebrew University
of Jerusalem
Jerusalem 91904
Israel

B.A. Plamenevskij
Department of Math. Physics
Faculty of Physics
St.-Petersburg State University
Uljanova 1
Stary Petershof
198904, St.-Petersburg
Russia

B.–W. Schulze
Max-Planck-Arbeitsgruppe
"Partielle Differentialgleichungen
und Komplexe Analysis"
Universität Potsdam
Postfach 60 15 53
14415 Potsdam
Germany

Yu.A. Semenov
Department of Mathematics
University of Toronto
Toronto
Ontario M5S 1A1
Canada

V.N. Senichkin
Department of Mathematics
St.-Petersburg University for
Telecommunications
Mojka 61
191065, St.-Petersburg
Russia

Advances in Partial Differential Equations

Editor in chief: B.–W. Schulze

Advances in Partial Differential Equations is a series originating from the work of the Max Planck Research Group for Partial Differential Equations and Complex Analysis at the University of Potsdam.

Editor in chief is B.–W. Schulze. Current coeditors are M. Demuth, E. Schrohe, and J. Sjöstrand.

The volumes will contain research expository papers as well as short communications. Contributions can be submitted to any one of the editors.

Research expository papers should be of lasting interest to a larger public. They are expected to be selfcontained, and should both contain new results and give an overview of the relevant area. Preferably, manuscripts should be between 60 and 110 pages long.

Papers to be published as short communications should not exceed 20 pages.

Advances in Partial Differential Equations will appear in the series **Mathematical Topics** of Akademie Verlag, Berlin. The manuscripts will be reproduced photomechanically and therefore must be prepared very carefully. Please contact the editors for detailed instructions.

Boundary Value Problems, Schrödinger Operators, Deformation Quantization

– Advances in Partial Differential Equations –

Edited by Michael Demuth, Elmar Schrohe, and Bert-Wolfgang Schulze (editor-in-chief)

(Mathematical Topics, Volume 8)

1995. 353 pages
170 mm x 240 mm
Hardcover DM/sFr 130,– / öS 1014,–
ISBN 3-05-501699-8

The analysis of boundary value problems has a long tradition in mathematics. Understanding the criteria for solvability and the structure of the solutions is of central interest both for theory and applications. Boundary value problems on manifolds with singularities present an additional challenge. They exhibit a wealth of analytic and algebraic structures, also under the aspect of index theory.

In the first contribution to this volume, boundary value problems without the transmission condition are interpreted as particular problems on manifolds with edges; it deals with the new effects caused by variable and branching asymptotics. In the second paper, a pseudo-differential calculus is constructed for boundary value problems on manifolds with conical singularities. A concept of ellipticity is introduced that allows a parametrix construction and entails the Fredholm property in weighted Sobolev spaces. Moreover, this approach lays the foundations for treating boundary value problems on manifolds with edges. Two further contributions deal with deformation quantization, an important topic of Mathematical Physics. The first one gives a complete proof of the index theorem in deformation quantization, while the other one treats trace densities. The final article in this volume, also from the area of Mathematical Physics, presents new results on the spectrum of perturbed periodic Schrödinger operators.

Contents: